Climate Change, Drinking Water Security, and Public Health

Azizur R. Molla • Seth H. Frisbie
Mohsen Sherif • Vijay P. Singh
Editors

Climate Change, Drinking Water Security, and Public Health

Global Challenges and Solutions

Springer

Editors
Azizur R. Molla
College of Health Professions, School of Interdisciplinary Health, Public Health Program
Grand Valley State University
Grand Rapids, MI, USA

Seth H. Frisbie
Department of Chemistry and Biochemistry, College of Arts and Sciences
Norwich University
Northfield, VT, USA

Mohsen Sherif
Civil and Environmental Engineering Department
United Arab Emirates University
Al Ain, United Arab Emirates

Vijay P. Singh
Department of Biological & Agricultural Engineering and Zachry Department of Civil & Environmental Engineering
Texas A&M University
College Station, TX, USA

ISBN 978-3-032-12073-1 ISBN 978-3-032-12074-8 (eBook)
https://doi.org/10.1007/978-3-032-12074-8

This Springer imprint is published by the registered company Springer Nature Switzerland AG
The registered company address is: Gewerbestrasse 11, 6330 Cham, Switzerland

Preface

The physical sciences easily explain how human activities cause climate change. However, some consequences of climate change are hard to understand. This book focuses on two of these consequences: (1) how climate change affects the amount and safety of drinking water and (2) how this climate-impacted drinking water affects human health.

In more detail, the Sun warms the Earth; some of this warmth is emitted from the Earth's surface into the atmosphere as infrared radiation. Atmospheric gases with three or more atoms, such as water vapor (H_2O), carbon dioxide (CO_2), methane (CH_4), nitrous oxide (N_2O), ozone (O_3), chlorofluorocarbons (CFCs), hydrofluorocarbons (HFCs), and others, absorb and then re-emit this infrared radiation in all directions. If this re-emitted infrared radiation is directed back toward the Earth, the heat is trapped, and our planet warms. Human activities are increasing the concentration of carbon dioxide and other greenhouse gases in the atmosphere; this must contribute to global warming and climate change. These events are the irrefutable consequences of thermodynamics and quantum mechanics, the rules that govern our universe.

These changes in our atmosphere must change our climate. These changes to our climate will change the quantity and quality of our drinking water, which will also impact public health.

The observed increase in global surface temperature, expansion of deserts, melting of glaciers, rising sea levels, and other consequences of climate change are decreasing our supply of readily available drinking water. For example, melting glaciers in the Himalayas and rising sea levels are decreasing the drinking water supply in South and Southeast Asia. A lack of drinking water on a warming planet can cause dehydration, heat injury, urinary tract infections, kidney stones, kidney failure, seizures, and low blood volume (hypovolemic) shock; some of these conditions are life-threatening.

The microbiological consequences of climate change can degrade the quality of drinking water. Microbial pathogens in drinking water include bacteria, viruses, protozoans, and parasites. In addition, microbially produced toxins, such as cyanobacterial toxins, threaten human health. Increases in extreme weather events such as heavy rainfall, floods, severe winter storms, droughts, and heatwaves impact microbial communities and human health. For example, both droughts and floods are associated with increased waterborne diseases. Microbiologically contaminated drinking water can cause diarrhea, cholera, dysentery, typhoid, polio, and other severe diseases. Currently, over 500,000 diarrheal deaths from contaminated drinking are expected each year.

The chemical consequences of climate change can degrade the quality of drinking water. As climate change progresses, sea levels will continue to rise, the area and duration of coastal floods will increase, and the underlying aquifer can become more saline and less oxygenated. This increase in the salinity of coastal drinking water aquifers will increase exposure to sodium ions (Na^+) and can increase the risk of hypertension, preeclampsia, and other adverse health effects. In addition, the salt effect and reduction are chemical processes associated with coastal flooding that can increase the release of many toxic chemical elements from sediments into drinking water and harm human health.

The physical consequences of climate change can degrade the quality of drinking water. For example, intense rainfall caused by climate change can erode soil into streams, rivers, ponds, and lakes. If the eroded soil has available phosphate, perhaps from over-fertilization, cyanobacteria may grow and release potent toxins into the water. Cyanobacteria toxins in drinking water can adversely affect the skin, liver, kidney, and central nervous system.

The authors of this book are pleased to share their research on these topics with you. Thank you for joining us on this journey.

Northfield, VT, USA   Seth H. Frisbie

Contents

About the Editors and Contributors

About the Editors

Azizur R. Molla, Ph.D., M.P.H. is a public health professor with over 20 years of experience in research and teaching. His research is focused on topics such as climate change, drinking water, disease and culture, environmental health, health policy, and community health. He holds both dual master's and Ph.D. degrees in medical anthropology and demography from the Pennsylvania State University, an M.P.H. from the University of North Carolina at Chapel Hill, and a B.S.S. from Dhaka University. Azizur is Professor of Public Health at Grand Valley State University in Michigan, USA. He has authored a book on drinking water and disease and co-authored book chapters and articles based on his public health research. He has conducted studies in the USA, Bangladesh, Ghana, and Haiti. His research and publications focus on diseases caused by water contamination and air pollution, as well as health disparities and public health policy. He has received the Faculty of Distinction Award and served on the steering committee of the National Academies of Sciences, Engineering, and Medicine. He has also designed several community-resilient environmental health projects. Azizur has held various leadership roles in environmental health, including with the American Public Health Association and with the Bangladesh Environment Network.

Seth H. Frisbie, Ph.D. has over 35 years of experience in public health. He earned his B.S. from the University of Massachusetts, his M.S. from Cornell University, and his Ph.D. from Cornell University. He is Professor Emeritus of Chemistry and Biochemistry at Norwich University in Vermont, USA. He is the principal or co-author of numerous articles in refereed journals, chapters in scientific books, and government and United Nations reports. His team has worked on drinking water and public health in Asia, Africa, North America, South America, and Europe. They made the first national-scale maps of arsenic, manganese, lead, nickel, and chromium-affected drinking water in Bangladesh. In addition, they published extensively on the risks of neurodevelopmental impairment due to excess manganese in infant formula and young child nutritional beverages sold in the USA. He has won numerous awards for outstanding research and teaching.

Mohsen Sherif, Ph.D. holds the position of Deputy Vice Chancellor for Academic Affairs (Provost) and Professor of Water Resources at United Arab Emirates University, UAE. Prior to his current appointment, he held the positions of Director of Research and Sponsored Projects, Director of the National Water and Energy Center, Associate Provost for Academic Personnel, Acting Dean of the College of Engineering, and Department Chair—Civil and Environmental Engineering at UAE University. Sherif has published more than 200 journal and conference papers and book chapters and completed several large-scale water projects. He has three patents focusing on renewable energy and low energy desalination systems. Sherif has 35 years of teaching and research experience in groundwater flow and contamination, artificial recharge and management of groundwater systems, hydrology, hydraulics, fluid mechanics, numerical simulation, climate change, and applications of AI and ML in water resources. He is acting as Associate Editor of the *Journal of Hydrology* and the *Journal of Hydrologic Engineering* (ASCE). Sherif is a recipient of several national and international awards. He received two Fulbright Scholarships, and he was also awarded the designation of Diplomate, Water Resources Engineer, American Academy of Water Resources Engineers.

Vijay P. Singh, Ph.D., D.Sc., NAE is Distinguished Professor, Regents Professor, and the inaugural holder of the Caroline and William N. Lehrer Distinguished Chair in Water Engineering at Texas A&M University, USA. He received his B.Sc. & Tech., M.S., Ph.D., and D.Sc. degrees, all in engineering. He is a registered professional engineer, a registered professional hydrologist, a Board-Certified Water Resources Engineer (ASCE-AAWRE), and an Honorary Diplomate of the European Water Resources Association. He is a member of the National Academy of Engineering (NAE), a member of the Canadian Academy of Engineering, Academician of the Georgia Fazisi Academy, a member of the European Academy of Sciences and Arts, and a fellow/member of 11 other international science/engineering academies. He is Distinguished Member of ASCE, Honorary Distinguished Member of IWRA, Distinguished Fellow of AGGS, Honorary Member of AWRA, and Fellow of AGU, IAHR, EWRI, ASCE, IAH, ISAE, IWRS, and IASWC. He has published extensively in the areas of hydrology, irrigation engineering, hydraulics, groundwater, water quality, food–water–energy–environment nexus, water resources, entropy theory, copula theory, and climate change impacts with more than 1688 refereed journal articles, 41 books, 105 edited reference books, 126 book chapters, and 330 conference papers.

Contributors

Martin Agbalenyo Southwestern AHEC, Inc., Shelton, CT, USA

Dalal Alshamsi National Water and Energy Center, United Arab Emirates University, Al Ain, United Arab Emirates

Department of Geosciences, College of Science, United Arab Emirates University, Al Ain, United Arab Emirates

Chad Cox GZA GeoEnvironmental, Inc., Brookline, MA, USA

Kahwa Douoguih William Lasky School of Business, Norwich University, Northfield, VT, USA

Abdel Azim Ebraheem National Water and Energy Center, United Arab Emirates University, Al Ain, United Arab Emirates

Boris Faybishenko Lawrence Berkeley National Laboratory, Berkeley, CA, USA

Seth H. Frisbie Department of Chemistry and Biochemistry, College of Arts and Sciences, Norwich University, Northfield, VT, USA

Oksana Halych Department of Hydropower, Thermal Power and Hydraulic Machines, Institute of Automation, Energy and Water Engineering, National University of Water and Environmental Engineering, Rivne, Rivne Region, Ukraine

Libby Hsu MIT D-Lab, Massachusetts Institute of Technology, Cambridge, MA, USA

Holger Janes Better Life Laboratories, Inc., Tallinn, Estonia

Volodymyr Korniichuk Department of Hydrotechnical Engineering and Hydraulics, Institute of Automation, Energy and Water Engineering, National University of Water and Environmental Engineering, Rivne, Rivne Region, Ukraine

Tara S. Kulkarni David Crawford School of Engineering, Norwich University, Northfield, VT, USA

Rebecca L. Larson School of Public Health, Augusta University, Augusta, GA, USA

William F. Lyons Jr. College of Graduate & Continuing Studies, Norwich University, Northfield, VT, USA

Ellie M. Madson Center for Climate Change and Health Equity, Department of Epidemiology and Environmental Health, School of Public Health and Health Professions, University at Buffalo, Buffalo, NY, USA

Yevgen Matseliuk Water and Sanitation Laboratory of the Department of Water Resources, Kyiv Agrarian University of the National Academy of Agrarian Sciences of Ukraine, Kyiv, Ukraine

Erika J. Mitchell Better Life Laboratories, Inc., East Calais, VT, USA

Azizur R. Molla College of Health Professions, School of Interdisciplinary Health, Public Health Program, Grand Valley State University, Grand Rapids, MI, USA

Susan Murcott D-Lab, Massachusetts Institute of Technology, Cambridge, MA, USA

Oleksandr Nechyporenko National Academy of Agrarian Sciences of Ukraine, National Scientific Centre, Institute of Agrarian Economics, Kyiv, Ukraine

Anuli U. Njoku Department of Public Health, College of Health and Human Services, Southern Connecticut State University, New Haven, CT, USA

Volodymyr Panteleyev Scientific Department, Institute of Water Problems and Land Reclamation of the National Academy of Agrarian Sciences of Ukraine, Kyiv, Ukraine

Fayzul K. Pasha Department of Civil and Geomatics Engineering, Lyles College of Engineering, California State University, Fresno, CA, USA

Afnan Rahman Department of Computational Mathematics, Science and Engineering, College of Natural Science, Michigan State University, East Lansing, MI, USA

Lakshmi Raghu Nagendra Prasad Rentachintala Department of Civil Engineering, Bapatla Engineering College, Bapatla, Andhra Pradesh, India

Mykhailo Romaschenko Institute of Water Problems and Land Reclamation of the National Academy of Agrarian Sciences of Ukraine, Kyiv, Ukraine

Kyiv Agrarian University of the National Academy of Agrarian Sciences of Ukraine, Kyiv, Ukraine

Roman Saydak Department of Agrarian Resources, Kyiv Agrarian University of the National Academy of Agrarian Sciences of Ukraine, Kyiv, Ukraine

Ahmed Sefelnasr National Water and Energy Center, United Arab Emirates University, Al Ain, United Arab Emirates

Ibrahim R. Shams WEB-AIS Project, University of Applied Sciences, Koeln, Germany

Mohsen Sherif Civil and Environmental Engineering Department, United Arab Emirates University, Al Ain, United Arab Emirates

Anatoliy Shevchenko Department of Water Resources, Institute of Water Problems and Land Reclamation of the National Academy of Agrarian Sciences of Ukraine, Kyiv, Ukraine

Stefanie Sobel Ukraine Water Association, Kyiv, Ukraine

Jacqueline Strenio William Lasky School of Business, Norwich University, Northfield, VT, USA

Donald K. Warne Center for Indigenous Health, Johns Hopkins Bloomberg School of Public Health, Johns Hopkins University, Baltimore, MD, USA

Matthew J. Weinburke Independent Public Health Consultant, Mariposa, CA, USA

Mykhailo Yatsuik Institute of Water Problems and Land Reclamation of the National Academy of Agrarian Sciences of Ukraine, Kyiv, Ukraine

Kyana R. L. Young Department of Engineering, Wake Forest University, Winston-Salem, NC, USA

Anwar Zahid Department of Disaster Science and Climate Resilience, University of Dhaka, Dhaka, Bangladesh

Bangladesh Water Development Board, Dhaka, Bangladesh

Part I
Climate Change and Impact on Drinking Water and Public Health

Chapter 1
Climate Change, Drinking Water, and Public Health: An Integrated Review

Azizur R. Molla, Ellie M. Madson, and Seth H. Frisbie

1 Learning Objectives

1. Discuss the microbial impacts of climate change on drinking water.
2. Addresses the chemical aspects of climate change and drinking water quality.
3. Share a public health discourse on climate change and water.

2 Introduction

Climate change, driven by increased greenhouse gas emissions, has become a critical issue affecting the environment and human health. Among its many impacts, climate change poses significant challenges to drinking water resources, with substantial implications for public health. Changes in precipitation patterns, rising temperatures, and more frequent extreme weather events affect water availability and quality, potentially leading to increased health risks. Climate change is now one of

A. R. Molla (✉)
College of Health Professions, School of Interdisciplinary Health, Public Health Program, Grand Valley State University, Grand Rapids, MI, USA
e-mail: mollaziz@gvsu.edu

E. M. Madson
Center for Climate Change and Health Equity, Department of Epidemiology and Environmental Health, School of Public Health and Health Professions, University at Buffalo, Buffalo, NY, USA
e-mail: elliemad@buffalo.edu

S. H. Frisbie
Department of Chemistry and Biochemistry, College of Arts and Sciences, Norwich University, Northfield, VT, USA
e-mail: sfrisbie@norwich.edu

A. R. Molla et al. (eds.), *Climate Change, Drinking Water Security, and Public Health*, https://doi.org/10.1007/978-3-032-12074-8_1

the biggest threats to the environment, and it significantly impacts humans, especially children's health (UNICEF 2024). Such a threat intensifies with water scarcity (Anawar 2013). Water security is essential to avoid waterborne diseases like diarrhea. For example, hot water distribution systems and water-cooling towers for air conditioners have been identified as the source of legionellosis outbreaks among hospital patients (Nelson and Williams 2014).

Microorganisms were the first and only form of life on Earth after it formed. Today, it is estimated that the Earth houses over 1 trillion species of microorganisms and that 99.999% of these taxa remain unknown to humans (Locey and Lennon 2016). Microbes provide the foundation of life on planet Earth, and only about 1400 have been determined to be human pathogens that can potentially lead to negative health outcomes (Balloux and Dorp 2017). While most of the known microbes are bacteria, there are also viruses, fungi, protozoans, algae, archaea, and helminths. The following content will mostly surround known human pathogens, with a small discussion about emerging and re-emerging pathogens and possible cross-species transmission.

The chemical aspect of this chapter is limited to inorganic contaminants in drinking water due to natural processes, such as arsenic (As), manganese (Mn), and uranium (U). Therefore, this study includes the 76 chemical elements and ions in the Earth's crustal rocks (Greenwood and Earnshaw 1984). These 76 elements and ions dissolve from the Earth's crustal rocks into surface and underground drinking water supplies; therefore, these are the 76 chemical contaminants we expect to find in drinking water. These 76 elements are listed in Chap. 9 of this book. This discussion does not include anthropogenic contaminants, such as industrial solvents, petroleum products, per- and polyfluoroalkyl substances, and the like. From a public health perspective, ingesting anthropogenic contaminants in drinking water cannot be justified.

3 Impact of Climate Change on Water Resources

3.1 Climatic Weather Events

About 70% of our planet Earth is covered with water, of which only 3% is freshwater; however, two-thirds of this freshwater is unavailable for use due to glaciers and other factors. As a result, about 1.1 billion, or 13% of world's population lack access to water. It is a fact that climate change affects precipitation patterns, which leads to increased droughts and more intense rainfall or floods. Climate change intensifies drought in the United States and many parts of the world (US EPA 2024a). Meanwhile, extreme rain can lead to flooding, which may overwhelm water treatment facilities and lead to contamination (Johnson and Lee 2024). The risk of flooding has been increasing due to climate change, precisely due to factors like sea level rise, wildfires, and anomalies in precipitation patterns (FEMA 2024).

Climate change causes long-term shifts in temperature and weather patterns; such shifts may be linked to variations in the solar cycle. Since the 1800s, anthropogenic factors have been the main driver of climate change, precisely due to burning fossil fuels like oil, gas, and coal (UN 2024). Greenhouse gases like water vapor, carbon dioxide, methane, and nitrous oxide absorb heat energy and emit it in all directions, keeping the Earth's surface and lower atmosphere warm (The Royal Society 2024).

The global surface temperature has been on the rise since the 1980s and the highest temperature was in 2023 since the recordkeeping started in 1880; our planet is 2.45 °F or 1.36 °C warmer in 2023 (NASA 2024). This trend will lead to more heat-related illnesses across the globe.

3.2 Water-Related Pathogens and Disease Transmission

Water-related diseases are often divided into four different groups: waterborne, water-washed, water-based, and water-related vectors. A waterborne pathogen is any microorganism responsible for negative health implications in humans or animals and is spread via contaminated water through ingestion, inhalation, or contact. Water-washed diseases are diseases that are caused by a lack of water or poor hygiene practices. Trachoma is a well-known example of a water-washed disease that leads to blindness due to lack of sanitation. Water-based diseases are caused by parasites that live in the water for part of their lifecycle and can infect hosts by either contact or ingestion. Water-related vectors result in diseases from vectors that breed in water, such as mosquitos. Water-related pathogens are commonly split into four groups: bacterial, viral, protozoan, and parasitic. A list of these pathogens can be found in Table 1.1. While this list is not comprehensive, the table includes some of the most common pathogens and their respective diseases.

Water-related illnesses often have a high burden of disease and can cause an array of health complications such as diarrhea, vomiting, fever, and dehydration, among many others (Funari et al. 2011). To decrease the overall burden of water-related illnesses, it is key to understand the routes of transmission and how climate change is impacting the transmission dynamics of waterborne pathogens. Ingestion of organisms by drinking contaminated water is one of the primary routes of exposure. While *Legionella* spp. is commonly spread by inhaling contaminated aerosolized water. Other pathogens, such as *Aeromonas* or *Naegleria,* can infect a host by contact. This can include through the mucous membranes or open wounds. Drinking contaminated water is a primary concern, but contamination of water used for bathing or recreational purposes also needs to be acknowledged.

Climate change is leading to increases in extreme weather events such as heavy rainfall, floods, severe winter storms, droughts, and heatwaves. Each of these events impacts the dynamics of microbial communities and how microbes adapt to circumstances. Droughts have been associated with increased waterborne diseases (WHO 2015). Floods have also been associated with increased waterborne and vector-borne diseases (Nijhawan and Howard 2022). Extensive droughts lead to a chain of

Table 1.1 Types of waterborne pathogens

Bacterial Pathogens	Viral Pathogens	Protozoan Pathogens	Parasitic Pathogens
Escherichia coli (E. coli) *Salmonella* spp. *Shigella* spp. *Vibrio cholerae* (causes cholera) *Legionella* spp. (causes Legionnaires' disease) *Campylobacter jejuni* *Pseudomonas aeruginosa* *Leptospira* spp. (causes leptospirosis) *Yersinia enterocolitica* *Aeromonas* spp. *Mycobacterium avium complex* (MAC) *Burkholderia pseudomallei* (causes melioidosis) *Chlamydia trachomatis* (causes trachoma) *Clostridium botulinum* (causes botulism)	Hepatitis A virus Hepatitis E virus Norovirus Rotavirus Adenovirus Enterovirus Astrovirus Sapovirus	*Giardia lamblia* (Giardia intestinalis) *Cryptosporidium parvum* *Entamoeba histolytica* (causes amoebiasis) *Naegleria fowleri* (causes primary amebic meningoencephalitis) *Acanthamoeba* spp. *Cyclospora cayetanensis*	*Schistosoma* spp. (causes schistosomiasis) *Dracunculus medinensis* (Guinea worm) *Toxoplasma gondii* *Strongyloides stercoralis* *Fasciola* spp. (liver flukes) *Necator americanus* (hookworm) *Ascaris lumbricoides* (roundworm) *Trichuris trichiura* (whipworm)

negative impacts. Drought conditions have been associated with increased levels of *E. coli* concentrations and cyanobacteria in water sources (Mosley 2015; Walter et al. 2018). Water quality is impacted in the short term, but droughts also have long-term impacts. Extensive droughts can lead to a decrease in absorbance and recharge of soil. When precipitation does occur, minimal water is absorbed back into the ground, leading to decreases in not only surface water but groundwater levels as well. This precipitation collects trash, debris, and pathogens and continues to transport those items to the remaining water source, leading to increased contamination (WHO 2015). Flooding events work in a similar way: When flooding occurs, it not only overloads sewer systems and contaminates surface waters but also can have long-term impacts on groundwater resources. There can also be widespread transport of microorganisms depending on the severity of the flood.

3.3 The Emergence of New Pathogens and Re-emergence of Diseases

Society has continued to evolve throughout history by creating new technologies and ways to advance human impact on the planet. Due to this behavior, many current drivers are leading to new emerging and re-emerging diseases: increased travel, poverty, behavioral changes, population density, and climate change. The drivers listed above are intertwined but are all associated with climate change. This is when clarification is needed for emerging and re-emerging diseases. Most emerging pathogens are often known to science for long periods of time but then are considered emerging once they are known to cause disease. For example, *Cryptosporidium* was first discovered in 1907, but it was not until 1976 that it was found to cause disease in humans, and it took the next two decades to determine exposure routes, transmission pathways, and pathogenicity (Thompson et al. 2016). A similar trend goes for re-emerging diseases. The disease is known and thought to be under control, and then it adapts.

Cholera, a debatable re-emerging disease, is a key example of a disease that is not only adaptable but is believed to have a worsening impact associated with climate change (Christaki et al. 2020; Polgreen and Polgreen 2017; Rodó et al. 2002). Cholera has been associated with at least seven pandemics in recent history, and the reason for its recurrence is due to the emergence of different strains. The first pandemic spread from the Bay of Bengal in 1817 from *Vibrio cholerae* O1 (Siddique and Cash 2014). Since the 1960s, cholera groups have been oscillating in their infectivity. Once a population is immune to one serogroup, the other re-emerges (WHO 2015). Today, the leading cause of cholera infections is due to drinking contaminated water. This event is often exacerbated during natural disasters, mass migrations, or times of infrastructure crisis, which are all happening with increased frequency.

Increases in temperature and rainfall give many microorganisms, which were not previously concerning, prime opportunities to become harmful infective agents. While most attention goes to bacterial or viral diseases, an area of concern should surround the emergence of fungal pathogens. It is estimated that there are over 6.5 million cases of severe fungal infections every year, with 3.8 million resulting in death (Denning 2024). Many fungal infections that can be transmitted through water are due to opportunistic pathogens, such as aspergillosis or pneumocystis pneumonia. They will not harm most people but are highly impactful for the ill or immunocompromised. There are many fungal infections associated with freshwater, with the most common being fungal keratitis, otitis externa, and tinea pedis (Baumgardner 2017). Fungal pathogens are predicted to increase in abundance and ways of distribution due to an increase in favorable growing temperatures (Nnadi and Carter 2021; Salazar-Hamm and Torres-Cruz 2024). An increase in severe fungal infections is concerning due to there being no vaccines presently available, and most current fungal treatments are designed for topical cases, instead of more severe respiratory infections.

3.4 Algal Blooms and Cyanobacteria

Cyanobacteria, also commonly known as blue-green algae, are naturally occurring toxic organisms that can have deadly implications even in low doses (Chorus and Welker 2021). They thrive in warm, shallow, stagnant waters, such as lakes, ponds, and some slow-moving rivers. Cases of cyanobacteria are usually associated with recreational water use, but growing concern has emerged around contaminated drinking water, especially sources that rely on surface water reservoirs. Cyanobacteria have been associated with a range of health issues, including but not limited to gastrointestinal issues, skin and eye irritation, respiratory issues, neurological impairment, liver and kidney disease, muscular weakness, and death (CDC 2024). The type of illness that is caused by cyanobacteria is highly variable based on exposure route—ingestion, inhalation, or contact—dose, concentration of bacteria, and individual characteristics. Due to warming water temperatures and fluctuating precipitation patterns creating more shallow waterways, the presence of cyanobacteria is increasing.

Most cyanobacteria are not toxic, but whenever algae are present, there is always the risk of toxins. Cyanobacterial toxins are broken down into categories based on their target organs, and the most common are hepatotoxins, neurotoxins, dermatotoxins, and cytotoxins. Examples of each of these toxins can be found in Table 1.2. All these toxins can be found from other sources not related to cyanobacteria, so this list only includes common cyanobacterial toxins. It is important to note that cyanobacteria toxins are species-specific, meaning that certain toxins are a result of specific species and will only grow under certain conditions.

The harms from cyanobacteria have inadvertently caused humans to start considering the One Health implications of certain organisms. This is due to increasing livestock and domestic animals and their contact with various bodies of water. Animals commonly enter or drink from a body of water that a human may be more hesitant to. Cyanobacteria is just as toxic to animals as it is to humans. A 2021 report from the Centers for Disease Control and Prevention (CDC) examined the incidence and severity of illnesses related to harmful algal bloom (HAB) events. This report compiled data from 16 states in the United States and determined that 368 HAB events resulted in 117 human cases and at least 2715 animal cases of illness (Thuneibat et al. 2021). The previous numbers exclude 2000 cases of illness in bats in Washington, as well as reports from many coastal states, resulting in a severe underestimation of cases. The animal cases were a mix of livestock, domestic pets, and wildlife. Of the 2715 reported animal cases, 2489 (92%) of the animals died. The disease burden for cyanobacteria may not be reflected properly in the human population, but it is severe and needs to be addressed with a One Health mindset.

Table 1.2 Common cyanobacterial toxins

Hepatotoxins	Neurotoxins	Dermatotoxins	Cytotoxins
• Microcystins • Nodularin	• Anatoxin-a • Homoanatoxin-a • Saxitoxins • Kalkitoxin	• Lyngbyatoxins • Aplysiatoxins	• Cylindrospermopsin

3.5 Need for One Health-Based Thinking in Relation to Climate Change Impacts

Water is needed to sustain all life on Earth. Water not only fuels humans and animals but also plant life. As the impacts of climate change on water become more apparent, most focus has gone to water availability and accessibility to humans. The downstream impacts of all organisms that need water are often overlooked. This is the same story for microorganisms. Microorganisms that impact humans are often well-understood. Decades of research have been conducted to understand the growing conditions of these organisms, how humans encounter them, how the organisms attack internal body systems, and how to prevent contact from ever happening. The same cannot be said for animal and plant counterparts. For example, *Aeromonas* is a commonly known pathogen that can infect humans via ingestion of contaminated water or food. The specific species known for these illnesses include *A. hydrophila, A. caviae, and A. veronii biovar sobria.* Many additional species of *Aeromonas* have been associated with diseases in fish and other animals, yet the impacts are not thoroughly understood. Humans must monitor microorganisms that have negative impacts on plants and animals, not simply to eliminate spillover and possible zoonotic transmission, but to address planetary health.

Humans also need to address the microbial changes and natural cycle alterations caused by human impact on the environment. For example, the nitrogen cycle is commonly referred to in agriculture when atmospheric nitrogen is converted into a form of nitrogen that can then be used by plants through nitrogen fixation. Humans are impacting the natural nitrogen cycle to increase yields and profit. It has been proven that fertilizers impact concentrations of certain nitrite-oxidizing bacteria such as *Nitrobacter* and *Nitrospira* (Han et al. 2018), leading to higher levels of these bacteria in soil and water.

As issues continue to arise with water scarcity and water quality on a global scale, biodiversity loss and disease emergence will worsen. Humans must look to the upstream causes of these broad impacts and work to make lasting changes to improve the health of planet Earth. A healthy environment leads to a healthy ecosystem for all organisms, from the smallest bacteria to the top animal in the food chain.

4 Water Quality and Public Health

4.1 Microorganism and Chemical Pollutants

The impact of climate change on water quality directly affects public health, particularly through the prevalence of waterborne diseases. Changes in water temperature and quality can increase the incidence of diseases such as cholera, giardiasis, and cryptosporidiosis. For example, higher temperatures can enhance the survival and growth of pathogens in water supplies. Climate-induced changes in water quality can also lead to the presence of harmful chemical contaminants. Increased runoff from intense rainfall can carry pollutants into water supplies, affecting water safety.

Microorganisms play a vital role in human disease, but many chemical pollutants are also a concern. Chemicals such as pesticides and heavy metals may enter drinking water sources, posing risks to human health.

4.2 Chemical Contaminants

This discussion is limited to inorganic chemical contaminants in drinking water due to natural processes, such as arsenic, manganese, and uranium. Therefore, this study includes the 76 chemical elements and ions in the Earth's crustal rocks (Greenwood and Earnshaw 1984). These 76 elements and ions dissolve from the Earth's crustal rocks into surface and underground drinking water supplies; therefore, these are the 76 chemical contaminants we expect to find in drinking water. These 76 elements are listed in Chap. 9 of this book. This discussion does not include anthropogenic contaminants, such as industrial solvents, petroleum products, per- and polyfluoroalkyl substances, and the like. From a public health perspective, ingesting anthropogenic contaminants in drinking water cannot be justified.

4.2.1 The Effect of Increasing Sodium Ions in Drinking Water on Public Health

As climate change progresses, sea levels will continue to rise, the area and duration of coastal floods will increase, and the underlying aquifer will become more saline. This increase in salinity will increase the exposure to sodium ions (Na^+) from drinking water.

Increased exposure to sodium ions in drinking water is associated with hypertension (Khan et al. 2011; Scheelbeek et al. 2016). Relatedly, pregnant women exposed to excess sodium ions in drinking water are not only at increased risk of hypertension but also increased rates of preeclampsia and other adverse pregnancy outcomes (Khan et al. 2011; Scheelbeek et al. 2016). A case study of hypertension, preeclampsia, and other health effects associated with increased salinity and sea level rise in coastal Bangladesh is in Chap. 20 of this book.

4.2.2 The Release of Toxic Ions into Coastal Aquifers by Reduction and by the Salt Effect

As sea levels continue to rise and floods and extreme rainstorms increase in area, duration, and intensity, the underlying aquifer can become more reducing. The resulting floodwaters and saturated soils are a barrier to the diffusion of oxygen gas

($O_{2(g)}$), a powerful oxidizing agent, from the atmosphere into the drinking water aquifer. During flooding, heterotrophic organisms and abiotic processes can deplete the dissolved oxygen (DO; $O_{2(aq)}$) concentration in the floodwater and the drinking well water; this also helps to make a reducing environment. These reducing conditions will increase the release of selected toxic elements, such as arsenic, manganese, and iron (Fe), from sediments and into groundwater (Frisbie et al. 2024). If this groundwater is used for drinking, this increases exposure and is a threat to public health. This process is discussed in detail in Chap. 9 of this book.

In addition, rising sea levels can cause an increase in the salt concentration of the underlying aquifer. In Bangladesh, this has increased the release of arsenic from sediments into drinking well water by the salt effect. The salt effect describes the increase in the solubility of an ionic solid, such as a mineral when it is in a solution of an electrolyte that does not have any ions that are in common with the solid. For example, the solubility of solid barium sulfate ($BaSO_{4(s)}$) increases as the concentration of dissolved potassium nitrate ($K^{+}_{(aq)}$ + $NO_3^{-}{}_{(aq)}$) increases (Eqs. 1.1 and 1.2; Frisbie et al. 2024).

$$\text{Without common ions}: \mathrm{BaSO_{4(s)}} \qquad \mathrm{Ba^{2+}_{(aq)} + SO^{2-}_{4(aq)}} \tag{1.1}$$

$$\text{With common ions}: \mathrm{BaSO_{4(s)} + K^{+}_{(aq)} + NO^{-}_{3(aq)}}$$

$$\mathrm{Ba^{2+}_{(aq)} + SO^{2-}_{4(aq)} + K^{+}_{(aq)} + NO^{-}_{3(aq)}} \tag{1.2}$$

In short, other ions dissolved in water will make an ionic solid more soluble (Frisbie et al. 2024). This process is discussed in detail in Chap. 9 of this book.

4.3 The Effect of Increasing Temperature on the Solubility of Minerals

The effect of rising temperatures on the dissolution of solid minerals into potential drinking water sources is controlled mainly by thermodynamics. More specifically, the change in Gibbs free energy (ΔG) determines if a chemical reaction is thermodynamically favorable or not. If ΔG is negative, the reaction is favored thermodynamically and is expected to spontaneously go from reactants to products. If ΔG is positive, the reaction is not favored thermodynamically and cannot spontaneously go from reactants to products (McMurry and Fay 2004).

The change in Gibbs free energy (ΔG) of a reaction is the difference between the change in enthalpy (ΔH) and the product of the temperature (T) in kelvins (K) and the change in entropy (ΔS). Therefore, $\Delta G = \Delta H—T\Delta S$ (McMurry and Fay 2004).

If a reaction is endothermic, ΔH is positive, increasing the likelihood that ΔG will be positive and that the reaction will not spontaneously go from reactants to products (McMurry and Fay 2004). For example, the ΔH for the dissolution of solid sylvite (potassium chloride, KCl) in water is 17.584 kilojoules per mole (kJ/mol; Kilday 1980) as follows (Eq. 1.3).

$$KCl_{(s)} \quad K^{+}_{(aq)} + Cl^{-}_{(aq)} \quad \Delta H^{0} = 17.584 \frac{kJ}{mol} \tag{1.3}$$

This reaction takes more energy to break the ionic bonds in the solid than is released by the bonds made between the ions and water. Therefore, the system cools as the reaction progresses, reducing the energy available to break additional bonds in the solid, slowing the reaction, and reducing the likelihood that the reaction spontaneously goes from reactants to products (McMurry and Fay 2004).

In contrast, if a reaction is exothermic, ΔH is negative, increasing the likelihood that ΔG will be negative and that the reaction will spontaneously go from reactants to products (McMurry and Fay 2004). For example, the ΔH for the dissolution of solid gypsum (calcium sulfate dihydrate, CaSO4·2H2O) in water is −23.81 kJ/mol (Newman and Wells 1938) as follows (Eq. 1.4).

$$CaSO_4 \bullet 2H_2O_{(s)} \quad Ca^{+2}_{(aq)} + SO^{-2}_{4(aq)} + 2H_2O_{(l)} \quad \Delta H^{0} = -23.81 \frac{kJ}{mol} \tag{1.4}$$

This reaction takes less energy to break the bonds in the solid than is released by the bonds made between the products and solvent. Therefore, the system warms as the reaction progresses, increasing the energy available to break additional bonds in the solid, speeding the reaction, and increasing the likelihood that the reaction spontaneously goes from reactants to products (McMurry and Fay 2004).

In addition to ΔH, the dissolution of solid minerals into potential drinking water sources is controlled by T and ΔS. Again, the change in Gibbs free energy (ΔG) of a reaction is the difference between the change in enthalpy (ΔH) and the product of the temperature (T) in kelvins (K) and the change in entropy (ΔS). Therefore, ΔG = ΔH—TΔS (McMurry and Fay 2004).

The temperature in kelvin is always positive. Dissolving a solid releases its component ions or molecules into solution; this increases randomness and makes ΔS positive. Therefore, for a dissolving solid, a positive T times a positive ΔS makes − TΔS negative. This negative − TΔS contributes to a favorable ΔG. As the temperature increases, this favorable—TΔS contributes to an increasingly favorable ΔG (McMurry and Fay 2004).

In summary, a balance between ΔH and—TΔS determines that the dissolution of a solid mineral into potential drinking water sources is thermodynamically favorable (Table 1.3).

As a practical example, the linear regression equation for the concentration of arsenic in Bangladesh's groundwater versus temperature has a positive slope; however, this 9.31 microgram of arsenic per liter per degree Celsius (μg/L °C) slope is not statistically significant. A possible reason why this expected relationship was not statistically significant is that the range of groundwater temperatures in this

Table 1.3 The criteria for the spontaneous dissolution of a solid into water: ΔG = ΔH—TΔS. (Petrucci 1985)

Case	ΔH	ΔS	ΔG	Result	Example
1	+	+	+ –	Nonspontaneous at low temperatures Spontaneous at high temperatures	$KCl_{(s)} \quad K^{+}_{(aq)} + Cl^{-}_{(aq)}$
2	–	+	–	Spontaneous at all temperatures	$CaSO_4 \bullet 2H_2O_{(s)} \rightleftarrows Ca^{+2}_{(aq)} + SO^{-2}_{4(aq)} + 2H_2O_{(l)}$

study (24.3 °C to 32.7 °C) was too narrow to give a statistically significant increase in the concentration of arsenic with temperature (Frisbie et al. 2024).

4.4 Chronic Health Effects

Historically, chronic diseases have been rising, mainly in the global south. Waterborne diseases like cholera or diarrhea were a concern in the 1880s when there was a cholera outbreak in Mecca, and diarrhea is still one of the health threats (Azizi and Azizi 2010). Climate change causes and impacts chronic health conditions, water quality, and pollution, causing diseases like diarrhea, cryptosporidiosis, campylobacter, leptospirosis, and harmful algal blooms.

Climate change affects human health in multiple ways, such as frequency of illness, duration, and severity. Such effects could be unprecedented or unanticipated and threaten public health (Balbus et al. 2016). Climate change negatively impacts some physiological functions of cardiovascular and kidney diseases. The disease risks will likely increase further due to climate change concerning air pollution, malnutrition, and extreme weather events (Kjellstrom et al. 2010).

The United States Global Change Research Program (2016) states that bacteria, viruses, protozoa, harmful algae, cyanobacteria, and human-produced chemicals cause water-related illnesses. In most parts of the United States, climate change will influence fresh and marine water resources, enhancing people's exposure to water-related contaminants that ultimately cause illness (Fig. 1.1).

In addition to drinking water, climate change will also impact health due to air pollution. According to the United Nations Declaration of cleaner air, it is a universal human right to have access to a healthy environment. However, about 10% of global deaths are caused by air pollution (UNEP 2024). Air pollution causes asthma and cardiovascular disease. The pathways for climate-related exposures and potential non-communicable and chronic disease effects are shown in Fig. 1.2.

Climate change is already causing an increased number of waterborne and vector-borne illnesses, heat-related illnesses, mental health issues, and even community health issues. Multiple factors affect public health and disease, and climate change is one of them.

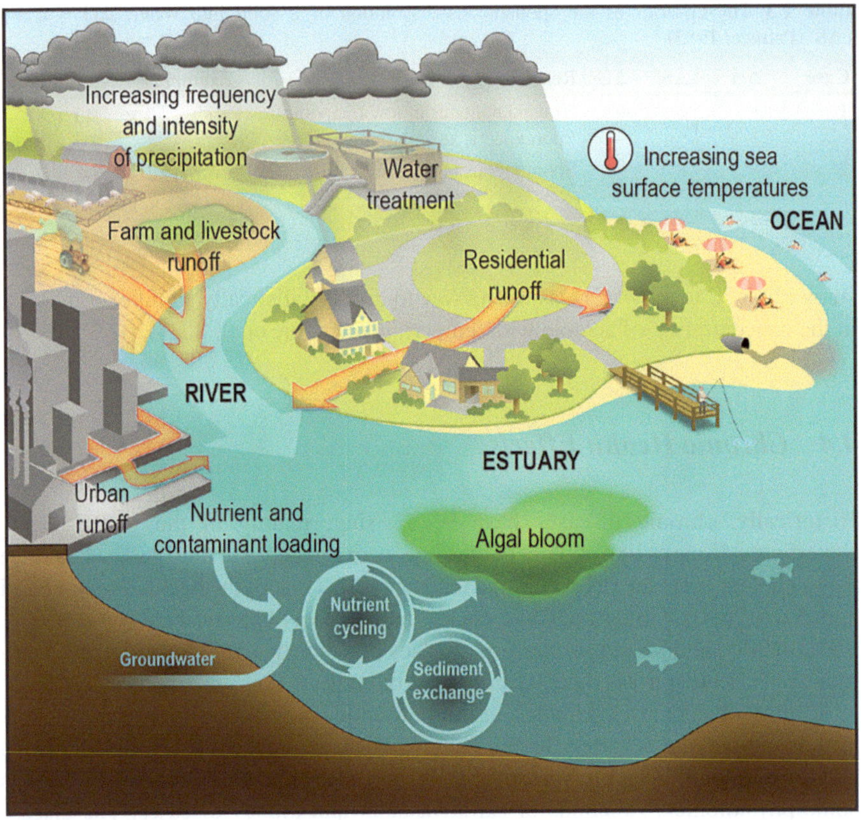

Fig. 1.1 The links between climate change, water quantity and quality, and human exposure to water-related illness. (Reproduced from the United States Global Change Research Program 2016)

5 Mitigation and Adaptation Strategies

5.1 Infrastructure Improvements

Improving water infrastructure is critical for managing the impacts of climate change. Harvesting water with a well-designed infrastructure is vital to protect biodiversity from climate risks like water shortage (Ojija and Nicholaus 2023). Green infrastructure is another option to protect costs. Such infrastructure will reduce erosion and flooding using plants, reefs, and natural barriers. It is proven that living shorelines are more durable than complex structures; this will ultimately help the ecosystem, including improving water quality (US EPA 2024b).

Investment in resilient infrastructure can help mitigate the effects of extreme weather events and ensure continued access to safe drinking water. Bangladesh, a country in the global south, uses technological options like dug wells, tubewells, and alternative water supply technologies, including shallow shrouded tube wells,

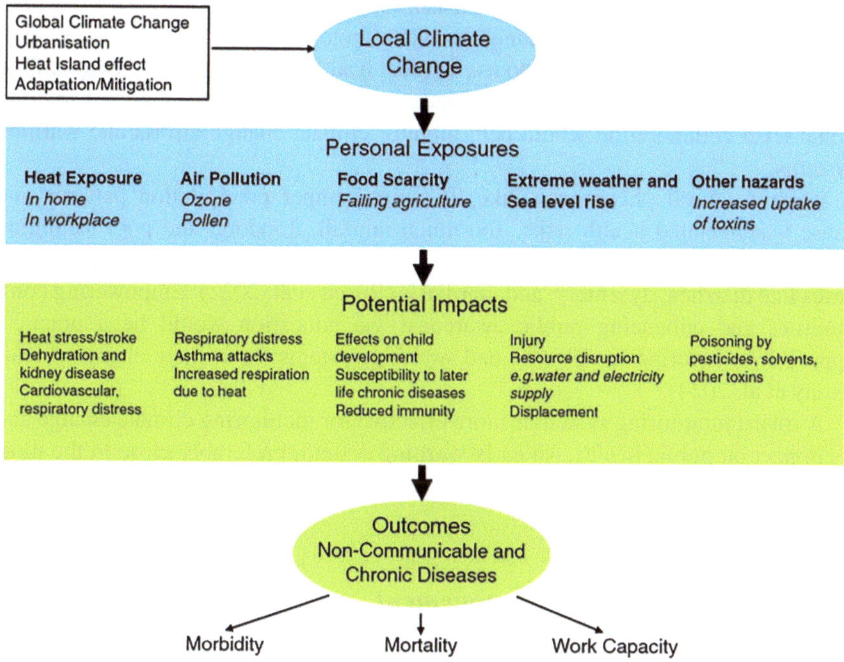

Fig. 1.2 The pathways for climate related exposures and potential non-communicable and chronic disease effects. (Kjellstrom et al. 2010) (Reproduced from Kjellstrom et al., Public health impact of global heating due to climate change: potential effects on chronic non-communicable diseases, *Int J Public Health*, Vol 55, p. 99, Figure 1, 2010, Springer Nature, https://doi.org/10.1007/s00038-009-0090-2)

very shallow shrouded tube wells, pond sand filters, household filters, solar desalination, and rainwater harvesting (Ahmed and Rahman 2000).

Considering the existing healthcare structure in many developing countries, local institutions like schools, mosques, and temples and their leaders can be used to educate people about improving sanitation so that poor sanitation and hygiene do not pollute water and cause diseases like diarrhea. It is vital to have a policy or policies to ensure that trained and certified healthcare personnel provide healthcare services; this will reduce the number of illnesses and public health catastrophes due to the adverse effects of climate change and environmental degradation (Molla 2008).

5.2 *Monitoring and Early Warning Systems*

Climate change is happening faster than we can imagine, while the increased number of illnesses across the globe is impacting public health. Climate change is undoubtedly a global issue; however, we need a more local approach. Individual climate actions concerning "reduce, reuse, recycle" are an easy-to-use mitigation approach. At the same time, technology like remote patient monitoring, wearables, and Bluetooth devices are easy to use to meet healthcare needs (Mian et al. 2024). In addition, community resilience, citizen governance, and national and international level collaboration would help monitor climate change efforts and warning systems.

Certain natural phenomena, like El Nino, hamper precipitation patterns and cause water-related health risks. Too much rainfall, flooding, and poor sanitation infrastructure contaminate water sources, ultimately contributing to waterborne diseases like diarrhea, dysentery, and giardiasis (Rusca et al. 2022). Empowering communities and enhancing public awareness via education would be a practical approach to ensuring monitoring and warning systems for climate change threats (Rony et al. 2024).

A robust monitoring system is more effective for monitoring climate change and its impact on public health. An early warning system, preferably close to the most likely affected community, would be a beneficial strategy and intervention.

5.3 Sustainable Water Management

Water will be the new petroleum, likely a costly life-saving item. As our human civilization moves on, there is no option except to keep our water safe for drinking, but it is also essential to sustain water supply and management. We must maintain our water resources to ensure easy access to safe drinking water. To manage sustainable water, we got to create a legal framework (Li et al. 2020; Wang et al. 2020).

It is also essential to integrate this framework with other stakeholders (Li et al. 2020). Such stakeholder engagement should remove administrative barriers to practical cooperation and coordination (Lestari et al. 2024). To ensure water resilience, it is essential to prevent water pollution, use water rationally and sustainably, eliminate existing wastewater management problems, strengthen cooperation, and encourage educational intervention among people (Ismayılov and Suleymanov 2024). We also need to take into consideration that sustainability relies more on the mind, not money; it demands collectivity and collaboration; it is now to be achieved with an individual effort or using a silo vision instead and a well-integrated comprehensive approach to be more effective.

5.4 Public Health Initiatives

Extremely warm temperatures coupled with anthropogenic climate change are worsening drought conditions. This situation ignites climate injustice in many parts of the world, especially among vulnerable communities (Sum 2024). A community-based approach to testing healthy water can prevent exposure to arsenic and other heavy metals in vulnerable communities. Other interventions would be significant public health strategies in rural states, such as homeowner awareness about the health risks of arsenic and other toxic metals and solving the problem through education, testing, and mitigation (Taylor et al. 2024).

In many parts of the world, it is women who take care of water needs for the household. Interestingly, women have become domestic water managers for collecting and using water (Medeiros 2024). The policymakers must prioritize more research, funding, and technical assistance as very small and private Community Water Systems are driving disproportionate drought effects on minorities like Latino communities. It is crucial to provide drinking water assistance for safe and affordable drinking water (Sum 2024). Community-focused research enhances community-engaged collaborations, initiates reciprocity in the community, and breeds community resilience (Stoll et al. 2024).

6 Conclusion

Climate change poses significant challenges to drinking water resources and public health. The impacts of altered precipitation patterns, saltwater inundation, increased temperatures, and extreme weather events on water quality and availability are substantial. Addressing these challenges requires a comprehensive approach that includes improving infrastructure, enhancing monitoring systems, adopting sustainable practices, and implementing public health initiatives. Continued research and adaptive strategies are essential to safeguard drinking water and protect public health in an increasingly variable climate.

7 Discussion Questions

1. Explain common misconceptions about emerging and re-emerging diseases. Provide examples of each and how climate change is impacting the presence of both.
2. Why does the solubility of an ionic solid, such as a mineral, not always increase with increasing temperature?
3. Explain how climate change is affecting drinking water quality. Discuss this in reference to a specific community or country situation.

References

Ahmed MF, Rahman MM (2000) Water supply and sanitation — rural and low income urban communities. ITN — Bangladesh, Center for Water Supply and Waste Management. Bangladesh University of Engineering and Technology (BUET), Dhaka. https://itn.buet.ac.bd/web/publications/water-supply-and-sanitation-for-rural-and-low-income-urban-communities/

Anawar HM (2013) Impact of climate change on acid mine drainage generation and contaminant transport in water ecosystems of semi-arid and arid mining areas. Phys Chem Earth Parts A/B/C 58–60:13–21. https://doi.org/10.1016/j.pce.2013.04.002

Azizi MH, Azizi F (2010) History of cholera outbreaks in Iran during the (19th) and (20th) centuries. Middle East J Dig Dis 2(1):51–55. https://pmc.ncbi.nlm.nih.gov/articles/PMC4154910/

Balbus J, Crimmins A, Gamble JL, Easterling DR, Kunkel KE, Saha S, Sarofim MC (2016) Chapter 1: introduction: climate change and human health. In: The impacts of climate change on human health in the United States: a scientific assessment. U.S. Global Change Research Program, Washington, DC, pp 25–42. https://doi.org/10.7930/J0VX0DFW

Balloux F, Dorp LV (2017) Q&a: what are pathogens, and what have they done to and for us? BMC Biol 15(91) https://bmcbiol.biomedcentral.com/counter/pdf/10.1186/s12915-017-0433-z.pdf

Baumgardner DJ (2017) Freshwater fungal infections. J Patient-Center Res 4(1):5. https://pmc.ncbi.nlm.nih.gov/articles/PMC6664362/pdf/jpcrr-4.1-32.pdf

Centers for Disease Control and Prevention (CDC) (2024) Symptoms of illnesses caused by freshwater harmful algal blooms. https://www.cdc.gov/harmful-algal-blooms/signs-symptoms/symptoms-freshwater-harmful-algal-blooms.html. Accessed 10 Dec 2023

Chorus I, Welker M (2021) Toxic cyanobacteria in water: a guide to their public health consequences, monitoring and management. Taylor & Francis Group, London. https://doi.org/10.1201/9781003081449

Christaki E, Dimitriou P, Pantavou K, Nikolopoulos GK (2020) The impact of climate change on cholera: a review on the global status and future challenges. Atmosphere 11(5):449. https://www.mdpi.com/2073-4433/11/5/449

Denning DW (2024) Global incidence and mortality of severe fungal disease. Lancet Infect Dis 24(7):e428–e438. https://www.thelancet.com/journals/laninf/article/PIIS1473-3099(23)00692-8/abstract

Federal Emergency Management Agency (FEMA) (2024) Climate change increases risk of flooding. https://www.floodsmart.gov/climate-change. Accessed 20 Dec 2024

Frisbie SH, Mitchell EJ, Molla AR (2024) Sea level rise from climate change is expected to increase the release of arsenic into Bangladesh's drinking well water by reduction and by the salt effect. PLoS One 19(1):e0295172. https://doi.org/10.1371/journal.pone.0295172

Funari E, Kistemann T, Herbst S, Rechenburg A (eds) (2011) Technical guidance on water-related disease surveillance. World Health Organization (WHO), Regional Office for Europe, Copenhagen Ø. https://www.who.int/europe/publications/i/item/9789289002417

Greenwood NN, Earnshaw A (1984) Chemistry of the elements. Pergamon Press, New York, pp 31–33. 1496

Han S, Zeng L, Luo X, Xiong X, Wen S, Wang B, Chen W, Huang Q (2018) Shifts in Nitrobacter- and Nitrospira-like nitrite-oxidizing bacterial communities under long-term fertilization practices. Soil Biol Biochem 124:118–125. https://doi.org/10.1016/j.soilbio.2018.05.033

Ismayılov R, Suleymanov F (2024) Water resilience under climate change in Azerbaijan. Geo J Tourism Geosites 53(2):677–686. https://doi.org/10.30892/gtg.53231-1243

Johnson P, Lee H (2024) Extreme weather and water quality: a critical analysis. Environ Sci Pol 48:23–31. https://doi.org/10.1016/j.envsci.2024.01.002

Khan AE, Ireson A, Kovats S, Mojumder SK, Khusru A, Rahman A, Vineis P (2011) Drinking water salinity and maternal health in coastal Bangladesh: implications of climate change. Environ Health Persp 119(9):1328–1332. https://doi.org/10.1289/ehp.1002804

Kilday MV (1980) The enthalpy of solution of SRM 1655 (KCl) in H2O. J Res Natl Bur Stand 85(6):467–482. https://doi.org/10.6028/jres.085.027

Kjellstrom T, Butler AJ, Lucas RM (2010) Public health impact of global heating due to climate change: potential effects on chronic non-communicable diseases. Int J Public Health 55:97–103. https://doi.org/10.1007/s00038-009-0090-2

Lestari I, Herdiansyah H, Nuraeni (2024) Water governance: urban water conservation as a response to climate change. Int J Conserv Sci 15(1):627–644. https://doi.org/10.36868/IJCS.2024.01.17

Li X, Wang X, Guo H, Ma W (2020) Multi-water resources optimal allocation based on multi-objective uncertain chance-constrained programming model. Water Resour Manag 34:4881–4899. https://doi.org/10.1007/s11269-020-02697-z

Locey KJ, Lennon JT (2016) Scaling laws predict global microbial diversity. Proc Natl Acad Sci 113(21):5970–5975. https://pmc.ncbi.nlm.nih.gov/articles/PMC4889364/pdf/pnas.201521291.pdf

McMurry J, Fay RC (2004) Chemistry, 4th edn. Pearson Education, Upper Saddle River, pp 721–762. 618–623

Medeiros I (2024) Access to safe drinking water: a global women's issue. Student paper and spotlight feature. J N Engl Water Works Assoc March 2024:39–44

Mian MQ, Ayman SA, Ali B, Ullah U (2024) Unraveling the impact of climate change on health dynamics in district swat, Pakistan: a pathway to sustainable resilience. Int J Soc Sci Arch 7(3):201–212. https://www.ijssa.com/index.php/ijssa/article/view/422

Molla AR (2008) Water, sewage, and disease in Bangladesh—a medical anthropology. The Edwin Mellen Press, New York, pp 86–87

Mosley LM (2015) Drought impacts on the water quality of freshwater systems; review and integration. Earth Sci Rev 140:203–214. https://doi.org/10.1016/j.earscirev.2014.11.010

National Aeronautics and Space Administration (NASA) (2024) Global temperature. https://climate.nasa.gov/vital-signs/global-temperature/?intent=121. Accessed 20 Dec 2024

Nelson KE, Williams C (2014) Infectious disease epidemiology: theory and practice, 3rd edn. Jones & Bartlett Learning, Burlington

Newman ES, Wells LS (1938) Heats of hydration and transition of calcium sulfate. J Res Natl Bur Stand 20:825–836. https://nvlpubs.nist.gov/nistpubs/jres/20/jresv20n6p825_A1b.pdf

Nijhawan A, Howard G (2022) Associations between climate variables and water quality in low- and middle-income countries: a scoping review. Water Res 210:117996. https://doi.org/10.1016/j.watres.2021.117996

Nnadi NE, Carter DA (2021) Climate change and the emergence of fungal pathogens. PLoS Pathog 17(4):e1009503. https://doi.org/10.1371/journal.ppat.1009503

Ojija F, Nicholaus R (2023) Impact of climate change on water resources and its implications on biodiversity: a review. East Afr J Environ Natl Resour 6(1):15–27. https://doi.org/10.37284/eajenr.6.1.1063

Petrucci RH (1985) General chemistry, 4th edn. Macmillan Publishing Company, New York, p 484

Polgreen PM, Polgreen EL (2017) 4 - emerging and re-emerging pathogens and diseases, and health consequences of a changing climate. Infect Dis 1:40–48.e2. https://doi.org/10.1016/B978-0-7020-6285-8.00004-6

Rodó X, Pascual M, Fuchs G, Faruque A (2002) ENSO and cholera: a nonstationary link related to climate change? Proc Natl Acad Sci 99(20):12901–12906. https://pmc.ncbi.nlm.nih.gov/articles/PMC130557/pdf/pq2002012901.pdf

Rony MKK, Wahiduzzaman M, Rahman MM, Bala SD (2024) Impact of El Niño on public health and its preparedness measures. Bull Natl Res Cent (Online) 48(1):1–10. https://doi.org/10.1186/s42269-023-01160-4

Rusca M, Gulamussen NJ, Weststrate J, Nguluve EI, Salvador EM, Paron P, Ferrero G (2022) The urban metabolism of waterborne diseases: variegated citizenship, (waste)water flows, and climatic variability in Maputo, Mozambique. Ann Am Assoc Geogr 112(4):1159–1178. https://doi.org/10.1080/24694452.2021.1956875

Salazar-Hamm P, Torres-Cruz TJ (2024) The impact of climate change on human fungal pathogen distribution and disease incidence. Curr Clin Micro Report 11:140–152. https://doi.org/10.1007/s40588-024-00224-x

Scheelbeek PFD, Khan AE, Mojumder S, Elliott P, Vineis P (2016) Drinking water sodium and elevated blood pressure of health pregnant women in salinity-affected coastal areas. Hypertension 68(2):464–470. https://doi.org/10.1161/HYPERTENSIONAHA.116.07743

Siddique A, Cash R (2014) Cholera outbreaks in the classical biotype era. Curr Top Microbiol Immunol 379:1–16. https://doi.org/10.1007/82_2013_361

Stoll N, Macaluso F, Canaly C, James KA (2024) Navigating community-engaged research to understand how drought affects water quality. J Environ Health 87(2):14–18. https://pubs.neha.org/view/900636073/14/

Sum S (2024) Impact of droughts on served drinking water disparities in California, 2007-2020. Am J Public Health 114(9):935–945. https://doi.org/10.2105/AJPH.2024.307758

Taylor A, Garretson A, Bieluch KH, Buckman KL, Lust H, Bailey C, Farrell AE, Jackson BP, Lincoln R, Arneson E, Hall SR, Stanton BA, Disney JE (2024) A mixed methods approach to understanding the public health impact of a school-based citizen science program to reduce arsenic in private well water. Environ Health Perspect 132(8). https://doi.org/10.1289/EHP13421

The Royal Society (2024) The basics of climate change. https://royalsociety.org/news-resources/projects/climate-change-evidence-causes/basics-of-climate-change/. Accessed 20 Dec 2024

Thompson RA, Koh WH, Clode PL (2016) Cryptosporidium—what is it? Food Waterborne Parasitol 4:54–61. https://doi.org/10.1016/j.fawpar.2016.08.004

Thuneibat M, Espinosa R, Vigar M, Jennings T, Jacobi A, Roberts VA (2021) Summary report—One Health Harmful Algal Bloom System (OHHABS), United States, 2021. https://www.cdc.gov/ohhabs/data/summary-report-united-states-2021.html. Accessed 10 Dec 2023

United Nations (UN) (2024) What is climate change? https://www.un.org/en/climatechange/what-is-climate-change. Accessed 20 Dec 2024

United Nations Children's Fund (UNICEF) (2024) Impact of climate change on water availability and quality. World Water Day 2024. https://www.unicef.org/serbia/en/impact-of-climate-change-on-water-availability-and-quality. Accessed 17 Dec 2024

United Nations Environment Programme (UNEP) (2024) After landmark UN declaration, hope for cleaner air. https://www.unep.org/news-and-stories/story/after-landmark-un-declaration-hope-cleaner-air. Accessed 20 Dec 2024

United States Environmental Protection Agency (US EPA) (2024a) Climate adaptation and source water impacts. https://www.epa.gov/arc-x/climate-adaptation-and-source-water-impacts#:~:text=During%20a%20drought%2C%20water%20utilities,need%20for%20drinking%20water%20treatment. Accessed 20 Dec 2024

United States Environmental Protection Agency (US EPA) (2024b) Green infrastructure: protect Coasts. https://www.epa.gov/green-infrastructure/protect-coasts. Accessed 15 Dec 2024

United States Global Change Research Program (2016) The impacts of climate change on human health in the United States: a scientific assessment. In: Crimmins A, Balbus J, Gamble JL, Beard CB, Bell JE, Dodgen D, Eisen RJ, Fann N, Hawkins MD, Herring SC, Jantarasami L, Mills DM, Saha S, Sarofim MC, Trtanj J, Ziska L (eds) U.S. global change research program, Washington, DC. https://doi.org/10.7930/J0R49NQX. 312 pp, Accessed 22 Dec 2024

Walter JM, Lopes FA, Lopes-Ferreira M, Vidal LM, Leomil L, Melo F, De Azevedo GS, Oliveira RM, Medeiros AJ, Melo AS (2018) Occurrence of harmful cyanobacteria in drinking water from a severely drought-impacted semi-arid region. Front Microbiol 9:176. https://pmc.ncbi.nlm.nih.gov/articles/PMC5835534/pdf/fmicb-09-00176.pdf

Wang H, Huang J, Zhou H, Deng C, Fang C (2020) Analysis of sustainable utilization of water resources based on the improved water resources ecological footprint model: a case study of Hubei Province, China. J Environ Manag 262(15):110331. https://doi.org/10.1016/j.jenvman.2020.110331

World Health Organization (WHO) (2015) Climate change and health in the Western Pacific region: synthesis of evidence, profiles of selected countries and policy direction. WHO Regional Office for the Western Pacific, Manila. https://www.who.int/publications/i/item/9789290617372

Chapter 2
Drinking Water Status Around the World and Its Effect on Health

Libby Hsu

1 Learning Objectives

1. Understand the current statistics on water availability and need.
2. Learn the definition of a safely managed drinking water source, and what percentage of people currently have access to water that meets this standard, across regions and levels of development.
3. Explore how climate change is expected to affect these statistics, including in low-income areas.

2 Water on Earth

Planet Earth contains approximately 1.386 billion cubic kilometers of water in solid, liquid, and gaseous states, including that in oceans, ice, lakes, rivers, the ground, and the atmosphere, as well as the water in living organisms (known as biological water). Most of this water is saline, with 96.54% of all water on Earth contained in its oceans, seas, and bays (Oki et al. 1999; USGS 2024).

Estimates of the freshwater distribution on Earth vary, but one study breaks down the world's freshwater distribution according to Table 2.1. This table shows that most freshwater is confined to sources that are not easily accessible to humans, including in ice, glaciers, and the ground. Most humans obtain some or all of their water from rivers, yet rivers contain just 0.006% of Earth's freshwater (Shiklomanov 1994; USGS 2024).

L. Hsu (✉)
MIT D-Lab, Massachusetts Institute of Technology, Cambridge, MA, USA
e-mail: slhsu@mit.edu

A. R. Molla et al. (eds.), *Climate Change, Drinking Water Security, and Public Health*, https://doi.org/10.1007/978-3-032-12074-8_2

Table 2.1 Estimate of global freshwater distribution. (Shiklomanov 1994)

Type of water	Water volume, in thousand km^3	Percent of total storage
Glaciers and permanent snowpacks	24,064	68.70
Fresh underground water	10,530	30.06
Underground ice in permafrost zone	300	0.86
Soil moisture	16.5	0.05
Freshwater lakes	91.0	0.26
Water in swamps	11.47	0.03
Water in river channels	2.12	0.006
Water in atmosphere	12.9	0.04
Total	*35,028*	*100*

Reproduced with permission from Igor A. Shiklomanov, World Water Resources: Assessment and Prediction. In: *Innovation, Resources and Economic Growth.* Table 1, p. 253, 1994, Springer Nature. https://link.springer.com/chapter/10.1007/978-3-642-78855-0_11

The global water cycle circulates the Earth's water through phases (solid, liquid, and gas) and locations over time. Although most water is stored in the locations listed in Table 2.1, processes including precipitation, evaporation, evapotranspiration, and runoff circulate small percentages of this water between storage locations. At a high level, water precipitates from the atmosphere to the Earth's surface and is available to humans when stored as either surface water or groundwater. Evaporation is the process that occurs when water changes from a liquid to a gas, while evapotranspiration is defined more specifically as the return of water to the atmosphere from the surface of the Earth through a combination of: a) the evaporation of water from the soil and b) the transpiration of plants in the soil (Oki et al. 1999; Brown 2000).

3 Human Uses and Demand for Water

Human beings used 3.9 trillion cubic meters (m^3) of water in 2014, and that number has continued to rise annually. 70% of humanity's current freshwater use is for agriculture, while approximately 12% is used in domestic and municipal settings, and 18% is used in industrial settings. Humanity's demand for freshwater has increased by about 1% per year since 1980. This increase is mostly from urban municipal sources, concurrent with the rapid rise in urban populations in that time period (Ritchie and Roser 2024; UNESCO 2024).

Adult human beings cycle approximately 140 liters (L) of water-containing blood through their kidneys daily. Most of this water is reabsorbed in the body, but a person doing moderate work in a temperate climate will lose 3 liters of water per day on average, mostly through urine and sweat. This loss is replaced by drinking water and eating food containing water. Therefore, humans need a minimum of 3 liters of water per day to survive (LeWine 2023; NIH NIDDK 2018).

Of course, humans need water for much more than personal consumption. The quantity of water needed for human activities per day varies widely among cultures and regions, but can be summarized according to short-term, medium-term, and long-term needs. The World Health Organization (WHO) has proposed a hierarchy of human water needs to assess water demand in emergencies or humanitarian crises. Water for drinking and cooking are the most critical needs for human survival in the short term. In the medium term, humans need additional water for tasks like bathing, sanitation, washing clothes, cleaning their surroundings, and growing food. In more stable long-term communities, humans need access to more water to support economic activities and food production, and they may be able to use water for recreational activities or other personal preferences. Generally, longer-term water uses do not require as high a level of treatment as short-term uses (for example, one does not need water treated to drinking standards to apply to agricultural crops) (Sphere Organization 2018).

In situations of water scarcity, short-term needs must be prioritized, and the WHO states that humans need a minimum of 7.5–15 L of water per day for basic survival. Of this, 2.5–3 L are needed for direct water intake, 2–6 L are needed for hygiene practices, and 3–6 L are needed for basic cooking. The exact volume depends on a person's physiology, health, gender, age, ability, culture, religious practices, and other variables. Additional water beyond this baseline may be needed if the person uses a toilet that requires water to function. Many populations around the world directly consume much more than 15 liters of water per person per day, through the addition of water uses beyond these basic needs. More importantly, the water used directly by an individual in their daily life does not represent that person's total water consumption per capita (Sphere 2018; Ritchie and Roser 2024).

To understand human water use, it is necessary to calculate the water that has been used in agricultural and industrial processes for products that an individual uses, and add this volume to the individual's water use. Figure 2.1 shows the per capita water withdrawal per year, including for agricultural, industrial, and municipal uses, around the world. Although data is not available for all countries, the figure shows major differences in water withdrawal among countries and regions. As a comparison, 200 m^3 of water use per year equates to 548 liters per day, which highlights how much water contributes to an individual's daily consumption through the inclusion of food production, consumer products, energy use, and other industrial processes in this calculation. An average person in the United States consumes more than 1000 m^3 of water withdrawal per year, which is equivalent to 2740 L per day (Ritchie and Roser 2024).

Data on total freshwater withdrawal is much more complete. Figure 2.2 shows freshwater withdrawal by country in 2020. Population density, as well as agriculture and industry, contribute to a country's water withdrawals. Three countries—China, India, and the United States—each withdrew more than 300 billion cubic meters of water that year, with India withdrawing the most, at 647.5 billion m^3 (Ritchie and Roser 2024).

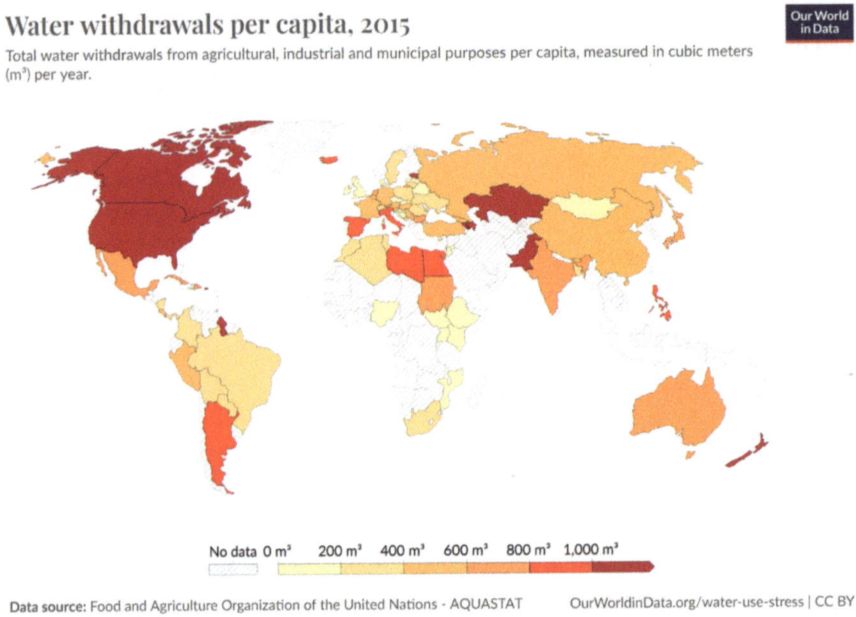

Fig. 2.1 Water withdrawals per capita in 2015. (Ritchie and Roser 2024) (https://ourworldindata.org/water-use-stress, licensed under the terms of the Creative Commons Attribution 4.0 International License (http://creativecommons.org/licenses/by/4.0/))

4 Water Availability

When assessing the availability of water for human and environmental needs, we must consider both the supply of water—including its quantity, quality, and the timing of its availability—and the demand for water, from humans and from the environment in which the water is located. Figure 2.3 provides one summary of the factors that are considered in assessing water supply and demand (USGS 2024).

The quantity of water available is assessed by calculating the water available in different steps of the water cycle. Hydrologists have methods for estimating the volume of water stored in locations such as lakes, rivers, reservoirs, ice, and in the ground—both within soil and in underground aquifers. They also have to estimate the water that enters and exits these sources as it flows through the water cycle, such as through the process of evapotranspiration from plants, when ice melts, and when water from rivers or the ground flows into the ocean.

Understanding the flux between water storage locations in this way is important because water is easier to obtain in some locations than others. As stated previously, many humans obtain water from surface locations such as rivers, where contact directly with the water source is easy to facilitate. Groundwater must be accessed through wells or with mechanical assistance from pumps, and rainwater can be collected with help from human-made collection and storage mechanisms.

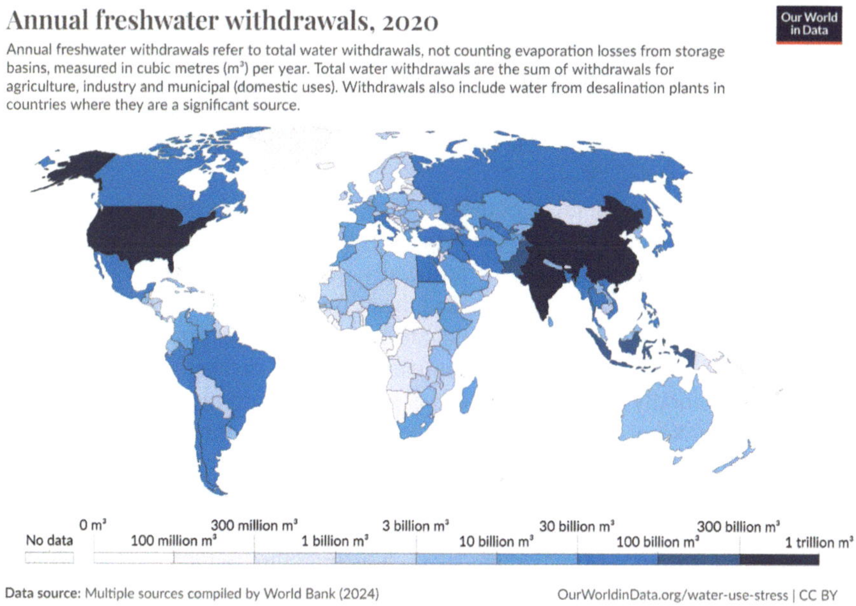

Fig. 2.2 Freshwater withdrawals by country in 2020. (Ritchie and Roser 2024) (https://ourworldindata.org/water-use-stress, licensed under the terms of the Creative Commons Attribution 4.0 International License (http://creativecommons.org/licenses/by/4.0/))

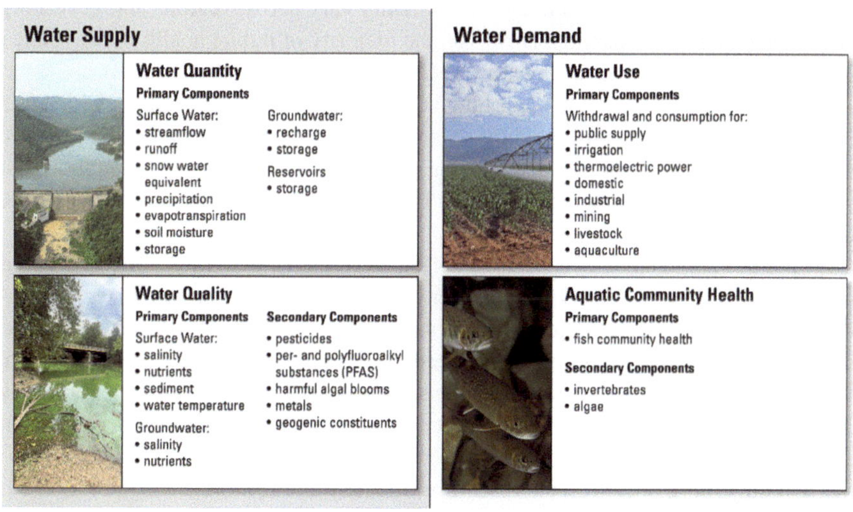

Fig. 2.3 Variables used by the United States Geological Service to assess water availability, which is driven by both supply and demand. (USGS 2023)

Technological progress has made it possible to obtain water from additional sources, such as the ocean and other saline bodies of water (through the use of desalination technology) or the air (through fog collection technology)—but these processes drive up complexity and cost, and can be difficult to scale. Water can also be recycled for repeated use, to eliminate the need to obtain additional water from the surrounding environment.

Accessing water is one piece of the puzzle. Another piece is the condition it is in once it has been obtained. Water can vary in its biological, chemical, and physical contents, which can be naturally or artificially occurring. The range of potential contaminants and their origins is vast, and their effects extend from inconvenient to fatal. Biological contaminants can include bacteria, viruses, protozoa, and helminths, among others. Chemical contaminants may be organic or inorganic, and range from naturally occurring—like the presence of arsenic, a carcinogen, in groundwater in Nepal and Bangladesh—to the byproducts of human activity, such as fertilizers in agricultural runoff—which can lead to excessive nitrates in water, causing conditions such as "blue baby syndrome," the evocative lay term for methemoglobinemia. Chemical components of water also include the salinity of the water, as well as the presence of radionuclides such as uranium, radium, and radon. The physical properties of water often have more aesthetic (also known as secondary) implications for human consumption, such as when sediments in the water affect its appearance, odor, or taste (WHO 2022).

The biological, chemical, and physical properties of water are too numerous to assess exhaustively in any location, and must be prioritized to minimize public health risk. Standards for drinking water quality vary across regions and governing bodies, and change frequently. However, the majority of public health issues related to water are microbiological in nature (Mihelcic et al. 2009; WHO 2022).

Finally, the availability of water may be affected by the time frame in which it is used. The water cycle is affected by seasonal temperature and weather conditions. A water source may not be adequate for a population consistently throughout the year based on these variables. As we will see, climate change is also affecting the timing of water availability.

Water demand is assessed by considering the types of activities humans want to use it for and the needs of the aquatic environment from which the water is withdrawn. Water is used by humans for resource extraction, energy generation, industrial processes, and agriculture. It also needs to be stored for domestic consumption, which may be on a regional, municipal, neighborhood, or household scale. Not only does a population need a certain quantity of water to be available in principle—it also needs that water to be available at the time it is desired. This is why municipalities and governing bodies create reservoir infrastructure to store water, so it is available on demand. Despite this planning, water may not be available in sufficient volume at times of peak human demand or in extreme climate conditions.

Lastly, human society must consider aquatic environmental needs in planning its water extraction. Human activity, including the extraction of water from aquatic environments, changes the quality of water in these environments. This affects the health and sustainability of aquatic environments, and can also affect the properties of the water, which may in turn change the processes that must be used to treat the water for human consumption in the future. Changes to the water cycle affect all components of Earth's ecosystems, including humans (USGS 2024).

5 Access to Safely Managed Drinking Water

The impact of clean water on public health is clear. Water treatment technology had a major impact on health in United States cities in the early twentieth century. Typhoid cases decreased sharply upon the introduction of centralized filtration and chlorination infrastructure in Baltimore, Chicago, Cincinnati, Cleveland, Detroit, and Jersey City between 1900 and 1935 (Cutler and Miller 2005).

5.1 Definition of Safely Managed Drinking Water

Definitions of water fit for human consumption have evolved over the decades, and today, the World Health Organization classifies drinking water sources according to its definition of "safely managed drinking water." The World Health Organization/United Nations International Children's Emergency Fund (WHO/UNICEF) Joint Monitoring Programme for Water Supply, Sanitation, and Hygiene (JMP) is a monitoring organization "affiliated with UN-WATER and is responsible for global monitoring of progress toward Sustainable Development Goal 6 (SDG6) targets for universal access to safe and affordable drinking water and adequate and equitable sanitation and hygiene services." Thus, understanding the JMP's definitions and methods is critical to engaging with water availability statistics in the twenty-first century (WHO and UNICEF 2021).

The JMP's "service ladder" is shown in Fig. 2.4 and is used to define levels of water access that can be compared across time and countries. Five service levels are specified, with the highest level, safely managed, required to meet the global indicator for SDG Target 6.1: "By 2030, achieve universal and equitable access to safe and affordable drinking water for all" (UN 2024; WHO and UNICEF 2021).

Safely managed drinking water meets supply standards for quality, quantity, and timing: It must be accessible in the household where it is being used—either indoors or outdoors; it must be accessible at the time the user desires it—if not continuously, then for at least 12 h per day or 4 days per week; and it must meet quality standards that are concentrated on the absence of fecal contamination (assessed by testing for *E. coli*), arsenic, and chloride. Because it would be difficult to test water for every possible contaminant, the JMP has identified these significant indicators of contamination to assess whether water is safely managed. The drinking water must come from an "improved source," defined as piped water, a borehole, a protected dug well, protected springs, rainwater, or packaged water that has been tested for quality (WHO and UNICEF 2021).

SERVICE LEVEL	DEFINITION
SAFELY MANAGED	Drinking water from an improved source that is accessible on premises, available when needed and free from faecal and priority chemical contamination
BASIC	Drinking water from an improved source, provided collection time is not more than 30 minutes for a round trip, including queuing
LIMITED	Drinking water from an improved source, for which collection time exceeds 30 minutes for a round trip, including queuing
UNIMPROVED	Drinking water from an unprotected dug well or unprotected spring
SURFACE WATER	Drinking water directly from a river, dam, lake, pond, stream, canal or irrigation canal

Fig. 2.4 The Joint Monitoring Programme for Water Supply, Sanitation, and Hygiene (JMP) service ladder for drinking water availability. (WHO and UNICEF 2021) (Reproduced from WHO and UNICEF 2021, Figure 25, p. 28. https://data.unicef.org/wp-content/uploads/2022/01/jmp-2021-wash-households_3.pdf, licensed under the terms of the Creative Commons Attribution-NonCommercial-ShareAlike 3.0 IGO (CC BY-NC-SA 3.0 IGO) licence (https://creativecommons.org/licenses/by-nc-sa/3.0/igo). Used with permission)

The next two rungs on the ladder, basic and limited, have the same standards for improved water sources as the safely managed rung. However, they differ in the location and timing of water availability. An improved drinking water source meets the "basic" standard if a user can collect water from it using no more than 30 min of travel and wait time. Similarly, the water source is classified as "limited" if the travel and wait time add up to more than 30 min. These classifications reflect the time and physical labor costs of not having access to an improved water source on the premises of one's own household (WHO and UNICEF 2021).

The final two rungs on the ladder, unimproved and surface water, are used to classify unimproved water sources that do not meet drinking water quality standards. Unimproved water sources, which comprise unprotected wells and springs, may have naturally higher water quality, but still do not meet the drinking water standards of improved water sources. Protected wells or springs may have coverings that prevent humans, non-aquatic animals and insects, surface plant matter, and other natural or anthropogenic contaminants from entering the water source, while unprotected sources do not have such mechanisms. It should be noted that a body of water used by humans to bathe or wash clothing or cooking implements is not a safely managed drinking water source. Surface water is used to define any other completely unprotected water source, such as a river or lake, and cannot have any particular water quality standards applied to it (WHO and UNICEF 2021).

Fecal contamination is a major source of illness among populations that do not have access to safely managed drinking water. Figure 2.5 shows the results of water quality testing performed by the JMP in 33 countries, and indicates

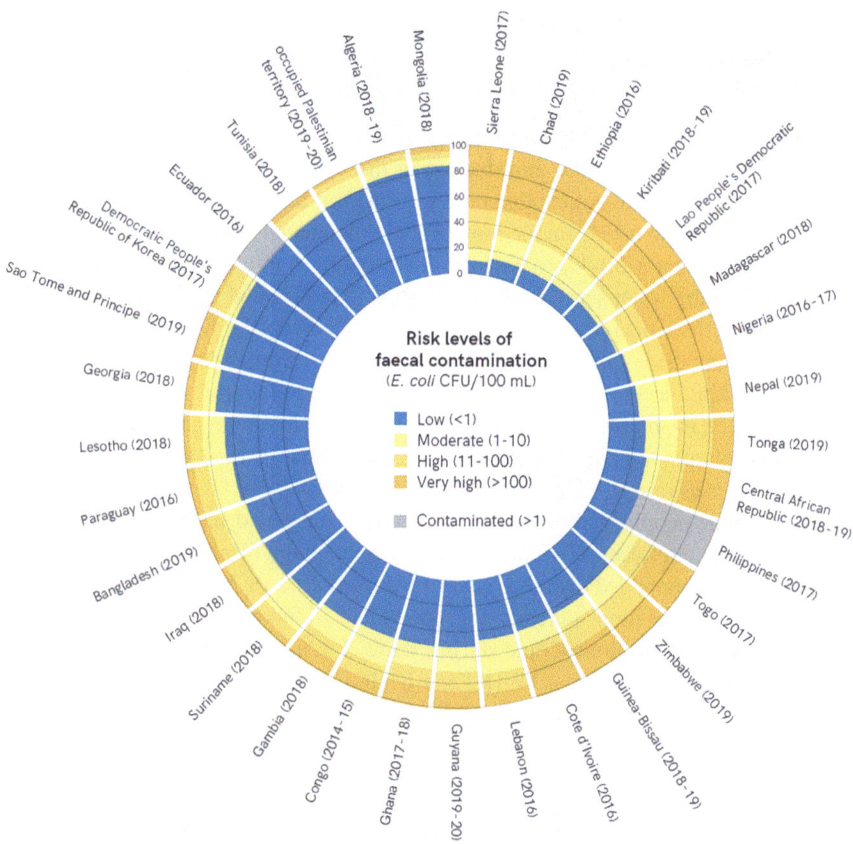

Fig. 2.5 Percentage risk of fecal contamination in drinking water sources measured in 33 countries, from selected surveys 2014–2020. (WHO and UNICEF 2021) (Reproduced from WHO and UNICEF 2021, Figure 37, p. 39. https://data.unicef.org/wp-content/uploads/2022/01/jmp-2021-wash-households_3.pdf, licensed under the terms of the Creative Commons Attribution-NonCommercial-ShareAlike 3.0 IGO (CC BY-NC-SA 3.0 IGO) licence (https://creativecommons.org/licenses/by-nc-sa/3.0/igo). Used with permission.)

high exposure to water contamination from human feces in many regions of the world. These results highlight the interconnected nature of water, sanitation, and hygiene in public health, and emphasize the importance of all six targets contained within SDG6. Safely managed drinking water cannot be realized worldwide without also addressing the need for access to safely managed sanitation systems and hygiene (Target 6.2), as well as wastewater treatment and recycling (Target 6.3), increased water-use efficiency (Target 6.4), and international water resource stewardship (Targets 6.5 and 6.6) (UN 2024; WHO and UNICEF 2021).

5.2 *Progress Toward Access to Safely Managed Drinking Water*

The WHO/UNICEF Joint Monitoring Programme has assessed access to safely managed drinking water since the introduction of the Sustainable Development Goals in 2015, and its publications include some information about improvements in access since 2000, thanks to data from the previously implemented Millennium Development Goals. SDG Indicator 6.1.1 is defined as the proportion of a population using safely managed drinking water services. It is important to realize that the availability of data varies around the world and can be difficult to collect. As the figures in this section reveal, the conclusions that can be made about water availability to meet Indicator 6.1.1 are lacking access to key data in some parts of the world. Between 2015 and 2020, however, the number of countries with such data available increased from 96 to 138, a major improvement. Figure 2.6 shows the results for Indicator 6.1.1 measured in these countries in 2020 (UN 2024; WHO and UNICEF 2021).

Two billion people gained access to safely managed drinking water between 2000 and 2020. Between 2015 and 2020, the percentage of global population with access to safely managed drinking water rose from 70% to 74%. However, because of concurrent population increase, the number of individuals with access actually *decreased* by 193 million people during these 5 years, and this burden has been disproportionate in rural areas (WHO and UNICEF 2021).

Thus, two billion people remain without full access to safe drinking water as of 2020. Of these, 1.2 billion have access to basic services (an improved water source requiring less than 30 min to obtain from), 282 million have limited services (requiring more than 30 min to obtain from), 367 million use unimproved services, and 122

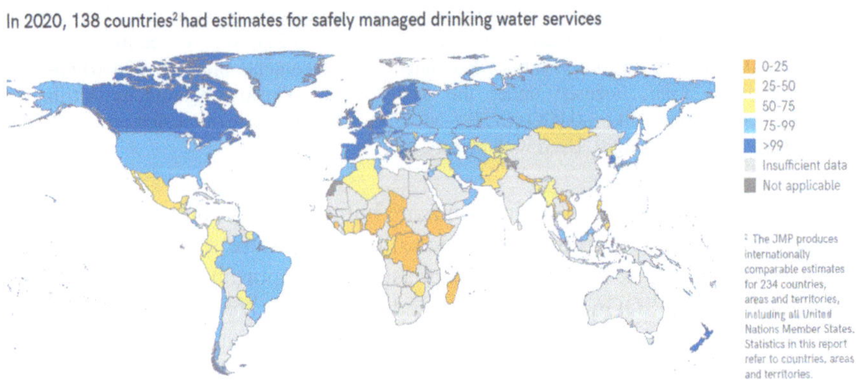

Fig. 2.6 Percentage of population using safely managed drinking water services in 2020, as defined by SDG Indicator 6.1.1. (WHO and UNICEF 2021) (Reproduced from WHO and UNICEF 2021, Figure 4, p. 8. https://data.unicef.org/wp-content/uploads/2022/01/jmp-2021-wash-households_3.pdf, licensed under the terms of the Creative Commons Attribution-NonCommercial-ShareAlike 3.0 IGO (CC BY-NC-SA 3.0 IGO) licence (https://creativecommons.org/licenses/by-nc-sa/3.0/igo). Used with permission.)

million drink surface water. Put another way, just under 1.5 billion of these individuals *do* have access to water that has been treated to international standards, albeit not in their own homes; while over 500 million do not have access to treated drinking water in any form (WHO and UNICEF 2021).

The time and labor burdens of procuring water outside the household, whether treated or not, should not be underestimated. Women and girls over age 15 bear the responsibility for procuring water in 70% of households without access to water on the premises, and girls under 15 are nearly twice as likely to have water-collecting responsibilities as boys under 15. This task wastes time that could be used for education, work, and rest, and the physical burden of carrying water manually has negative health effects. UNICEF estimated in 2016 that women and girls spend 200 million hours a day collecting water (UNICEF 2016; WHO and UNICEF 2023).

Of the 500 million people without access to treated water, over one million of them die from water-related diseases per year, and death rates related to untreated water are much higher in low-income countries. Figure 2.7 shows these death rates by country in 2021, clearly indicating higher rates in many sub-Saharan African and Asian countries (Ritchie et al. 2024).

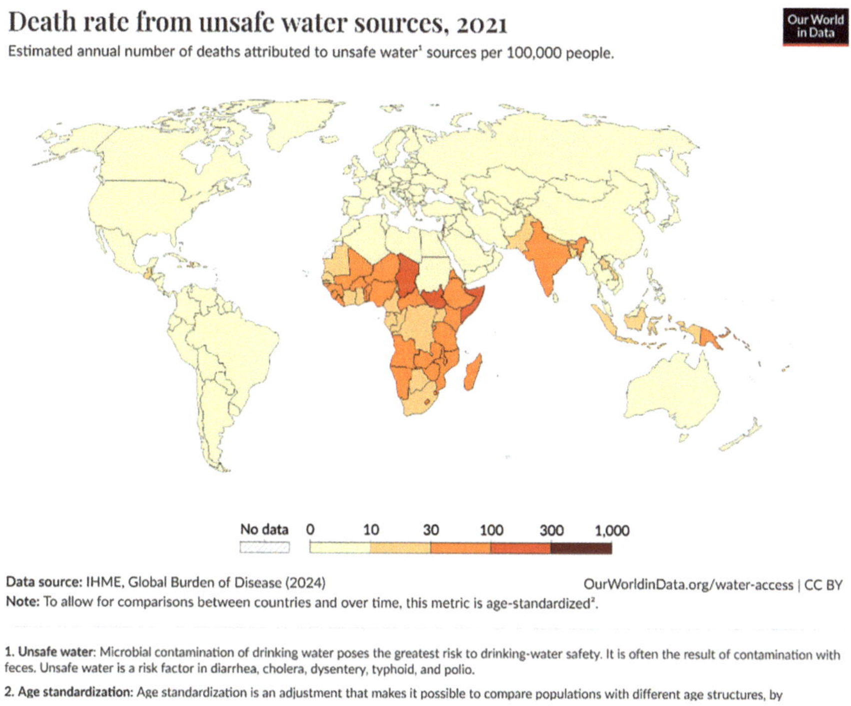

Fig. 2.7 Death rate by country from diseases related to the consumption of untreated water in 2021. (Ritchie et al. 2024) (https://ourworldindata.org/clean-water, licensed under the terms of the Creative Commons Attribution 4.0 International License (http://creativecommons.org/licenses/by/4.0/))

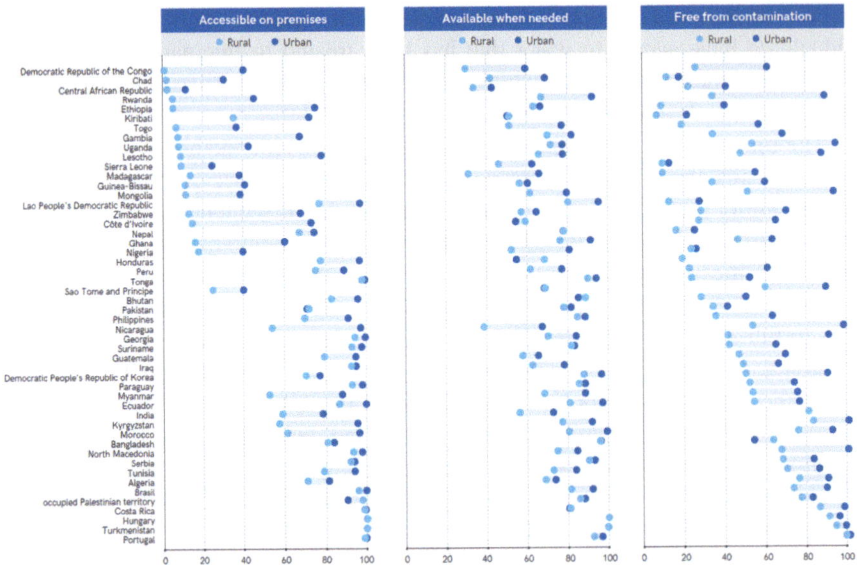

Fig. 2.8 Percentage of rural and urban populations using improved sources available on premises, available when needed, and free from contamination, in 50 countries in 2020. (WHO and UNICEF 2021) (Reproduced from WHO and UNICEF 2021, Figure 33, p. 35. https://data.unicef.org/wp-content/uploads/2022/01/jmp-2021-wash-households_3.pdf, licensed under the terms of the Creative Commons Attribution-NonCommercial-ShareAlike 3.0 IGO (CC BY-NC-SA 3.0 IGO) licence (https://creativecommons.org/licenses/by-nc-sa/3.0/igo). Used with permission.)

There can be a significant difference in access to safely managed water between urban and rural areas in the same country, highlighting the fact that water infrastructure may be more restricted to urban areas. Some countries have begun disaggregating their data between urban and rural areas. Figure 2.8 shows the gaps between rural and urban access in 50 countries that make such data available.

The JMP stated in 2021 that, in order to meet SDG Target 6.1 by 2030—universal access to safely managed drinking water—humanity would need to implement these services at four times the rate it did from 2015 to 2020. In contrast, if the current rate of implementation continues, there will still be 1.6 billion people without adequate access in 2030. The data makes clear that this disparity will continue to disproportionately affect poorer, more rural people around the world (WHO and UNICEF 2021).

6 Effects of Climate Change on Water Access

UNICEF has developed an Extreme Water Vulnerability Index to help assess water scarcity risk for populations around the world. The index combines data on physical water scarcity—including water stress, time variability, drought, and lowering of groundwater tables—with data from the 2021 JMP report on access to drinking water, to understand where humans live with extreme water vulnerability. The index concludes that at least 1 billion people live in areas of high water vulnerability and at least 489 million people live in areas of extreme water vulnerability. Much of this population is concentrated in Africa and Asia (WHO and UNICEF 2021).

Data from the Food and Agriculture Organization of the United Nations complements the Extreme Water Vulnerability Index by highlighting the stress on freshwater withdrawals by country, as shown in Fig. 2.9. A country's internal freshwater resources are considered to be under greater stress when that country is drawing more of its freshwater from non-renewable resources, such as underground aquifers, or obtaining freshwater through resource-intensive means such as desalination. This definition of water stress was high or critical in many Middle Eastern and North African countries in 2020. Figure 2.10 presents additional data on the decrease over time in renewable water resources in several high-population countries, and in the world as a whole (Ritchie and Roser 2024).

Climate change, in addition to human activity, will continue to increase strain on renewable water resources and availability of freshwater because its effects intensify the global water cycle, contributing to more extreme weather events, including floods and droughts. Between 2002 and 2021, 1.4 billion people were affected by drought around the world, and the United Nations Educational, Scientific and Cultural Organization (UNESCO) has estimated that about half of the world's population is now affected by water scarcity at some point each year (UNESCO 2024).

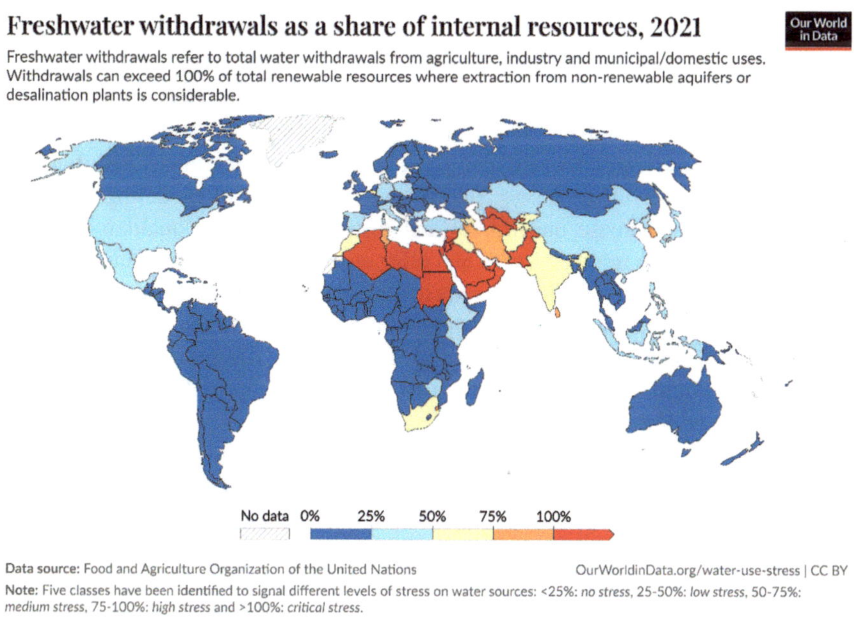

Fig. 2.9 Stress on freshwater withdrawal by country in 2021. (Ritchie and Roser 2024) (https://ourworldindata.org/water-use-stress, licensed under the terms of the Creative Commons Attribution 4.0 International License (http://creativecommons.org/licenses/by/4.0/))

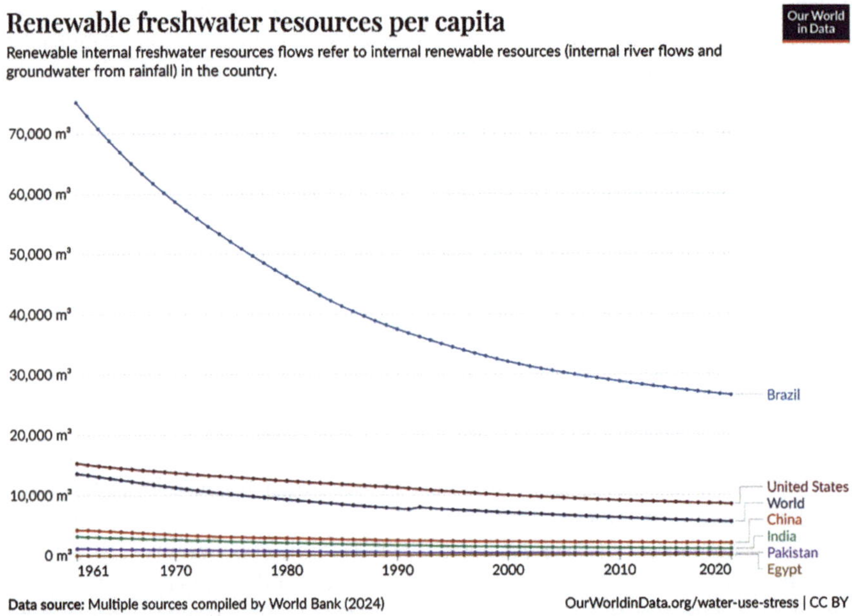

Fig. 2.10 Renewable freshwater resources per capita in selected countries and worldwide. (Ritchie and Roser 2024) (https://ourworldindata.org/water-use-stress, licensed under the terms of the Creative Commons Attribution 4.0 International License (http://creativecommons.org/licenses/by/4.0/))

In its 2014 report, the Intergovernmental Panel on Climate Change (IPCC) predicted severe decreases of 10–30% in surface water runoff in parts of the world that were already experiencing water scarcity. The IPCC also acknowledged that such predictions are difficult to assess, due to the complexity of modeling the water cycle, which can increase uncertainty and restrict analysis to qualitative assessments. Smaller, more refined models can offer more clarity for a particular region of the world, but often only indicate trends toward greater water scarcity and extreme events. For example, a study of India noted major changes in temperature and precipitation from 2000 to 2015 compared to the previous 100 years, suggesting that these trends would continue as a result of climate change (IPCC 2014; Garrote 2017; Goyal and Surampalli 2018).

Africa is particularly vulnerable to the effects of climate change, including increased water scarcity. Analysis of weather patterns in Southern Africa has indicated that increased events such as droughts, cyclones, and floods can be attributed to climate change and will continue to worsen. Increased temperature leads to more evaporation and rainfall, and changing patterns in rainfall and groundwater storage; and temperature increases have been more extreme in the southern hemisphere on the African continent. This will have implications for agriculture as well as household water availability as the changes to the water cycle become more noticeable and extreme, which in turn will continue to affect the health of ecosystems in which humans live (Kusangaya et al. 2014; Garrote 2017).

The impacts of climate change on water resources will require water supply and demand management, particularly in the context of agricultural irrigation. Water users will have to adapt to increased restrictions on the time and location of water availability, recycle more water for additional uses, and practice greater efficiency in their use of water. Governments maintaining water infrastructure will have to promote these practices, reduce leakage in water distribution systems, and adopt new technologies to provide additional water supply, such as through desalination (Garrote 2017).

7 Conclusions

Earth's water resources are available to humans in many forms, but freshwater makes up only a fraction of that water. The water cycle is a system that moves water between the atmosphere, oceans, surface, and ground in gas, liquid, and solid forms, affecting where, when, and how easily humans can access freshwater for their needs.

Water availability is affected by supply and demand, and humans need both a minimum amount of freshwater to survive as well as additional freshwater for a variety of activities, including household, agricultural, and industrial use. The water used per capita varies greatly across geographic regions and economies, and has contributed to water scarcity in some parts of the world.

Clean water is essential for survival, health, and economic activity, and the WHO has provided a definition of "safely managed drinking water" that is being used in the Sustainable Development Goals, which have a target of achieving universal access to safely managed drinking water by 2030. There is currently a major divide between urban and rural access to safely managed drinking water around the world. Water scarcity is being exacerbated by climate change, which affects the water cycle and contributes to changing weather patterns that include more extreme events like flooding and drought. However, the effect of climate change on water resources is difficult to model, which limits the predictions currently available.

8 Discussion Questions

1. Explain how water is distributed across Earth's surface and how much freshwater is available to humans.
2. Discuss how much water humans need and for what purposes.
3. Define safely managed drinking water and discuss why it is important.
4. Explain how climate change is affecting access to safely managed drinking water around the world.

References

Brown P (2000) Basics of evaporation and evapotranspiration. University of Arizona, Tucson. https://repository.arizona.edu/handle/10150/146968

Cutler D, Miller G (2005) The role of public health improvements in health advances: the twentieth-century United States. Demogr 42(1):1–22. https://doi.org/10.1353/dem.2005.0002

Garrote L (2017) Managing water resources to adapt to climate change: facing uncertainty and scarcity in a changing context. Water Resour Manag 31:2951–2963. https://doi.org/10.1007/s11269-017-1714-6

Goyal MK, Surampalli RY (2018) Impact of climate change on water resources in India. J Environ Eng 144(7). https://doi.org/10.1061/(ASCE)EE.1943-7870.0001394

Intergovernmental Panel on Climate Change (IPCC) (2014) Climate change 2014: impacts, adaptation, and vulnerability. Part a: global and sectoral aspects. In: Contribution of working group II to the fifth assessment report of the intergovernmental panel on climate change. Cambridge University Press, Cambridge and New York

Kusangaya S, Warburton ML, Archer van Garderen E, Jewitt GPW (2014) Impacts of climate change on water resources in southern Africa: a review. Phys Chem Earth 67-69:47–54. https://doi.org/10.1016/j.pce.2013.09.014

LeWine HE (2023) How much water should you drink? Harvard Health Publishing. https://www.health.harvard.edu/staying-healthy/how-much-water-should-you-drink. Accessed 17 Oct 2024

Mihelcic JR, Fry LM, Myre EA et al (2009) Field guide to environmental engineering for development workers. ASCE Press, Reston, pp 14–27

National Institutes of Health (NIH) National Institute of Diabetes and Digestive and Kidney Functions (2018) Your kidneys & how they work. NIH. https://www.niddk.nih.gov/health-information/kidney-disease/kidneys-how-they-work. Accessed 17 Oct 2024

Oki T, Entekhabi D, Harrold TI (1999) The global water cycle. Global Energy and Water Cycles 10(27)

Ritchie H, Roser M (2024) Water use and stress. Our world in data. https://ourworldindata.org/water-use-stress. Accessed 10 Oct 2024

Ritchie H, Spooner F, Roser M (2024) Clean water. Our world in data. https://ourworldindata.org/clean-water. Accessed 17 Oct 2024

Shiklomanov I (1994) World water resources: assessment and prediction. Innovation, resources and economic growth. Springer Berlin Heidelberg, Berlin, Heidelberg

Sphere Association (2018) The sphere handbook: humanitarian charter and minimum standards in humanitarian response, 4th edn. Sphere. www.spherestandards.org/handbook. Accessed 10 Oct 2024

United Nations Department of Economic and Social Affairs (UN DESA) (2024) Goal 6. UN. https://sdgs.un.org/goals/goal6#targets_and_indicators. Accessed 17 Oct 2024

United Nations Educational, Scientific and Cultural Organization (UNESCO) (2024) The United Nations world water development report 2024: water for prosperity and peace. UNESCO. https://unesdoc.unesco.org/ark:/48223/pf0000388948. Accessed 10 Oct 2024

United Nations International Children's Emergency Fund (UNICEF) (2016) Collecting water is often a colossal waste of time for women and girls. UNICEF. https://www.unicef.org/press-releases/unicef-collecting-water-often-colossal-waste-time-women-and-girls. Accessed 17 Oct 2024

United States Geological Survey (USGS) (2024) How much water is there on, in, and above the Earth? https://www.usgs.gov/special-topics/water-science-school/science/how-much-water-there-earth. Accessed 10 Oct 2024

United States Geological Survey (USGS) Water Resources Mission Area (2023) IWAA water availability components. https://www.usgs.gov/media/images/iwaa-water-availability-components-0. Accessed 10 Oct 2024

World Health Organization (WHO) (2022) Guidelines for drinking-water quality, 4th edn. WHO, Geneva

World Health Organization, United Nations Children's Fund (WHO, UNICEF) (2021) Progress on household drinking water, sanitation and hygiene 2000-2020: five years into the SDGs. WHO and UNICEF, Geneva

World Health Organization, United Nations International Children's Emergency Fund (WHO, UNICEF) (2023) Progress on household drinking water, sanitation and hygiene 2000-2022: special focus on gender. WHO and UNICEF, Geneva

Chapter 3
Consumption Patterns of Drinking Water and Its Impact on Public Health

Anuli U. Njoku and Martin Agbalenyo

1 Learning Objectives

1. Describe the impacts of water injustice on health in the United States.
2. Identify spatial, racial, and socio-economic disparities in access to water in the United States.
3. Highlight efforts made to address water insecurities and disparities.
4. Discuss recommendations to address disparities in access to water.

2 Introduction

2.1 Global Perspective

Access to water has been a major concern to humanity over the years. The availability of water has influenced settlement and migration for ages. Water scarcity has been identified as a global crisis (UNESCO 2024; Mohanty 2003). Water is a basic nutrient necessary for life, as it is essential for food digestion, transportation, and removal of waste from the body (Eneh 2007). Although access to potable water is considered a fundamental human right, many people around the globe are faced

A. U. Njoku (✉)
Department of Public Health, College of Health and Human Services, Southern Connecticut State University, New Haven, CT, USA
e-mail: njokua3@southernct.edu

M. Agbalenyo
Southwestern AHEC, Inc., Shelton, CT, USA
e-mail: martin.agbalenyo@ynhh.org

A. R. Molla et al. (eds.), *Climate Change, Drinking Water Security, and Public Health*, https://doi.org/10.1007/978-3-032-12074-8_3

with its scarcity (Ehinomen et al. 2018). In recent years, the global water demand has surged at twice the rate of the world's population growth (Benavides-Muñoz et al. 2024; Connor 2015). It has been predicted that demand for water around the globe will rise by 55% from 2000 to 2050 (Wang and Zimmerman 2016). According to Ercin and Hoekstra (2014), the world is facing water scarcity due to an increase in demand for water and a decline in water quality and quantity. Over the last century, demand for water has increased by 600% (Wada et al. 2016). Factors such as population growth, economic growth, and increased preference for commodities such as meat and sugar have intensified the competition for this precious resource (Ercin and Hoekstra 2014). Improving access to water will not only promote physical health, but also enhance the socio-emotional wellbeing of the people. In the developing world, it is estimated that every 1 dollar invested in water and sanitation will yield a return from $5 to $28 (Connor 2015).

Globally, about 4600 km^2 of groundwater is used per year, which is close to the maximum sustainable level (Gleick and Palaniappan 2010). Groundwater withdrawals in China, Pakistan, Iran, and the United States constituted 67% of the global usage in the 2010s (Wada et al. 2016). Groundwater withdrawal is expected to increase by 39% by 2050 (Wada et al. 2016). The increasing demand for water and the rapid depletion of water resources may cause a breaking point before 2050 (Boretti and Rosa 2019).

Although access to drinking water has increased over the years and efforts are being made globally to solve the menace of water shortage, a great number of people are still facing water shortage. According to UNICEF and WHO (2018), an estimated 844 million people did not have access to drinking water in 2015. A survey conducted in 2018 among 78 countries revealed that 50% of healthcare facilities in middle-income and low-income countries do not have access to potable water (Cronk and Bartram 2018). People, especially children, die daily due to using water from unsafe sources in the developing world (WHO 2018).

For most parts of the developing world where the cultural roles of women and children involve providing water for their households, the lack of potable water poses a tremendous challenge to these groups of people. This can cause stress and physical burden on these individuals. A study conducted by Adeleye et al. (2014) revealed that women and children in the Kpakungu area of Nigeria had to spend several hours to secure water for their households. This situation compelled some of them to walk long distances in search of water, which may not be wholesome for consumption.

2.2 *The United States: Access to Water*

The challenge of water scarcity is not just confined to developing countries. Both the developed and the developing world are faced with the problem of water injustice (Meng 2023). Although the developed world has improved access to drinking water, there are social and spatial disparities. According to Vanderslice (2011), even though 99% of Americans have access to safe drinking water, the United States faces income and racial disparities in access to water.

Although there are numerous studies on access to water, Vanderslice (2011) revealed that there is a dearth of studies on disparities. Previous studies revealed the impacts of water insecurities, but there is limited research on comprehensive interventions to address the impacts of water insecurities (Heidari et al. 2021). Therefore, this chapter aims to: (a) describe the impacts of water injustice on health in the United States; (b) identify spatial, racial, and socio-economic disparities in access to water the United States; (c) highlight efforts made to address water insecurities and disparities; (d) discuss recommendations to address disparities in access to water.

More than 400,000 homes in the United States do not have complete indoor plumbing (Sterling III et al. 2023). In 2013, 26.5 million people in the United States obtained their water from 9800 public water systems that failed to comply with the standards set by the Safe Drinking Water Act (US EPA 2015). A different source revealed that the majority of Americans depend on about 150,000 public water systems (PWS) for their water needs. The PWS serves about 94% of the population of the United States, while 13% of the people get water from private wells (Dieter and Maupin 2017). A section of the population uses both private wells and PWS for their water needs (Dieter and Maupin 2017).

Other groups of people depend on alternate sources such as bottled water, as PWS and wells are not readily available to them. Bottled water is the most consumed beverage since 2016 (Jaffee 2024). With an annual growth rate of 5%, 15.7 million gallons of bottled water were consumed by Americans in the year 2021 (Jaffee 2024). However, bottled water should not be perceived as a safer option, as it is more likely to contain microplastic compared to tap water (Mason et al. 2018). Until recently, bottled water was considered a luxurious good consumed by high-income earners (Szasz 2007). Jaffee (2024) challenged the assertion that bottled water is an optional commodity for higher-income earners by revealing that the product is rather a necessity for many low-income earners and racial minorities such as the Black and Hispanic populations due to distrust and avoidance of tap water.

Regions in the United States that are experiencing population growth and drought are more likely to face water insecurities (Warziniack and Brown 2019). Some states in the United States have implemented measures such as rationing of water to curb water shortage in the short-term. Droughts in Texas had compelled more than 100 public water providers to regulate water use in 2018 (Warziniack and Brown 2019). To mitigate the impacts of drought in California, the state has passed a law that restricts the daily use of water by each person to 50 gallons by 2030 (Warziniack and Brown 2019).

2.3 Environmental Justice and Health Disparities in Access to Water

"Environmental justice embraces the principle that all people and communities are entitled to equal protection of environmental and public health laws and regulations" (Bullard 1996, p. 493). Our environment is not just the physical components, as there are social elements that influence our day-to-day lives. Mohai et al. (2009) revealed that the definition of environmental justice should not only include the natural surroundings, but also the cultural environments. According to Vaz et al. (2017), environmental injustices are structural social challenges that adversely affect vulnerable groups. Social determinants of health factors affecting water disparities include racism, physical location, ethnicity, and socioeconomic status (Balazs and Ray 2014; VanDerslice 2011). There is a direct relation between equal access to basic needs and the health of the people. Balazs and Ray (2014) revealed that disparities in basic needs adversely affect the health of the people. Unequal access to safe drinking water exacerbates racism and environmental injustice (Jones 2001).

It has been observed that domestic water usage during COVID-19 had increased by 14% as safety required frequent washing of hands (Nemati and Tran 2022). Although handwashing was very expedient during the pandemic, the economic cost of water limited the amount of water used in low-income households (Nemati and Tran 2022).

The San Joaquin Valley is a region in California's Central Valley that is known for its agriculture and considered the most productive agricultural region in the United States. The area is predominantly rural but also has densely populated urban centers in San Joaquin Valley. Demographically, area counties have had some of the highest poverty rates in the United States. Balazs and Ray (2014) stated that complex dimensions such as natural, built, social, and political factors at various levels national, state, community, and household levels influence water access disparities.

According to the analysis of the water disparity framework by Balazs and Ray (2014), although there was no explicit correlation between race and disparity in San Joaquin Valley, racism is historically embedded in natural factors, built environments, and socio-political factors that drive disparities in access to water. For example, inequitable access to infrastructure and poorly enforced drinking water regulations and building codes fuel unequal access to safe drinking water. A multifaceted research and intervention agenda is needed to address the disproportionate environmental burdens by race, ethnicity, and socioeconomic status and to alleviate the burden of drinking water vulnerability within these communities.

San Joaquin is also exposed to nitrate. According to the Permits, Inspections, Compliance Monitoring, and Enforcement (PICME) system (2008), three out of four nitrate violation cases in 2007 in California can be traced to San Joaquin Valley. Harter et al. (2012) revealed that this problem can be attributed to agriculture, as some farmers in the Valley use nitrate. Reliance on groundwater for drinking water can make area communities susceptible to nitrate contamination of groundwater. Two types of vulnerability are brought about by groundwater nitrate contamination: the risks to the public's health and the financial burden of mitigating those risks through treatment, source reduction, remediation, or alternative water sources. Moreover, the area has many impoverished communities that cannot afford drinking water treatment or alternative water supplies.

According to Rosinger et al. (2022), African Americans, Hispanics, and low-income earners in the United States are less likely to access tap water because they are more likely to experience shut-offs due to the economic cost of water. This situation may compel them to resort to alternatives such as bottled water and sugar-sweetened beverages which may be unfavorable due to higher average costs and lack of fluoride, an essential protective mineral (Rosinger et al. 2022). The demand for bottled water has increased among Black Americans, Asians, and Hispanics, and this can increase their economic burdens due to the extra bottled water levies (Rosinger et al. 2022). There are also health risks linked to inadequate access to safe drinking water. For example, Black communities in Jackson, Mississippi have experienced significantly higher public health diseases than majority-white communities in the region due to White flight, urban population decline, poverty concentration, lack of tax revenue, social and racial stratification and segregation, and deterioration of water infrastructure (Meng 2023).

Undocumented immigrants are other groups of people who are susceptible to water insecurity. Lack of citizenship hampers community participation and discourages residents from requesting support in times of difficulty (Egge and Ajibade 2021). According to Egge and Ajibade (2021), undocumented immigrants in East Porterville feared reaching out for assistance during a period of water scarcity, as they did not want their immigration status to be known by the authorities. The perception of immigrants toward water in their countries of origin can influence their choices of water in the United States. Immigrants from countries where tap water is not safe are less likely to have trust in tap water in the United States. This situation may force them to look for alternative water sources such as bottled water (Patel and Schmidt 2017).

2.4 *Human Health Impacts of Water Injustice*

The strong correlation between access to safe drinking and health cannot be overemphasized. Access to safe water is an effective means to curb infectious diseases (Motoshita et al. 2011). Even though they are living close to community water systems, Black residents living in peri-urban neighborhoods in Wake County of North Carolina have challenges obtaining safe drinking water compared to other residents in adjacent neighborhoods (Stillo and MacDonald Gibson 2017). This situation compelled them to resort to unregulated wells. Stillo and Gibson (2018) further revealed that this situation may lead to health problems such as elevated levels of lead in the blood of children who consume water from unsafe sources. The situation made these vulnerable residents more susceptible to gastrointestinal issues that required emergency response by 22%. Out of 171 private wells assessed in Wake district, Stillo and MacDonald Gibson (2017) revealed that 50 of them representing 29.2% samples tested positive for total coliform bacteria while only 0.556% of water samples from community water services tested for the same bacteria. The research further revealed that 6.43% of samples from private wells tested positive for Escherichia coli compared to 0.00850% of samples from municipal services. Unregulated private wells can cause several health risks to children. Robertson et al. (2003) revealed that lack of safe drinking water can increase the incidence of diseases such as diarrhea, schistosomiasis, ascariasis, hookworm, and trichuriasis. Access to clean water, improved sanitation, hygiene, and sustainable use of water resources can reduce the prevalence of diseases by 10% (Prüss-Üstün et al. 2008). From 2007 to 2008, the Centers for Disease Control and Prevention (CDC) recorded 36 drinking water-related outbreaks in the United States, including subsequent illness in over 4000 people (Brunkard et al. 2011).

According to Schaider et al. (2019), Hispanic populations are more likely to consume water containing higher levels of nitrate and arsenic compared to other ethnic groups. Prolonged exposure to water sources containing nitrate levels above 5 mg/L increases susceptibility to negative health outcomes such as cancer, birth defects, and preterm birth (Schaider et al. 2019).

Nguyen Viet and Vu (2013) stated that 80% of diseases in the developing world are caused by poor sanitation and lack of safe water.

Access to potable water has an impact on infant mortality. A study conducted by Gibson and Mace (2006) revealed a negative correlation between access to safe drinking water and infant mortality. According to Whetten (2018), 44% of the world's population obtains water outside their homes. Women and children who have the primary responsibility to provide water mostly spend more than 1 h per trip obtaining water. People who walk long distances to secure water are more likely to use less water than they need (Whetten 2018). Proximity to water sources also influences health behavior, as people who have access to water are more likely to practice good hygiene such as handwashing and cleaning and disinfecting frequently touched surfaces (Whetten 2018).

In the United States, Balazs and Ray (2014) identified several challenges related to access to water, such as elevated levels of impurities in the San Joaquin Valley, inadequate water facilities in the US-Mexico Colonias, and unsupervised chemical pollution in the Navajo Nation.

3 Case Studies

3.1 Case Study 1: Flint Water Crisis

One major environmental injustice in U.S. history is the Flint Water Crisis, which occurred in Flint, Michigan. Flint switched its water source from Lake Huron to the Flint River without adding anti-corrosive inhibitors (Olson et al. 2017). This situation led to increased levels of lead in the water. The Safe Drinking Water Act was violated due to excessive trihalomethane levels (Pauli 2020). Changing the water source has resulted in adverse health outcomes, including low fertility (Grossman and Slusky 2019). Despite public concerns, residents were assured that the water was safe for consumption (Mohai 2018). A growing mistrust emerged due to the authorities' failure to address the problem (Rosinger et al. 2022). Groups disproportionately affected included Black residents and those with lower socioeconomic status (Butler et al. 2016; Njoku 2021). Children exposed to lead face a higher risk of developmental issues (Edwards et al. 2009). Lead exposure among children can lead to problems with verbal reasoning and aggression control (Cecil et al. 2008). Additionally, children exposed to lead are more likely to experience neurostructural issues in adulthood (Cecil et al. 2008). The government's failure to prioritize water supply was a key component of the crisis. Although faulty infrastructure was identified well before the crisis, repairs were delayed (Katner et al. 2018). Other social determinants also contributed to health issues. Factors behind the water crisis include historical and structural disparities, as well as government inefficiency in addressing the issue (Highsmith 2018). According to Sadler and Highsmith (2016), this represented an environmental injustice, as the concentration of people of color and those living in poverty in the area was the result of policies enacted by authorities (Sadler and Highsmith 2016).

Liévanos et al. (2021) revealed that individuals with lower levels of income and education were disproportionately affected. Another group of people who were disproportionately affected were women. According to Highsmith (2019), there has been a historical and sustained issue of "feminization of poverty" since the 1970s, with women representing 54% of people living below the federal poverty line in 2010. Flint, the birthplace of General Motor Company (GMC), historically provided more opportunities to males than females. Highsmith (2019) revealed that the automobile industry discriminated against women by denying them high-paying jobs.

The Flint Water Crisis did not only affect people directly exposed to the water. It also affected the socio-emotional well-being of people not directly in contact with the water. According to Kilpatrick et al. (2023), the crisis has disproportionately intensified the psychological stress among Black women residing within and outside Flint who perceived the incident as a deliberate attempt to marginalize the Black race. Overall, the Flint Water Crisis has highlighted the need for research on the experiences of environmental justice communities, including racial minorities and low-income households (Statman-Weil et al. 2020).

3.2 *Case Study 2: Colonias*

Colonias are unincorporated rural and peri-urban communities along the US-Mexico border (Núñez-Mchiri 2012). Residents in the colonias have faced socio-economic and environmental injustices over the years (Núñez-Mchiri 2012). Zheng et al. (2022) identify socio-environmental injustice, water scarcity, and lack of sanitation and hygiene resources as the root causes of water and sanitation insecurity in the colonias.

More than 2000 colonias can be found in four border states: Arizona, New Mexico, Texas, and California (Patwoary 2021). Some scholars have suggested that the term "colonias," which is derived from Spanish meaning "neighborhood," should be replaced with "disadvantaged unincorporated communities" (DUC) due to the stigma associated with the residents, particularly among Latinx individuals and Spanish speakers (Patwoary 2021). Colonias have limited access to essentials such as housing, potable water, sewage systems, and transportation (Lopez 2020). They are characterized by lower household incomes and inadequate amenities (Huntoon and Becker 2001). Residents face water insecurity, which adversely impacts their physical health and psychological well-being (Wutich et al. 2022). Zheng et al. (2022) revealed that water scarcity, lack of sanitation and hygiene resources, as well as social and environmental injustice, were the main causes of water and sanitation insecurities in the colonias. Zheng et al. (2022) developed a model illustrating the causes and impacts of water and sanitation insecurities within the colonias. The model is presented in Fig. 3.1.

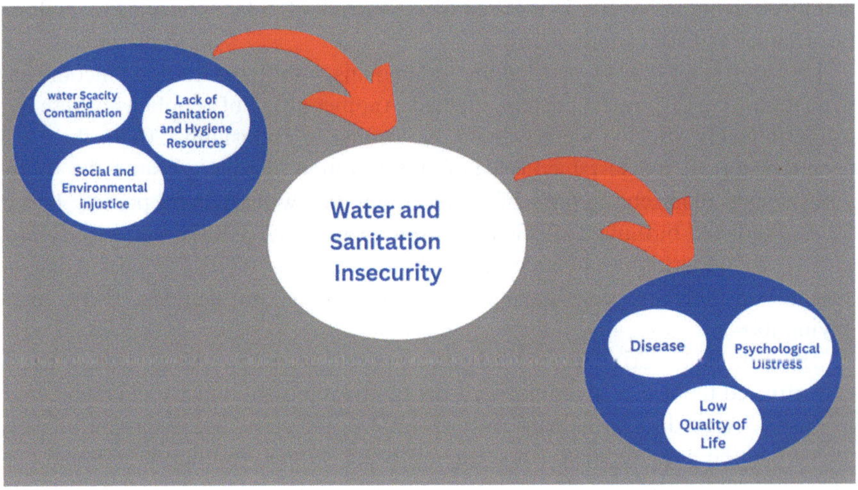

Fig. 3.1 Causes and impacts of water and sanitation insecurities. *Source*: Zheng et al. (2022) (Reproduced from Zheng et al. 2022, graphical abstract. Minor modification was made. https://doi.org/10.2166/wh.2022.085, licensed under the terms of the Creative Commons Attribution 4.0 International License (http://creativecommons.org/licenses/by/4.0/))

Zheng et al. (2022) identified poor quality of life, diseases, and psychosocial distress are the tragic consequences of water and sanitation insecurity in the colonias.

Efforts by the federal government and states to provide water infrastructure in these areas are severely lacking, which forces many residents to rely on vendor machines for their daily water needs. Due to financial constraints, some residents cannot afford water from these vending machines (Jepson and Brown 2014). Others turn to polluted irrigated water for hygiene and other domestic purposes (Jepson and Brown 2014). Approximately 840,000 people, or 30% of the population, lack access to drinking water in the colonias (Gu et al. 2023).

3.3 *Case Study 3: Navajo Nation*

The Navajo reservation, covering over 27,000 square miles, is located in the Southwestern part of the United States (Navaho Nation 2011). The Navajos represent the largest population of Native Americans (Lowe et al. 2018). There are more than a quarter of a million Navajos, with approximately 44% of them being children under 19 years (Navaho Nation 2011). For more than 500 years, Native Americans, including the Navajos, have faced poor health outcomes (Jones 2001). Health disparities have been a persistent issue among the Navajos. According to Lowe et al. (2018), the Navajos have suffered from mistreatment, trauma, and environmental injustice. Economic and political interests have been identified as contributors to health disparities among the Navajos (Jones 2001). Inadequate water systems, environmental pollution, and historical neglect have increased the vulnerability of the Navajos (Ed 2024). Accessing water can be difficult for these communities. According to Bell (2024), the Navajo people have been fighting for their rights to basic necessities such as water. Four out of five individuals in the Navajo Nation obtain their water from distant sources (Lyttle 2010). Although authorities are making efforts to address the issue, Lamoureux (2012) characterized the interventions as "build first, ask questions later." This indicates that the efforts made by policymakers do not focus on the actual needs of the people.

In some instances, potable water was not available to approximately two out of five Navajos, a situation which made many of them resort to trucks for water supply (Ogunwole 2006). Hoover et al. (2017) stated that the quality of water in the Navajo Reserve is threatened by 500 abandoned uranium sites in the Southern parts of the country. Due to exposure to uranium, the Dine people of Navajo are more susceptible to cancers of the kidney, stomach, and gallbladder when compared to their white neighbors (Ed 2024). Elevated levels of uranium have also been found among children in the reserve (Morales 2016).

The Navajo reserve has been created as tribal land since June 1868 (Greene 2024). Although the land is recognized as tribal, they still compete with the government over the control of natural resources such as water bodies. Even though authorities are not able to meet the water needs of the Navajo Nation, the government still controls a lot of water bodies in the reserve (Greene 2024). The Navajo Nation is facing water colonialism, a situation where the government uses water as a control mechanism by taking it away from the Indigenous people (Greene 2024).

For centuries, the Navajos have been persistently fighting for their right to water resources. Although they won a legal battle that gave them the right to the San Juan River Basin, this tremendous achievement was inadequate to address the problem of water disparity in the reserve (Lamoureux 2012).

In 2023, the Supreme Court ruled in favor of the Government of the United States, stating that it is not the duty of the government of the United States to provide water infrastructure to the Navajo Nation (Greene 2024). Moreover, inadequate access to water has worsened the social and economic disparities among the Navajo Nation (Ed 2024).

3.4 Case Study 4: Wake County, North Carolina

Although access to safe drinking water has improved over the years, several minority groups living on the outskirts of communities in North Carolina still face water insecurities (Gibson et al. 2014). Wake County, located in North Carolina, is one of the areas experiencing uneven access to water. Black residents are disproportionately affected by the water crisis in Wake County (Clonch 2019). Research conducted by Gibson et al. (2014) in Wake County revealed a 3.8% decrease in access to the community water system as the population of Black residents increased by 10%. This issue is deeply rooted in racial segregation (Gibson et al. 2014). Some residents obtain water from wells that are unsafe for consumption. Those who rely on well water are more vulnerable to health risks due to exposure to contaminants such as fecal matter (Stillo and MacDonald Gibson 2017). The poor quality of water can be attributed to factors like inadequate monitoring, lack of maintenance, and residents' inability to afford septic systems (Stillo and MacDonald Gibson 2017). Although the North Carolina General Assembly has passed a law requiring all wells to be tested, this legal requirement exempts wells constructed before July 1, 2000, and those that have been tested at least once before (North Carolina General Assembly 2016). Access to potable water in North Carolina is an environmental health issue that disproportionately impacts African Americans and individuals from lower socioeconomic backgrounds (Sohns 2023). According to DeFelice et al. (2016), unsafe water from wells accounts for 99% of emergency room visits for gastrointestinal illnesses in North Carolina. Approximately $39.9 million is spent annually on emergency department visits due to consuming water from private wells in North Carolina (DeFelice et al. 2016).

4 Efforts to Address Environmental Injustice and Disparities in Water

Until recently, most of the water policies in the United States were based on impromptu initiatives in times of conflicts. They were policies by default formulated and implemented to achieve short-term goals and aspirations (Deason et al. 2001).

The United States Safe Drinking Water Act gives the mandate to the Environmental Protection Agency (EPA) to regulate the activities of Public Water Systems. EPA delegates its regulatory roles to authorities such as states, territories, and tribes which may raise the standard higher than those set by the EPA (Patel et al. 2020). However, wells are not regulated in most states. Although about 43 million Americans depend on water from wells, there are no federal policies to regulate them (Bowen et al. 2019). Also, there are discrepancies in the current laws that regulate various wells because they are made by individual states (Bowen et al. 2019). Although the federal government has set standards for underground water quality, the states have the power and autonomy to make laws and implement policies to regulate underground water resources (Megdal et al. 2015).

The US Public Health Services set the microbiological drinking water quality standard in 1914 to regulate providers who supply water to interstate transporters. To ensure that people got access to safe drinking water, the US Public Health Service was regulating 28 contaminants by 1962 (Baum et al. 2015). In 2002, the EPA revealed that about 90 contaminants were being regulated (EPA 2002).

The United States has been investing in water at national and international level. The United States has been making efforts to improve access to water and sanitation at national and global levels for years. In 2010, $ 953 million was invested in Water and Sanitation by the United States at a global level (Salaam-Blyther 2012).

In addition to investing in water infrastructure, efforts are made to manage water bodies and other natural resources to ensure sustainability. Numerous programs in the United States focus on protecting watersheds across the country. Barnett et al. (2013) revealed that there were 93 watershed programs in the United States with a total investment of $383 million in 2013. Protecting watersheds not only safeguards the natural environment but also enhances water system performance, reduces operational costs, and increases predictability regarding future water production (Gartner et al. 2013). In 2007, the Central Arkansas Water Project was established to protect the watersheds of Lake Winna and Lake Maumelle, serving a population of 400,000 (Gartner et al. 2017). With a funding of $27.7 million, 2500 conservation lands were acquired (Gartner et al. 2017). Additionally, to tackle significant issues like ecological threats and water quality degradation, the Colorado-Big Thompson Headwaters Partnership was established in 2012 (Gartner et al. 2017). The Thompson Headwaters Partnership initiated efforts such as providing new water infrastructure and developing a 20-year Municipal Watershed Plan (Gartner et al. 2017).

As noted earlier, low-income households are disproportionately affected by water insecurity. To address this issue among this group of people, the Eastern

Kentucky Personal Responsibility in a Desirable Environment (PRIDE) was implemented to promote environmental sustainability, public education, and mobilization of local resources. The PRIDE interventions include promoting water quality by installing septic systems for residents whose income levels are below 55% of federal poverty guidelines (Arcipowski et al. 2017).

Attempts are also made to provide financial support to low-income households that are disproportionately affected by water insecurities. The Supplemental Nutrition Assistance Program (SNAP) is being utilized to increase access to safe drinking water. Beneficiaries of SNAP can receive a gallon of water as part of their monthly benefits (Yaktine and Caswell 2013). However, this is sorely inadequate, particularly for individuals facing water insecurity. To enhance access to water in six communities in Kern County that are disproportionately impacted by water insecurity, the State of California allocated $5 million for water provision through SNAP in 2017 (Wilking et al. 2022). By March 1, 2022, beneficiaries would receive $50 per month, which would halve their water expenditures (Wilking et al. 2022).

To increase access to potable water and promote trust in the providers of piped water, several agencies have taken the initiative to provide safe drinking water in public places such as schools, libraries, and parks. The Agua4All project in California provided water bottle-filling stations in schools, libraries, and parks around areas where access to potable water is limited (Patel and Schmidt 2017). Also, public schools in Baltimore and Boston provided bottled water to students when their communities were faced with elevated levels of lead in their water and needed access to drinking water while the lead in the water was remediated (Patel and Schmidt 2017).

As water resources are increasingly becoming more competitive, institutions and private individuals are taking various initiatives to improve access to potable water. Efforts are being made to concentrate on other non-traditional sources such as desalination, reuse, and rainwater harvesting. A typical example is the Prescott Valley which uses wastewater treated for agricultural purposes (Gleick 2010). Another example is the Palo Verde nuclear power plant which depends on reused wastewater for its cooling system (Gleick 2010).

5 Recommendations

A multidisciplinary approach is required to tackle water insecurity because it is an environmental injustice issue, with multiple drivers such as social, political, economic, and environmental factors. A study conducted by Onufrak et al. (2014) revealed that about 20% of young people between the ages of 9 and 19 years perceived tap water as unsafe for drinking. Blacks, Hispanics, and people from lower socio-economic backgrounds are more likely to hold the belief that pipe water is unsafe (Onufrak et al. 2014). The provision of safe drinking water alone is not enough. The negative perceptions toward pipe water need to be improved through public education. The perceptions of individuals influence their health behaviors.

Hubbart and Gootman (2021) stated that water-related public programs should include strategies to educate individuals about the quality of the water they consume.

Projects like PRIDE in Eastern Kentucky, which provides economic assistance to low-income earners for installing septic systems, should be encouraged nationwide. Low-income individuals are more likely to rely on well water. Installing septic systems will help reduce waterborne diseases among these populations.

As stated earlier, one of the main reasons for the low consumption of tap water is distrust. According to Patel and Schmidt (2017), reports on water testing are too technical and beyond the understanding of most of the residents. To promote trust in tap water, technical results should be simplified. The simplified result should be translated into the preferred languages of the consumers.

Testing private wells should be mandatory. Private wells serve about 13% of the population of America (Dieter and Maupin 2017). The EPA should pass a law that requires periodic testing of wells in order to determine their qualities. This practice will enhance the safety of water to the 13% of Americans who depend on private wells.

As established earlier, those who are disproportionately affected by water are racial minorities and low-income earners. These groups of people may not have the advocacy skills to fight for themselves. Stakeholders such as non-governmental organizations should partner with community-based organizations for advocacy. Advocacy can influence policies in favor of the vulnerable faced with water insecurity.

Problems with water infrastructure should be solved immediately before they become worse. Concerning the Flint crisis, Katner et al. (2018) revealed that a faulty machine was detected before the crisis, but the authorities were not proactive enough to solve the problem in time. Lessons from the Flint Water Crisis have taught us that addressing problems promptly will not only enhance the well-being of the people but also save the authorities a lot of economic resources. Timely repairs of broken parts and periodic maintenance of water infrastructure should be encouraged.

6 Conclusions

Unequal access to safe drinking water has intensified the issues of social and environmental injustices in the United States. While efforts are underway to enhance access to water, the issue of disparity also needs to be tackled. Simply increasing access does not necessarily resolve the problem of disparity. Equity must guide the provision of water. According to the United Nations World Water Development Report 2015, "*the principle of equity, perhaps more than any technical recommendation, carries with it the promise of a more water-secure world for all*" (Connor 2015).

There is the need to intensify research in order to address water scarcity and disparity. Jackson et al. (2001) revealed that the increasing pressure on freshwater resources requires immediate research on water resource management. Research should also focus on both conventional and less known means sources such as desalination, reuse, and recycling. As noted earlier, efforts are being made to broaden access to water by exploring non-conventional means. However, only a few institutions are currently exploring these means. More research is needed to determine the viability of these lesser-known interventions to make them more widely available and accessible.

7 Discussion Questions

1. Based on the examples discussed in this chapter, explain at least five human health impacts of water injustice and unequal access to safe drinking water.
2. Describe at least three events of environmental injustice and disparate access to water that have occurred in the history of the United States.
3. Based on the recommendations described in this chapter, what are at least three strategies to prevent or mitigate instances of water injustice and unequal access to safe drinking water?

References

Adeleye B, Okelola O, Medayese S (2014) Problems of water supply and sanitation in Kpakungu area of Minna (Nigeria). Glocalism (1-2). https://doi.org/10.12893/gjcpi.2014.1-2.9

Arcipowski E, Schwartz J, Davenport L, Hayes M, Nolan T (2017) Clean water, clean life: promoting healthier, accessible water in rural Appalachia. J Contemp Water Res Educ 161(1):1–18. https://doi.org/10.1111/j.1936-704X.2017.3248.x

Balazs CL, Ray I (2014) The drinking water disparities framework: on the origins and persistence of inequities in exposure. Am J Public Health 104(4):603–611. https://doi.org/10.2105/AJPH.2013.301664

Barnett J, O'Neill S, Waller S, Rogers S (2013) Reducing the risk of maladaptation in response to sea-level rise and urban waterscarcity. In: Moser S, Boykoff M (eds) Successful Adaptation to Climate Change (pp 37-49). Routledge.

Baum R, Amjad U, Luh J, Bartram J (2015) An examination of the potential added value of water safety plans to the United States national drinking water legislation. Int J Hyg Environ Health 218(8):677–685. https://doi.org/10.1016/j.ijheh.2014.12.004

Bell MM (2024) A Nation Without Water: An Analysis of the Navajo Nation's Water Crisis

Benavides-Muñoz HM, Lapo-Pauta M, Martínez-Solano FJ, Quiñones-Cuenca M, Quiñones-Cuenca S (2024) Global events and surge in residential water demand: exploring possible hydraulic scenarios. Water 16(7):956. https://doi.org/10.3390/w16070956

Boretti A, Rosa L (2019) Reassessing the projections of the world water development report. NPJ Clean Water 2(1):15. https://doi.org/10.1038/s41545-019-0039-9

Bowen K, Krishna T, Backer L, Hodgins K, Waller LA (2019) State-level policies concerning private wells in the United States. Water Policy 21(2):428–435. https://doi.org/10.2166/wp.2019.205

Brunkard JM, Ailes E, Roberts VA, Hill V, Hilborn ED, Craun GF, Rajasingham A, Kahler A, Garrison L, Hicks L, Carpenter J (2011) Surveillance for waterborne disease outbreaks associated with drinking water—United States, 2007–2008. Morbidity and Mortality Weekly Report. MMWR Surveill Summ, 60(12):38–68.

Bullard RD (1996) Environmental justice: it's more than waste facility siting. Soc Sci Q 77(3):493–499. https://www.jstor.org/stable/42863495

Butler LJ, Scammell MK, Benson EB (2016) The Flint, Michigan, water crisis: a case study in regulatory failure and environmental injustice. Environ Justice 9(4):93–97. https://doi.org/10.1089/env.2016.0014

Cecil KM, Brubaker CJ, Adler CM, Dietrich KN, Altaye M, Egelhoff JC, Wessel S, Elangovan I, Hornung R, Jarvis K, Lanphear BP (2008) Decreased brain volume in adults with childhood lead exposure. PLoS Med 5(5):e112. https://doi.org/10.1371/journal.pmed.0050112

Clonch A (2019) Annexation and Water Utility Extensions in Wake County, NC: The Role of Race, Income, and Other Demographic Characteristics. 10.17615/2qb3-5f55. https://doi.org/10.1007/978-3-319-71389-2_5

Connor R (2015) The United Nations world water development report 2015: water for a sustainable world, vol 1. UNESCO publishing

Cronk R, Bartram J (2018) Environmental conditions in health care facilities in low- and middle-income countries: coverage and inequalities. Int J Hyg Environ Health 3:409–422. https://doi.org/10.1016/j.ijheh.2018.01.004

Deason JP, Schad TM, Sherk GW (2001) Water policy in the United States: a perspective. Water Policy 3(3):175–192. https://doi.org/10.1016/S1366-7017(01)00011-3

DeFelice NB, Johnston JE, Gibson JM (2016) Reducing emergency department visits for acute gastrointestinal illnesses in North Carolina (USA) by extending community water service. Environ Health Perspect 124:1583–1591. https://doi.org/10.1289/EHP160

Dieter CA, Maupin MA (2017) Public supply and domestic water use in the United States, 2015 (No. 2017–1131). US Geological Survey

Ed RC (2024) Caught between two worlds: a study of water access on the Navajo reservation and the educational fallout of the youth who thirst. Northern Arizona University. https://www.proquest.com/docview/3094572145?pq-origsite=gscholar&fromopenview=true&sourcetype=Dissertations%20&%20Theses

Edwards M, Triantafyllidou S, Best D (2009) Elevated blood lead in young children due to lead-contaminated drinking water: Washington, DC, 2001– 2004. Environ Sci Technol 43(5):1618–1623. https://doi.org/10.1021/es802789w

Egge M, Ajibade I (2021) A community of fear: emotion and the hydro-social cycle in East Porterville, California. J Political Ecol. https://doi.org/10.2458/jpe.2366. https://archives.pdx.edu/ds/psu/36156

Ehinomen C, Afolabi B, Agu O (2018) An empirical analysis of the implications of insufficient portable water supply on productivity: a case study of South-Western Nigeria. Eur J Soc Sci 57(3):347–361. ISSN 1450-2267. http://www.europeanjournalofsocialsciences.com/

Eneh OC (2007) Improving the access to potable water in Nigeria. Afr J Sci 8(2):1962–1971. https://www.researchgate.net/profile/Onyenekenwa-Eneh/publication/279885361_IMPROVING_THE_ACCESS_TO_POTABLE_WATER_IN_NIGERIA/links/559d2e0008ae04e365088c62/IMPROVING-THE-ACCESS-TO-POTABLE-WATER-IN-NIGERIA.pdf

EPA (2002) National primary drinking water regulations: long term 1 enhanced surface water treatment rule. Final rule. Fed Regist 67(9):1811–1844

Ercin AE, Hoekstra AY (2014) Water footprint scenarios for 2050: a global analysis. Environ Int 64:71–82. https://doi.org/10.1016/j.envint.2013.11.019

Gartner T, Mulligan J, Schmidt R, Gunn J (2013) Natural infrastructure: investing in forested landscapes for source water protection in the United States. World Resources Institute, Washington

Gartner T, Di Francesco K, Ozment S, Huber-Stearns H, Lichten N, Tognetti S (2017) Protecting drinking water at the source: lessons from US watershed investment programs. Journal (American Water Works Association) 109(4):30–41. https://www.jstor.org/stable/90005411

Gibson MA, Mace R (2006) An energy-saving development initiative increases birth rate and childhood malnutrition in rural Ethiopia. PLoS Med 3(4):e87. https://doi.org/10.1371/journal.pmed.0030087
Gibson JM, DeFelice N, Sebastian D, Leker H (2014) Racial disparities in access to community water supply service in Wake County, North Carolina. Am J Public Health 104(12):e45. http://uknowledge.uky.edu/frontiersinphssr/vol3/iss3/6?utm_source=uknowledge.uky.edu%2Ffrontiersinphssr%2Fvol3%2Fiss3%2F6&utm_medium=PDF&utm_campaign=PDFCoverPages
Gleick PH (2010) Roadmap for sustainable water resources in southwestern North America. Proc Natl Acad Sci 107(50):21300–21305
Gleick PH, Palaniappan M (2010) Peak water limits to freshwater withdrawal and use. Proc Natl Acad Sci 107(25):11155–11162
Greene MP (2024) A case of water: a rhetorical analysis of the legal and social effects of the Arizona V. Navajo Nation Supreme Court Case. https://scholarworks.umt.edu/etd/12332
Grossman DS, Slusky DJ (2019) The impact of the Flint water crisis on fertility. Demography 56(6):2005–2031. https://doi.org/10.1007/s13524-019-00831-0
Gu Z, Li W, Hanemann M, Tsai Y, Wutich A, Westerhoff P, Landes L, Roque AD, Zheng M, Velasco CA, Porter S (2023) Applying machine learning to understand water security and water access inequality in underserved Colonia communities. Comput Environ Urban Syst 102:101969. https://doi.org/10.1016/j.compenvurbsys.2023.101969
Harter T, Lund JR, Darby J, Fogg GE, Howitt R, Jessoe KK, Pettygrove GS, Quinn JF, Viers JH (2012) Addressing nitrate in California's drinking water. With a focus on Tulare Lake Basin and Salinas Valley groundwater. Report for the state water resources control board report to the legislature. UC Davis Center for Watershed Sciences, Davis. https://groundwaternitrate.ucdavis.edu/files/138956.pdf
Heidari H, Arabi M, Warziniack T (2021) Vulnerability to water shortage under current and future water supply-demand conditions across US river basins. Earth's Future 9(10):e2021EF002278. https://doi.org/10.1029/2021EF002278
Highsmith AR (2018) A poisonous harvest: race, inequality, and the long history of the Flint water crisis. J Law Soc 18:121. https://heinonline.org/HOL/LandingPage?handle=hein.journals/jls18&div=12&id=&page=
Highsmith AR (2019) Demolition means progress: Flint, Michigan, and the fate of the American metropolis. University of Chicago Press. https://doi.org/10.7208/9780226251080
Hoover J, Gonzales M, Shuey C, Barney Y, Lewis J (2017) Elevated arsenic and uranium concentrations in unregulated water sources on the Navajo Nation, USA. Expo Health 9:113–124. https://doi.org/10.1007/s12403-016-0226-6
Hubbart JA, Gootman KS (2021) A call to broaden investment in drinking water testing and community outreach programs. Challenges 12(2):32. https://doi.org/10.3390/challe12020032
Huntoon L, Becker B (2001) September. Colonias in Arizona: A changing definition with changing location. In Memoria of a Research Workshop "Irregular Settlement and Self-Help Housing in the United States" (p. 34). https://www.academia.edu/27570823/Building_Colonias_Land_Policy_from_the_Inside_Out
Jackson RB, Carpenter SR, Dahm CN, McKnight DM, Naiman RJ, Postel SL, Running SW (2001) Water in a changing world. Ecol Appl 11(4):1027–1045. https://doi.org/10.1890/1051-0761(2001)011[1027:WIACW]2.0.CO;2
Jaffee D (2024) Unequal trust: bottled water consumption, distrust in tap water, and economic and racial inequality in the United States. Wiley Interdiscip Rev: Water 11(2):e1700
Jepson W, Brown HL (2014) 'If no gasoline, no water': privatizing drinking water quality in South Texas colonias. Environ Plan A 46(5):1032–1048. https://doi.org/10.1068/a46170
Jones CP (2001) Invited commentary: "race," racism, and the practice of epidemiology. Am J Epidemiol 154(4):299–304. https://doi.org/10.1093/aje/154.4.299
Katner AL, Brown K, Pieper K, Edwards M, Lambrinidou Y, Subra W (2018) America's path to drinking water infrastructure inequality and environmental injustice: The case of flint,

Michigan. The Palgrave handbook of sustainability: Case studies and practical solutions, pp. 79–97

Kilpatrick SK, Bauer KW, Heard-Garris N, Malone AM, Abdou CM, Weeks HM, Clayson M, Allgood KL, Dokshina D, Needham BL (2023) Experiences of the Flint water crisis among reproductive-age Michigan women in communities outside of Flint: differences by race and ethnicity. J Racial Ethn Health Disparities 10(3):993–1005. https://doi.org/10.1007/s40615-022-01287-6

Lamoureux C (2012) Quenching the thirst of a Nation: returning water to the Navajo people. Arizona J Environ Law Policy

Liévanos RS, Evans CR, Light R (2021) An intercategorical ecology of lead exposure: complex environmental health vulnerabilities in the Flint water crisis. Int J Environ Res Public Health 18(5):2217. https://doi.org/10.3390/ijerph18052217

Lopez D (2020) Identity building in militarized U.S.-Mexico border Colonias: Testimonios of women of color resisting militarization. The University of Arizona

Lowe AA, Bender B, Liu AH, Solomon T, Kobernick A, Morgan W, Gerald LB (2018) Environmental concerns for children with asthma on the Navajo Nation. Ann Am Thorac Soc 15(6):745–753. https://doi.org/10.1513/AnnalsATS.201708-674PS

Lyttle SD (2010) The third world in the American southwest: the Navajo nation's water crisis and the failures of water law. Geo J Law Mod Critical Race Perspect 2:83. https://heinonline.org/HOL/Page?handle=hein.journals/gjmodco2&div=8&g_sent=1&casa_token=&collection=journals

Mason SA, Welch VG, Neratko J (2018) Synthetic polymer contamination in bottled water. Front Chem 6:407. https://doi.org/10.3389/fchem.2018.00407

Megdal SB, Gerlak AK, Varady RG, Huang LY (2015) Groundwater governance in the United States: common priorities and challenges. Groundwater 53(5):677–684. https://doi.org/10.1111/gwat.12294

Meng Q (2023) Urban water crisis and its relationship to health inequities against black communities in the USA: spatial analytics of the Jackson region in Mississippi. J Clean Prod 411:137356. https://doi.org/10.1016/j.jclepro.2023.137356

Mohai P (2018) Environmental justice and the Flint water crisis. Mich Sociol Rev 32:1–41. https://www.jstor.org/stable/26528595

Mohai P, Pellow D, Roberts JT (2009) Environmental justice. Annu Rev Environ Resour 34(1):405–430. https://doi.org/10.1146/annurev-environ-082508-094348

Mohanty KK (2003) The near-term energy challenge. Am Institute Chem Eng AIChE J 49(10):2454. https://www.proquest.com/docview/199446093?pq-origsite=gscholar&fromopenview=true&sourcetype=Scholarly%20Journals

Morales L (2016) For the Navajo Nation, Uranium Mining's Deadly Legacy Lingers. *NPR. April, 10*

Motoshita M, Itsubo N, Inaba A (2011) Development of impact factors on damage to health by infectious diseases caused by domestic water scarcity. Int J Life Cycle Assess 16(1):65–73

Navaho Nation (2011) Welcome to the Navajo Nation Government [accessed 2017 Sep 20]. http://www.navajo-nsn.gov/history.htm

Nemati M, Tran D (2022) The impact of COVID-19 on urban water consumption in the United States. Water 14(19):3096. https://doi.org/10.3390/w14193096

Nguyen Viet C, Vu T (2013) The impact of piped water on household welfare: evidence from Vietnam. J Environ Plan Manag 56(9):1332–1358. https://doi.org/10.1080/09640568.2012.720568

Njoku AU (2021) COVID-19 and environmental racism: challenges and recommendations. Eur J Environ Public Health 5(2):em0079. https://doi.org/10.21601/ejeph/10999

Núñez-Mchiri GG (2012) Housing, colonias, and social justice in the US-Mexico border region. In: Social justice in the US-Mexico border region. Springer Netherlands, Dordrecht, pp 109–125. https://doi.org/10.1007/978-94-007-4150-8_6

Ogunwole SU (2006) We the people: American Indians and Alaska natives in the United States, vol 28. US Department of Commerce, Economics and Statistics Administration, US Census Bureau

Olson TM, Wax M, Yonts J, Heidecorn K, Haig SJ, Yeoman D, Hayes Z, Raskin L, Ellis BR (2017) Forensic estimates of lead release from lead service lines during the water crisis in Flint, Michigan. Environ Sci Technol Lett 4(9):356–361. https://pubs.acs.org/doi/full/10.1021/acs.estlett.7b00226

Onufrak SJ, Park S, Sharkey JR, Merlo C, Dean WR, Sherry B (2014) Perceptions of tap water and school water fountains and association with intake of plain water and sugar-sweetened beverages. J Sch Health 84(3):195–204. https://doi.org/10.1111/josh.12138

Patel AI, Schmidt LA (2017) Water access in the United States: health disparities abound and solutions are urgently needed. Am J Public Health 107(9):1354–1356. https://doi.org/10.2105/AJPH.2017.303972

Patel AI, Hecht CE, Cradock A, Edwards MA, Ritchie LD (2020) Drinking water in the United States: implications of water safety, access, and consumption. Annu Rev Nutr 40(1):345–373. https://doi.org/10.1146/annurev-nutr-122319-035707

Patwoary NS (2021) The health impacts of water insecurity in U.S. Arizona State University, Colonias

Pauli BJ (2020) The Flint water crisis. Wiley interdisciplinary reviews. Water 7(3):e1420. https://doi.org/10.1002/wat2.1420

Permits Inspections Compliance Monitoring and Enforcement (PICME) (2008) Sacramento. California Department of Public Health, CA

Robertson W, Stanfield G, Howard G, Bartram, J (2003) Monitoring the quality of drinking water during storage and distribution. In: Dufour A, Snozzi M, Koster W, Bartram J, Ronchi E, Fewtrell L (eds) Assessing microbial safety of drinking water (p.179). IWA Publishing.

Rosinger AY, Patel AI, Weaks F (2022) Examining recent trends in the racial disparity gap in tap water consumption: NHANES 2011–2018. Public Health Nutr 25(2):207–213. https://doi.org/10.1017/S1368980021002603

Sadler RC, Highsmith AR (2016) Rethinking Tiebout: the contribution of political fragmentation and racial/economic segregation to the Flint water crisis. Environ Justice 9(5):143–151. https://doi.org/10.1089/env.2016.0015

Salaam-Blyther T (2012) Global access to clean drinking water and sanitation: US and international programs

Schaider LA, Swetschinski L, Campbell C et al (2019) Environmental justice and drinking water quality: are there socioeconomic disparities in nitrate levels in U.S. drinking water? Environ Health 18:3. https://doi.org/10.1186/s12940-018-0442-6

Sohns A (2023) Differential exposure to drinking water contaminants in North Carolina: evidence from structural topic modeling and water quality data. J Environ Manag 336:117600. https://doi.org/10.1016/j.jenvman.2023.117600

Statman-Weil Z, Nanus L, Wilkinson N (2020) Disparities in community water system compliance with the safe drinking water act. Appl Geogr 121:102264

Sterling CW III, Krometis LA, Pingel T, Winling L (2023) Connections between present-day water access and historical redlining. Environ Justice. https://doi.org/10.1089/env.2022.0115

Stillo F, Gibson JM (2018) Racial disparities in access to municipal water supplies in the American south: impacts on children's health. Int Public Health J 10(3)

Stillo F, MacDonald Gibson J (2017) Exposure to contaminated drinking water and health disparities in North Carolina. Am J Public Health 107(1):180–185. https://ajph.aphapublications.org/doi/abs/10.2105/AJPH.2016.303482

Szasz A (2007) Shopping our way to safety: how we changed from protecting the environment to protecting ourselves. University of Minnesota Press

The United Nations Educational, Scientific and Cultural Organization - UNESCO (2024) Imminent risk of a global water crisis, warns the UN World Water Development Report 2023. viewed 29 September 2024, https://www.unesco.org/en/articles/imminent-risk-global-water-crisis-warns-un-world-water-development-report-2023

U.S. Environmental Protection Agency (US EPA). Providing Safe Drinking Water in America: 2013 National public water systems compliance report. 305-R-15-001. Washington, D.C.:

Office of Enforcement and Compliance Assurance. p. 2015. https://www.epa.gov/sites/default/files/2015-06/documents/sdwacom2013.pdf
VanDerslice J (2011) Drinking water infrastructure and environmental disparities: evidence and methodological considerations. Am J Public Health 101(S1):S109–S114. https://doi.org/10.2105/AJPH.2011.300189
Vaz E, Anthony A, McHenry M (2017) The geography of environmental injustice. Habitat Int 59:118–125. https://doi.org/10.1016/j.habitatint.2016.12.001
Wada Y, Flörke M, Hanasaki N, Eisner S, Fischer G, Tramberend S, Satoh Y, Van Vliet MTH, Yillia P, Ringler CJGMD, Burek P (2016) Modeling global water use for the 21st century: the water futures and solutions (WFaS) initiative and its approaches. Geosci Model Dev 9(1):175–222. https://gmd.copernicus.org/articles/9/175/2016/
Wang R, Zimmerman J (2016) Hybrid analysis of blue water consumption and water scarcity implications at the global, national, and basin levels in an increasingly globalized world. Environ Sci Technol 50(10):5143–5153. https://pubs.acs.org/doi/full/10.1021/acs.est.6b00571
Warziniack T, Brown TC (2019) The importance of municipal and agricultural demands in future water shortages in the United States. Environ Res Lett 14(8):084036
Whetten J (2018) Water and healthcare access disparities: Impacts on health, wealth and education. https://digitalrepository.unm.edu/econ_etds/95/
Wilking C, Nink E, Cradock AL (2022) Drinking water and the supplemental nutrition assistance program: current policy and opportunities for improvement. J Public Health Manag Pract 28(3):218–221. https://doi.org/10.1097/PHH.0000000000001474
World Health Organization, United Nations Children's Fund (2018). Available: https://www.unicef.org/media/47746/file/Annual_Results_Report_2017_WASH.pdf
Wutich A, Jepson W, Velasco C, Roque A, Gu Z, Hanemann M, Hossain MJ, Landes L, Larson R, Li WW, Morales-Pate O (2022) Water insecurity in the global north: a review of experiences in US colonias communities along the Mexico border. Wiley Interdiscip Rev: Water 9(4):e1595. https://doi.org/10.1002/wat2.1595
Yaktine AL, Caswell JA (eds) (2013) Supplemental nutrition assistance program: examining the evidence to define benefit adequacy. National Academies Press (US), Washington (DC)
World Health Organization, United Nations Children's Fund (2018) Available: https://www.unicef.org/media/47746/file/Annual_Results_Report_2017_WASH.pdf
Zheng M, Wutich A, Brewis A, Kavouras S (2022) Health impacts of water and sanitation insecurity in the global north: a scoping literature review for US colonias on the Mexico border. J Water Health 20(9):1329–1342. https://doi.org/10.2166/wh.2022.085

Chapter 4
Climate Change and Its Impact on Water and Food Security

Rebecca L. Larson, Oleksandr Nechyporenko, and Donald K. Warne

1 Learning Objectives

1. List Sustainable Development Goals related to water, food security, and climate.
2. Define water security, food security, and nutrition security.
3. Identify components of a sustainable food system.
4. Explain the interactions between climate change and water, food, and nutrition security.
5. Discuss the relationship between climate change and conflict on water and food security.
6. Describe the intersection of climate change and water on Indigenous food systems.
7. Discuss possible solutions for improving sustainable food systems, and water, food, and nutrition security.

R. L. Larson (✉)
School of Public Health, Augusta University, Augusta, GA, USA
e-mail: rlarson@augusta.edu

O. Nechyporenko
National Academy of Agrarian Sciences of Ukraine, National Scientific Centre, Institute of Agrarian Economics, Kyiv, Ukraine
e-mail: nechyporenko@iae.kiev.ua

D. K. Warne
Center for Indigenous Health, Johns Hopkins Bloomberg School of Public Health, Johns Hopkins University, Baltimore, MD, USA
e-mail: dwarne1@jhu.edu

A. R. Molla et al. (eds.), *Climate Change, Drinking Water Security, and Public Health*, https://doi.org/10.1007/978-3-032-12074-8_4

2 Introduction

Billions of people experience water and food insecurity every year. Climate change influences food and water systems; at the same time, water and food systems contribute to climate change (FAO 2024b; Mbow et al. 2019). This chapter begins with a review of the Sustainable Development Goals related to water, food security, and climate, followed by definitions of water, food, and nutrition security. The modern food system is compared to sustainable food systems and impacts of climate change on water, food, and nutrition security are reviewed within the context of the food system. The relationship between climate change and conflict on water and food security and the intersection of climate and water on Indigenous food systems are also explored. This chapter concludes with solutions for and progress toward improving sustainable food systems and water, food, and nutrition security.

3 Sustainable Development Goals

The United Nations Sustainable Development Goals (SDGs) (Department of Economic and Social Affairs n.d.) outline 17 goals, several (Table 4.1) of which have obvious and direct significance to the discussion about water and food security and climate change.

The Intergovernmental Panel on Climate Change (IPCC) (Mbow et al. 2019) connected Goal 2 (no hunger) and Goal 13 (climate action) to other SDGs, except 14 and 16, based on data from 227 countries for the official indicators of SDGs. While insufficient data existed at the time of their research for SDGs 14 and 16, other researchers make the connections between SDG 2 and 13 and all other SDGs

Table 4.1 Sustainable Development Goals related to water and food security and climate change (Department of Economic and Social Affairs n.d.)

Goal 2: No hunger	End hunger, achieve food security and improved nutrition, and promote sustainable agriculture
Goal 3: Good health and well-being	Ensure healthy lives and promote well-being for all at all ages
Goal 6: Clean water and sanitation	Ensure availability and sustainable management of water and sanitation for all
Goal 13: Climate action	Take urgent action to combat climate change and its impacts
Goal 14: Life below water	Conserve and sustainably use the oceans, seas, and marine resources for sustainable development
Goal 15: Life on land	Protect, restore, and promote sustainable use of terrestrial ecosystems, sustainably manage forests, combat desertification, and halt and reverse land degradation and halt biodiversity loss

clear. A collaboration of interdisciplinary scientists (Food Systems Countdown Initiative 2025; Schneider et al. 2025) identified direct and indirect interconnections between 50 food system indicators important for reaching the SDGs, including several related to climate change and water, food, and nutrition security. Access to safe water is directly affected by four indicators, including pesticide use and nitrogen use efficiency. The share of the population experiencing food insecurity directly mutually (i.e., bidirectionally) directly affects 5 indicators, including cost of a healthy diet, share of the population that cannot afford a healthy diet, and percentage of the population that relies on extreme coping strategies to combat food insecurity. The share of the population experiencing food insecurity directly affects 7 other indicators, including the prevalence of undernourishment, while also being directly affected by 12 other indicators, including fruit and vegetable yield and availability in the food supply per capita per day, food supply variability, and food price volatility.

4 Water Security, Food Security, and Nutrition Security

4.1 Water Security

Water is essential for daily living; however, only 0.5% of water on Earth is available as freshwater and it is becoming increasingly scarce due to an increasing population, overextraction, contamination, poor management, conflict, extreme weather events, and climate change (United Nations Children's Fund [UNICEF] 2021; United Nations-Water 2025). Globally, 1.42 billion people, including 450 million children, live in areas with high or extremely high physical water scarcity risks, and 2.3 billion people live in countries that experience water stress (UNICEF 2021; United Nations Convention to Combat Desertification 2022). Children experiencing water scarcity and unsafe drinking water often develop diarrhea, which limits their ability to absorb nutrients, leading to physical and mental impairment and increasing the risk of death (UNICEF 2021). Not only do we need water to drink, food systems are dependent on freshwater. Approximately 72% of freshwater is used for agriculture purposes, and the demand for food is expected to rise 50% by 2050, thereby increasing the demand for freshwater (FAO & UN-Water 2024; United Nations-Water 2025).

The United Nations Educational, Scientific and Cultural Organization (UNESCO 2012) International Hydrological Programme defines water security as "the capacity of a population to safeguard access to adequate quantities of water of acceptable quality for sustaining human and ecosystem health on a watershed basis, and to ensure efficient protection of life and property against water related hazards—floods, landslides, land subsidence, and droughts." The United Nations (UN-Water 2013) adapted the UNESCO definition to state that water security is "the capacity of a population to safeguard sustainable access to adequate quantities of and acceptable quality water for sustaining livelihoods, human well-being, and socio-economic development, for ensuring protection against water-borne pollution and water-related disasters, and for preserving ecosystems in a climate of peace and political stability."

Other terms related to water security are water stress and water scarcity. Water stress exists when a region withdraws 25% or more freshwater resources than it can replenish, while water scarcity occurs when the demand for water exceeds the supply, when there is inadequate water infrastructure, and/or when institutional responses to water needs are not balanced (UN-Water n.d.). UNICEF (2021) defines water scarcity as existing "… where the demand for water exceeds supply and where available water resources are approaching or have exceeded sustainable limits. Water scarcity can either be physical or economic."

4.2 *Food Security*

Norman Borlaug (1970), known as the father of the green revolution, identified access to food as an essential component of social justice. Borlaug is credited with saving millions of people from starvation in Mexico, India, and Pakistan during the 1950s and 60s, through his research in increasing wheat, rice, and corn yields and as an international advisor on scaling up food production. However, in 2023, approximately 2.33 billion people, 28.9% of the global population, continued to face moderate or severe food insecurity, and 713–757 million people, or 9.1% of the population, experienced hunger (Food and Agriculture Organization of the United Nations [FAO] et al. 2024).

The 1996 World Food Summit Plan of Action (FAO 1996) defined food security as existing "when all people, at all times, have physical and economic access to sufficient, safe and nutritious food that meets their dietary needs and food preferences for an active and healthy life." The FAO food insecurity report (2001) recognized the cultural aspects of food and feeding and updated the World Food Summit definition to state that food security "exists when all people, at all times, have physical, *social* and economic access to sufficient, safe and nutritious food that meets their dietary needs and food preferences for an active and healthy life"; this definition is most commonly used today. The FAO definition highlights four dimensions of food security: availability, access, utilization, and stability. In 2020, the United Nations High Level Panel of Experts (HLPE 2020) on Food Security and Nutrition of the Committee on World Food Security sought to draw attention to two other aspects of food insecurity not always captured in food security conversations: agency and sustainability. The HLPE (2020) provides definitions for each of the six dimensions of food security (Table 4.2). The United States Department of Agriculture (USDA) (National Institute of Food and Agriculture [NIFA] 2025) defines household food security as "access by all members at all times to enough food for an active, healthy life" and similar to the FAO recognition of safety, nutrition, and cultural aspects of food security, it notes that food should be nutritionally adequate and safe and that individuals should be able to acquire food in socially acceptable ways (i.e., not scavenging, stealing, or relying on emergency supplies).

A second aspect of food security not readily apparent in the FAO definition is the level of severity of food security. The USDA Economic Research Service (ERS

Table 4.2 The six dimensions of food security

Availability	Having a quantity and quality of food sufficient to satisfy the dietary needs of individuals, free from adverse substances and acceptable within a given culture, supplied through domestic production or imports
Access (economic, social, and physical)	Having personal or household financial means to acquire food for an adequate diet at a level to ensure that satisfaction of other basic needs are not threatened or compromised; and that adequate food is accessible to everyone, including vulnerable individuals and groups
Utilization	Having an adequate diet, clean water, sanitation, and health care to reach a state of nutritional well-being where all physiological needs are met
Stability	Having the ability to ensure food security in the event of sudden shocks (e.g., an economic, health, conflict, or climatic crisis) or cyclical events (e.g., seasonal food insecurity)
Agency	Individuals or groups having the capacity to act independently to make choices about what they eat, the foods they produce, how that food is produced, processed, and distributed, and to engage in policy processes that shape food systems. The protection of agency requires socio-political systems that uphold governance structures that enable the achievement of FSN for all
Sustainability	Food system practices that contribute to long-term regeneration of natural, social, and economic systems, ensuring the food needs of the present generations are met without compromising the food needs of future generations

Source: HLPE. 2020. Food security and nutrition: Building a global narrative towards 2030, Rome. https://openknowledge.fao.org/server/api/core/bitstreams/8357b6eb-8010-4254-814a-1493faaf4a93/content. Reproduced with permission

2017) distinguishes between four levels food security severity, either at the individual or the household level, as measured by the U.S. Household Food Security Survey Module. Food security includes high and marginal food security, whereas food insecurity includes low food security and very low food security. High food security is defined as "no reported indications of food-access problems or limitations," and marginal food security is described as "one or two reported indications—typically of anxiety over food sufficiency or shortage of food in the house. Little or no indication of changes in diets or food intake" (ERS 2017). Low food security (previously labeled as food insecurity without hunger) is defined as "reports of reduced quality, variety, or desirability of diet. Little or no indication of reduced food intake," and very low food security (previously identified as food insecurity with hunger) is described as "reports of multiple indications of disrupted eating patterns and reduced food intake" (ERS 2017). The USDA makes the distinction between food insecurity and hunger; food insecurity is an economic and social measure of inadequate food accessibility, while hunger is a physiological condition. The FAO (n.d.) also sought to describe levels of food security severity, delineating three levels at either the individual or household level as measured by the Food Insecurity Experience Scale, derived in part from the U.S. Household Food Security Survey Module. Food security or mild food insecurity is defined as "people who are food secure have adequate access to food in both quantity and quality. They become mildly food insecure when facing uncertainty about continued ability to obtain adequate food," whereas moderate food insecurity is described as "people experiencing moderate food insecurity have been forced to decrease the quality and/or quantity of the food they consume," and severe food insecurity is defined as "people experiencing severe food insecurity have typically run out of food and, at worst, gone a day (or days) without eating" (FAO n.d.).

The Integrated Food Security Phase Classification (IPC n.d.), a global partnership initially formed to address the 2004 food crisis in Somalia, adds a third dimension to the definition of food security—time. Acute food insecurity is defined as "food insecurity found at a specific point in time and of a severity that threatens lives or livelihoods, or both, regardless of the causes, context or duration," whereas chronic food insecurity is "food insecurity that persists over time mainly due to structural causes, including intra-annual seasonal food insecurity" (e.g., poverty, marginalization, lack of access to basic services) (Integrated Food Security Phase Classification n.d.).

4.3 Nutrition Security

A closely related concept to food security described by several researchers and organizations (Blankenship and Blancato 2022; Mozaffarian et al. 2021; NIFA 2025; Tirado et al. 2013; United Nations Standing Committee on Nutrition 2010; World Bank 2006) is nutrition security. The United Nations Standing Committee on Nutrition (2010) defined nutrition security as existing "when food security is combined with a sanitary environment, adequate health services, and proper care and feeding practices to ensure a healthy life for all household members." The World Bank (2006) describes nutrition security as an outcome of multiple factors, including food security, good health, a healthy environment, and caring practices. Some scholars (Poblician et al. 2022) argue that the distinction between nutrition security and food security is unnecessary because access to nutritious food is already included in the definition of food security. However, other researchers (Tirado et al. 2013; World Bank 2006; Mbow et al. 2019) recognize that individuals can be food secure while still being nutrition insecure, due to a lack of knowledge, poor health, cultural practices, or even personal preferences. The World Bank (2006) states that "food security is therefore often a necessary but not sufficient condition for nutrition security." The FAO et al. (2024) also makes the distinction between food insecurity and the prevalence of undernourishment, which occurs when "habitual food consumption is insufficient to provide the amount of dietary energy required to maintain a normal, active and healthy life" and is a measure of malnutrition. Malnutrition results from a deficiency, excess, or imbalance of energy and/or nutrients, which may lead to underweight, stunting and wasting, macronutrient deficiencies (primarily protein deficiency, but fat deficiency is possible; carbohydrate is unlikely), micronutrient deficiencies (of vitamins and minerals; also known as 'hidden hunger'), and/or overweight and obesity (FAO et al. 2024). Similar to the different phases of food security, the IPC (n.d.) also makes a distinction between malnutrition and food insecurity and provides a description of different phases and response objectives for acute malnutrition (Table 4.5). Several other tools exist to measure food security and nutrition status; see the International Dietary Data Expansion Project (n.d.).

5 Food Systems

Swinburn et al. (2019) describe a global syndemic of obesity, undernutrition, dietary risks for non-communicable diseases, and climate change. Swinburn and colleagues conclude that climate change worsens malnutrition in all its forms. The individual pandemics in this syndemic, a synergy of epidemics, co-occur; interact with each other; produce negative outcomes on social equity, human health and wellbeing, and ecological health and wellbeing; and share common societal drivers, including food and agriculture, land use, transportation, and urban design.

5.1 The Modern Food System and Planetary Boundaries

Food systems include various geographies, actors, agricultural services, societal elements, and natural elements (FAO 2018). Food systems also encompass different levels of inputs, technological advancement, and production outputs, depending on geography and resources. Actors within the food system include those directly involved in food production, those that support the food system, and consumers of the food systems; various actors include input suppliers, producers (ranging from households with gardens to supplement their nutrition needs and livelihoods, subsistence farmers, small scale landowners, to large-scale agribusiness), aggregators, processors, wholesalers, retailers, foodservice, consumers, national and international financing institutions, governments, intergovernmental organizations, civil society, and research organizations (FAO 2024a; HLPE 2020; IPCC 2023; Mbow et al. 2019). In fact, 892 million people are directly involved in production and over a billion people are employed in some capacity throughout the broader food system (Davis et al. 2023; FAO 2024b). These actors are impacted differently by the various food system components; vulnerable populations, including small-scale farmers, women, children, Indigenous Peoples, and individuals with disabilities bear the greatest burdens (FAO 2024a; IPCC 2023). Food flows through several channels within food systems, including production, aggregation, processing, distribution, consumption, and disposal of waste (FAO 2018). Agricultural services that support this flow of food include access to natural resources, labor, inputs, knowledge, finance, and other services. Societal elements of the food system include governments and organizations; policies, laws, and regulations; infrastructure; and sociocultural norms. Natural elements of the food system include ecosystems, genetics, air, soil, water, and climate.

Modern agriculture has significantly altered earth's ecosystems, more so than any other human activity, yet soil health and biodiversity are neglected in the current system. Globally, approximately 40% of land is used for agricultural purposes (United Nations Convention to Combat Desertification 2022). Nearly 80% of this agricultural land is devoted to livestock production but it only produces about 20% of global calories; this is because animal production is inefficient at converting animal feed to edible human food compared crop production (Mekonnen and Gerbens-Leenes 2020; United Nations Convention to Combat Desertification 2022). Modern agriculture, with intensive fossil fuel intensive inputs, accounts for 70% of freshwater use and is responsible for 80% of global deforestation, as well as being responsible for approximately 30% of greenhouse gas emissions, 50% of freshwater biodiversity loss, 52% of agricultural land degradation, and 70% of terrestrial biodiversity loss (United Nations Convention to Combat Desertification 2022).

When assessing food's environmental impact using four indicators—greenhouse gas emissions, water stress, land use, and eutrophication potential—meat, fish, and cheese have a high environmental impact, whereas fruits, breads, and sugary beverages have a low impact (Clark et al. 2022). Animal products generally use the most water (L/kg product), beef in particular, compared to crop foods (Mekonnen and Gerbens-Leenes 2020). Beef requires approximately 15,000 L of water/kg, pork 6000 L/kg, chicken meat 4300 L/kg, egg 3200 L/kg, milk 1000 L/kg, almonds 8000 L/kg, lentils 5900 L/kg, beans 5500 L/kg, peanuts 2800 L/kg, wheat 1800 L/kg, rice 1700 L/kg, and maize 1200 L/kg. Currently, 22.5% of global cropland is equipped for irrigation (FAO 2024b). More area is equipped for irrigation in Asia, mainly for rice production (which is a significant producer of the greenhouse gas, methane), while more area in Africa and Oceania is rainfed. Those areas that are irrigated often must compete for water for other needs (community use, energy, industry), while areas not irrigated are more susceptible to weather variability and climate change.

Planetary health researchers report that global food production is the largest threat to climate stability and ecosystem resilience and is the largest cause of environmental degradation and crossing of planetary boundaries (EAT-Lancet Commission on Food, Planet, Health 2019). The concept of planetary boundaries was developed by Rockstrom and colleagues (2009) as a set of safe limits for the earth system. Researchers (Richardson et al. 2023) quantified nine interdependent boundaries and determined that six have now been crossed, thereby increasing the risk of large-scale detrimental environmental change; freshwater change, land system change, climate change, biochemical flows, biosphere integrity, and novel entities have all exceeded planetary boundaries, and only ocean acidification, atmosphere loading, and stratospheric ozone depletion remain within safe planetary limits.

Agrifood systems have contributed to approximately $10 trillion in environmental, socioeconomic, and health impacts (FAO 2024a). Hidden cost drivers occur across three main pathways: environmental, socioeconomic, and health. Hidden cost drivers in the environmental pathway include greenhouse gas emissions throughout the food system leading to climate change and agricultural losses, nitrogen emissions and runoff causing water and air pollution, use of blue water resulting

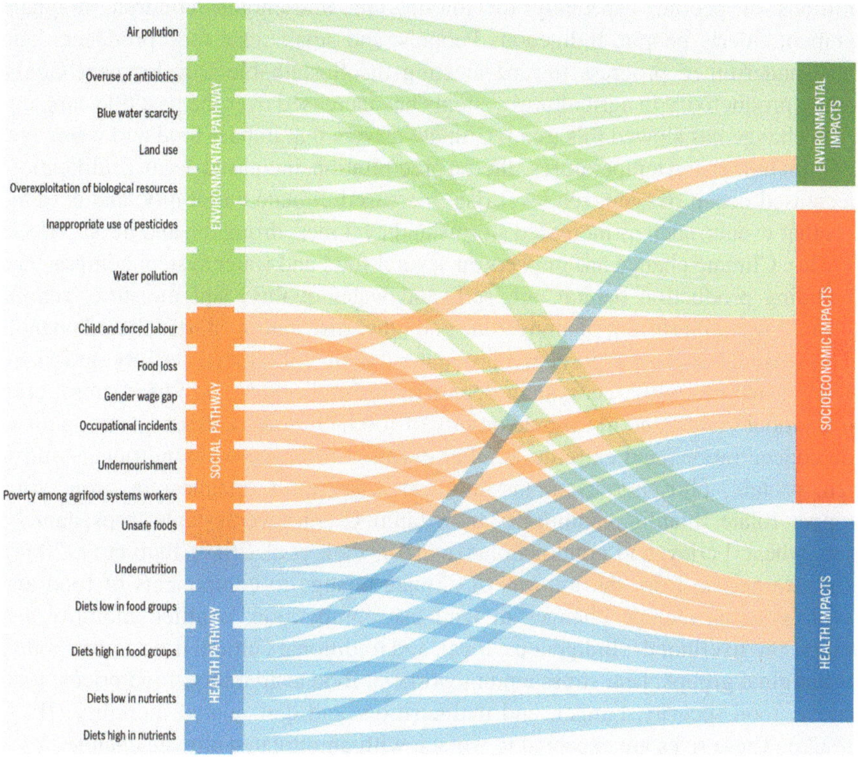

Fig. 4.1 Hidden cost drivers (of the agrifood system) along the environmental, social, and health pathways and their impact domains (FAO 2024a) (Reproduced from FAO 2024a, p. 7. https://doi.org/10.4060/cd2616en, licensed under the terms of the Creative Commons Attribution- 4.0 International licence (CC BY 4.0: https://creativecommons.org/licenses/by/4.0/legalcode.en))

in scarcity and subsequent labor and agricultural losses, and land use change causing ecosystem degradation (Fig. 4.1). Hidden costs of the socioeconomic pathway include failures in the distribution pathway that lead to undernourishment and poverty among farmworkers and other vulnerable populations leading to productivity losses. Hidden costs in the health pathway include consumption of unhealthy diets and malnutrition leading to productivity losses. In fact, unhealthy diets—low intake of whole grains, fruits, and vegetables—and related non-communicable disease contribute 70% of these hidden costs.

5.2 Climate Change Impact on Water and Food Security

The IPCC (2022a) recognizes with high confidence that increases in frequency and intensity of weather and climate extremes, losses in food production and food access, and decreased diet diversity have reduced food and water security for millions of people, especially for low-income households, children, pregnant women, elderly people, Indigenous Peoples, and small-scale food producers, and have undermined progress toward meeting the Sustainable Development Goals. While productivity in agricultural systems has increased over the past 50 years, climate change has slowed this trend. Climate drivers that impact food and water systems include rising temperatures, altered precipitation, increased ocean acidification, decreased ocean oxygen, sea level rise, increased climate variability, and extreme weather events, such as increased storms and cyclones, droughts, and floods (IPCC 2022a). Climate change has impacts on several food and water system components, including production habitat; air, soil, and water quality; soil moisture; runoff; streamflow; soil erosion; planting zone and other distribution shifts; altered phenology (seasonal changes); pests and pollinators; plant and animal fertility; grassland quality; tree mortality; ecosystem degradation; biodiversity loss; heat stress; crop and animal yields; and nutritional quality of food. Of note, increasing carbon dioxide concentrations and extreme weather events can decrease key nutrients—nitrogen, protein, phosphorous, potassium, calcium, sulfur, magnesium, iron, zinc, copper, folate, thiamin, vitamin B6, and vitamin C—in several staple crops, namely, rice, wheat, barley, and potato (Loladze 2014; Park et al. 2019; Taub et al. 2008). Risks to people from the impacts of climate change to components of food and water systems include changes in land use competition; conflict; mobility and migration; livelihoods; Indigenous, local, and traditional cultural uses; vulnerability of marginal groups; heat stress among workers; food availability; food prices; food safety; food security, hunger, and malnutrition; and diet related mortality (IPCC 2022a). These risks are expected to worsen with unmitigated climate change.

The impacts of climate change, food systems, and ecosystems are bidirectional (FAO 2022, 2024b; Mbow et al. 2019; Sutton et al. 2024). While climate change impacts ecosystems (land, water, oceans, biodiversity, soils) and food systems (production, processing, transport, retail, consumption, storage, loss, and waste) through altered temperatures, precipitation, extreme events, and other factors, ecosystems and food systems impact climate change through greenhouse gas emissions, pollutants, and other impacts. Additionally, ecosystems (i.e., water systems) and food systems impact each other. Both food security and insecurity are outcomes of the food system. Well-being is an outcome of food security, while malnutrition is an outcome of food insecurity. However, humans have the ability to impact greenhouse gas emissions and other ecosystem components through various mitigation and adaptation strategies.

The IPCC (2022a, 2023) described the negative impacts of climate change to food and water systems, and human health. Extreme weather events, such as extreme heat and drought, contribute to reduced soil moisture and health, reduced food yield and quality, and reduced labor capacity. Reduced soil moisture and health lead to food yield and quality losses. Losses in food yield and quality increase food prices, reduce household incomes, and reduce food security. Reduced household income contributes to reduced labor capacity and to decreased quality of life. Reduced labor capacity contributes to reduced household income, losses in food yield and quality, and decreased quality of life. Increased food prices also contribute to reduced food

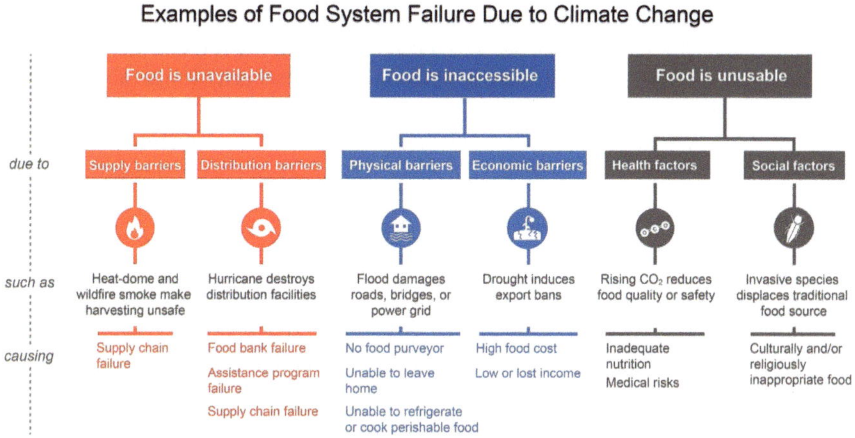

Fig. 4.2 Examples of food system failure due to climate change (Bolster et al. 2023) (Adapted from Chodur, G.M., Zhao, X., Biehl, E. et al. 2018, Figure 3. Assessing food system vulnerabilities: a fault tree modeling approach. *BMC Public Health* 18, 817. https://doi.org/10.1186/s12889-018-5563-x, licensed under the terms of the Creative Commons Attribution 4.0 International License (http://creativecommons.org/licenses/by/4.0/))

security. Reduced food security contributes to increased malnutrition and decreased quality of life. Increased malnutrition also contributes to decreased quality of life. The U.S. Fifth National Climate Assessment (Bolster et al. 2023) also provides a depiction of the interaction between climate change and the food system (Fig. 4.2). Long term trends and extreme weather events cause supply and distribution barriers which contribute to food being unavailable, physical and economic barriers contribute to food being inaccessible, and health and social factors contribute to food being unusable.

Population growth, increasing demand for food production, and shifting diets will drive future water requirements (IPCC 2019; Mekonnen and Gerbens-Leenes 2020). Additionally, climate change is projected to impact future water and food systems. Of note, half of grain production is projected to be impacted by severe water scarcity by 2050; this is especially concerning because approximately 50% of global calories come from grains (Joint World Health Organization [WHO]/FAO Expert Consultation 2005; United Nations Convention to Combat Desertification 2022).

The IPCC (Mbow et al. 2019) mapped current and future impacts of climate change to the four pillars of food security—availability, access, utilization, and stability. For example, reduced crop yields, reduced food quality, and disrupted food storage and transport networks from climate change may impact food availability. Disrupted infrastructure, price spikes, and changes in farmer livelihoods may lead to reduced food access. Increasing CO_2 and decreased nutritional quality in foods, increased food toxins, and increased diarrheal and infectious diseases can reduce food utilization. Increased food prices, unstable livelihoods, migration, and conflict may contribute to reduced stability. Combined systemic impacts include increasing environmental degradation and GHG emissions, limited biodiversity, competition for land and natural resource, and increasing food insecurity, undernourishment, obesity, and ill health.

5.3 *Sustainable Food Systems*

In light of the modern food system's inability to eliminate food insecurity and malnutrition in an ever-growing population, researchers have shifted attention to sustainable food systems that focus on the right to food and that recognize the interconnections of the food system to other supporting systems and interactions of the SDGs (HLPE 2020). The United Nations Brundtland Commission (United Nations Environment Program 1987) defines sustainable development as meeting "needs of the present without compromising the ability of future generations to meet their own needs." The FAO (2018) defines sustainable food systems as those that "deliver food security and nutrition for all in such a way that the economic, social and environmental bases to generate food security and nutrition for future generations are not compromised." The HLPE (2020) provides a framework to better identify components of a sustainable food system (Fig. 4.3). Sustainable food systems include food environments within larger food systems, as well as several drivers of

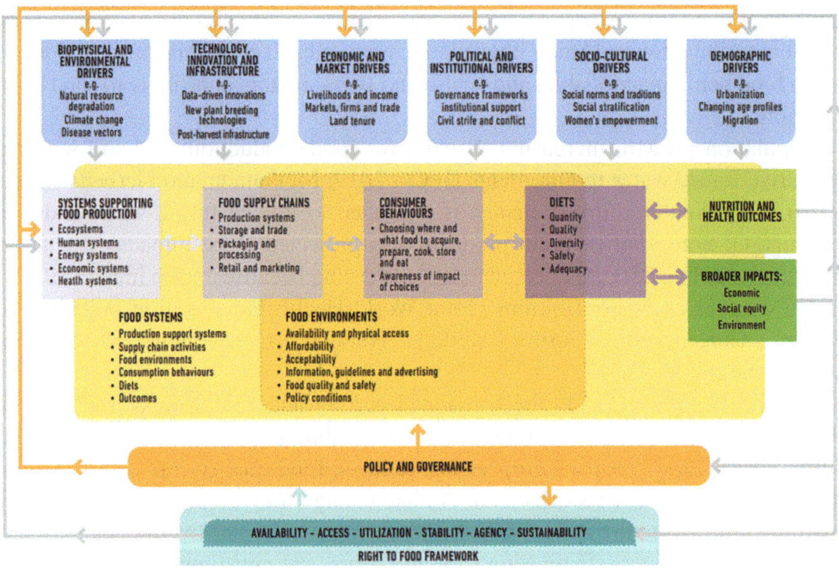

Fig. 4.3 Sustainable food system framework (Source: HLPE. 2020. Food security and nutrition: Building a global narrative towards 2030, Rome. https://openknowledge.fao.org/server/api/core/bitstreams/8357b6eb-8010-4254-814a-1493faaf4a93/content. Reproduced with permission)

food systems, policy and governance, and food security. Drivers of food systems are biophysical and environmental drivers, including climate change; technology, innovation, and infrastructure; economic and market drivers, including livelihoods and land tenure; political and institutional drivers, including conflict; sociocultural drivers, including social norms and traditions; and demographic drivers. Supporting systems to the food system, often seen as the beginning of the flow of food through the system, include ecosystems (water, land, air) and human, energy, economic, and health systems. The drivers also influence food supply chains (production, storage, trade, packaging, retail, marketing), consumer behaviors (choosing where and what food to acquire, prepare, cook, store, and eat), diets (quantity, quality, diversity, safety, adequacy), and outcomes (nutrition, health, economic, social equity, and environment). Several qualities of a sustainable food system are linked to the six pillars of food security: productive and prosperous (availability), equitable and inclusive (access), respectful and empowering (agency), resilient (stability), regenerative (sustainability), and healthy and nutritious (utilization) (HLPE 2020).

Sustainable food systems should help consumers eat a sustainable diet, a diet that combines concepts of both food security and sustainability. The FOA and WHO (2019) define sustainable healthy diets as "dietary patterns that promote all dimensions of individuals' health and wellbeing; have low environmental pressure and impact; are accessible, affordable, safe and equitable; and are culturally acceptable."

6 Climate Change, Conflict, and Water and Food Security

Water and food scarcity can be both a driver and a consequence of conflict. Countries and territories in protracted conflict or crisis may experience disruptions of water and food systems resulting in water and food scarcity (FAO 2024a; HLPE 2024; United Nations Children's Fund 2021). Other causes of food and water scarcity in areas of conflict occur from displacement, the use of withholding food and water as a weapon, destruction of food production and supply chains, interruption of humanitarian aid, currency devaluation, volatile food prices and price inflation, unemployment, loss of assets, and diminished coping strategies and support systems. Areas of conflict also experience significant social and environmental hidden costs throughout the food system, 18% and 20% of gross domestic product, respectively (FAO 2024a). These costs include limited agricultural support and financing; low productivity, fertilizer intensity, and rural electrification; few social protections, poverty, and underemployment; fragmented government structures and low government effectiveness; land degradation; loss of biodiversity; and increased vulnerability to food insecurity. Conscription or migration may also worsen food and water security in countries with protracted conflict and crisis. Climate change may act to intensify the effects of conflict on food and water security (Intergovernmental Panel on Climate Change 2023). One area particularly affected by conflict and climate change is Ukraine. See Case Study 4.1 for a deeper look at the intersection of climate change, conflict, and water.

Case Study 4.1 The impact of the war on food and water security in Ukraine

Author: Dr. Oleksandr Nechyporenko.

The negative impact of the consequences of the Russian Federation's military aggression against Ukraine, as well as the challenges of increasing climate change, have become the dominant causes of negative transformations in the economy of the agricultural sector of the state in recent years. Thus, if we compare the average annual results of production activities in agriculture in the 3 pre-war years with the average annual indicators of the 3 years of the war, it should be noted that the volume of gross agricultural output decreased by almost 17% and amounted to an average of UAH (Ukrainian monetary unit) 1048 billion during the full-scale invasion, while an almost equal decrease was observed in both crop production and animal husbandry. The overall structure of gross output hardly changed during this time: 80% is crop production, of which 13–14% is wheat, 16–18% is corn, 27–30% is oilseeds, and 20% is animal husbandry, of which 10%–11% is animal and poultry farming and 5%–6% of the total structure is milk. At the same time, households reduced gross output by 18.5%, compared to 14.7%, which characterizes, in addition to the aforementioned key factors, a decrease in peasants' motivation to engage in agriculture.

The limited resources, especially fixed assets, caused a significant drop in production in natural terms. Thus, the harvest of grain crops decreased by more than 26%, oilseed crops by almost 6%, and vegetables by 19%. At the same time, the average annual yield of the vast majority of crops, despite the high figures in 2021, increased during wartime. For example, grain and leguminous crops increased their average yield by 1.4 centners/ha, and sugar beets by almost 65 centners/ha, while the yield of oilseed crops remains at the prewar level—on an average, over 23 centners/ha. In animal husbandry, egg and wool production declined by more than 25%, total meat production by 10%, and milk production by almost 19%.

On average, over the 3 years of the war, the agricultural consumption has not been replenished to the pre-war level only for: milk (−19.4%), meat (−3.6%), vegetables and melons (−21.5%), as well as fruits, berries, grapes (−22%). However, per capita population remaining in the country, the provision of these products has remained almost at the pre-war level, meat has even increased 5.6%. At the same time, it is worth mentioning that these indicators are significantly less than the recommended consumption norms, and therefore motivation and resources are needed to focus the efforts of agrarian business to increase the production of these types of products.

It should be noted that in addition to military operations, the performance of the agricultural sector of the state's economy is significantly affected by climate change, the main negative factor of which is a decrease in the moisture supply of plants, especially during the growing season. Therefore, irrigation, as well as providing livestock and the rural population with water, is of great importance for Ukraine.

In the pre-war years, Ukraine had 63 thousand rivers, 40 thousand lakes, 1100 reservoirs, and 400 thousand ponds—one million 100 thousand hectares of freshwater area. At the request of the World Financial Bank, studies were conducted that put Ukraine in 125th place out of 180 countries in terms of the amount of drinking water per capita. That is, Ukrainians are somewhere between the African countries of Chad and Sudan. And in the Kherson region, there is less drinking water than in the central part of the country, more than twice as much. The main component of the water balance of Ukraine is the flow of rivers, which forms surface irrigation sources. It is determined that the average annual amount of precipitation was 609 mm or 366 cubic km (in different climatic zones, the amount ranges from 310 to 1350 mm), which is spent mainly on evaporation (526 mm, or 86%) and in a much smaller amount—on the formation of local flow (83 mm, or 14%). At the same time, the share of surface flow is 64 mm, and underground—19 mm, they form 60–80% of river flow and 10–20% of groundwater assets.

Irrigated agriculture is characterized by the largest irreversible water consumption, the value of which reached 75% of the total amount of water taken for irrigation. According to the State Agency of Water Resources of Ukraine, in the pre-war period, more than a third of the total water resources asset, 19.24 cubic km, was used annually to meet industrial, municipal, and agricultural needs (1.2 cubic km). Over the last two pre-war decades, the structure of water consumption has significantly shifted toward industry, which "consumed" 70% of the total volume of extracted, and the share of agriculture has decreased from 30% in the 1980s to 5%–6% in recent years, mainly due to a significant reduction in the area of irrigated land.

In the pre-war years, Ukraine had a total area of irrigated land of 2170.5 thousand hectares (1770 thousand hectares excluding the Autonomous Republic of Crimea) compared to 2.5 million hectares at the end of the 1990s. Every year, more than 1.2 million hectares of productive land in mainland Ukraine, previously equipped for artificial irrigation, are used without irrigation. Almost a third of them were privately owned and 68% were leased.

The non-use of almost 70% of the recorded irrigated areas for their intended purpose was mainly due to the violation of the technological integrity of land reclamation systems due to the hasty and ill-considered division of agricultural lands, as well as the inadequate condition, including unauthorized dismantling, of engineering infrastructure.

The efficiency of using actually irrigated land (400–500 thousand hectares), in addition to the need to modernize the outdated infrastructure, was also influenced by other factors. Thus, the profitability of crop production was significantly reduced by: the high cost of irrigation services and electricity; imperfection of crop cultivation and irrigation technologies; large losses of irrigation water, up to 40–50% in irrigation canals, silting and overgrowth of them with algae; violation of the design structure of crops; a general decrease in soil fertility; and a growing shortage of qualified personnel from year to year, especially middle-level land reclamation specialists.

As a result, at the beginning of the 1990s, the area of irrigated land in Ukraine was 8% of the total area of arable land then in the pre-war year 2021; this share decreased to 1.54% of all areas from which was harvested and during the war (as of 2023) it decreased to 0.4%. In the best years, the production of agricultural products on irrigated lands amounted to up to 20% of the total volume in the country. At the same time, in regions with developed irrigation this indicator was significantly higher. Thus, in Kherson region, the share of products obtained on irrigated lands (25.6% of arable land) was 46%, in Zaporizhia region (13.4% of arable land)—30%, Odessa region (11.2% of arable land)—29%, and Mykolaiv region (11.1% of arable land)—28%. In the Crimean Autonomous Region, 76% of fodder, 90% of vegetables, and about 70% of fruits were grown on irrigated lands (29.2% of arable land). However, already in the pre-war year 2021, agricultural enterprises produced only 1.78% of the country's gross agricultural output on irrigated lands, and as a result of the occupation and destruction of the Kakhovka hydroelectric power station dam, this share decreased to 0.48%, or more than 40 times less compared to the peak of irrigation development in Ukraine. During the war years, the area of irrigated land cultivated by enterprises decreased almost 5 times and amounted to only 84.5 thousand hectares in 2023, or 0.4% of the total area from which the harvest was harvested in the state. During this period, the production of grain and leguminous crops on these areas in Ukraine decreased almost five times, and oilseed crops decreased almost nine times. But even now, Ukraine produces several times more of these crops than is necessary to ensure food security. Thus, to meet domestic needs, only a third of the grown grain and about 8% of oilseeds are needed. However, the more than halving (to 0.53 million tons) of production on irrigated lands of vegetables and melon crops, which the country's population had previously largely been provided with through imports, was significant. The war greatly affected the total production of these products by Ukrainian agriculture. Thus, the total production of open ground vegetables during the war years decreased by almost 23% and amounted to 7.7 million tons in 2024, compared to more than 9.9 million tons in 2021. At the same time, the regions where military operations took place reduced production by half, although the yield in these territories is traditionally higher mainly due to irrigated areas. Significant production losses were caused by the explosion in June 2023 of the Kakhovka hydroelectric power station dam by Russian aggressors. This resulted in the loss of about 20% of Ukraine's vegetable production and 46% of its melon production. About 20,000 hectares of fields out of more than 300,000 hectares irrigated from the destroyed reservoir were occupied by vegetables.

The Kakhovka HPS provided annual regulation of the Dnipro River flow for electricity supply, irrigation, and water supply to arid regions of southern Ukraine from Kherson to Zaporizhia, as well as the Kakhovka, North Crimean canals and the Dnipro-Kryvyi Rih canal. A characteristic feature of the Kakhovka hydroelectric complex was the location of an earthen dam 30 m high on silt, which allowed, using a closed distribution device, to create the Kakhovka reservoir with a volume of 18.2 billion cubic meters of water and raise the water level in the Dnipro River to 16 meters. The Kakhovka reservoir is one of six large reservoirs in the cascade on the Dnipro River, is 240 km long, and is located in the Zaporizhia, Dnipropetrovsk, and Kherson regions.

Therefore, the negative consequences of this terrorist attack for wildlife will be manifested in an area of at least 5000 km^2 (flooding zone and drainage zone). Only the zone that has been covered by the reservoir waters for the past 70 years and is now exposed to the open air for many years will be larger than 1000 km^2. By blowing up the Kakhovka hydroelectric power station, the occupiers caused significant direct damage to Ukrainian territories and indirect damage that will manifest itself over time (Fig. 4.4). If we speak exclusively in the context of agriculture, in particular irrigated agriculture, then irrigation from the reservoir will be impossible in the region, according to experts, until 2–3 years after the dam is restored.

Before the explosion of the dam, it was estimated that the 1.5 m drop in the water level in the reservoir led to the loss of 14% of the export potential of

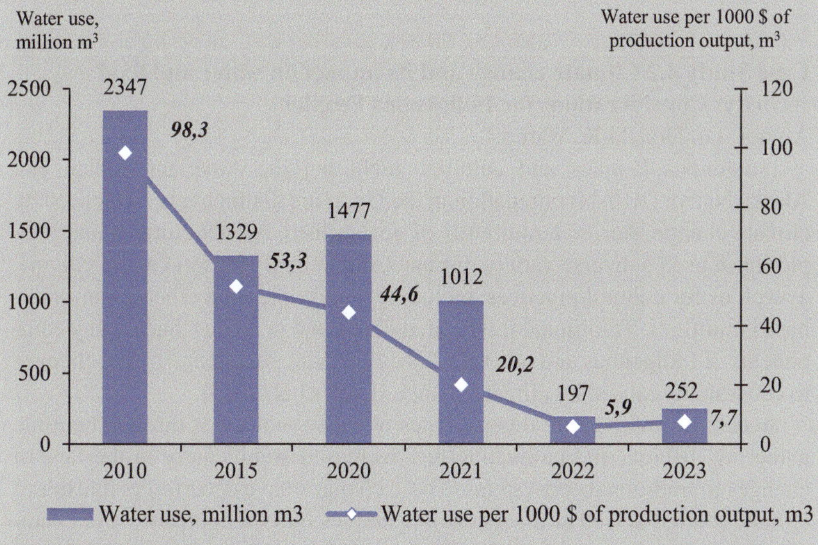

Fig. 4.4 Water usage in Ukraine (Source: author Nechyporenko, based on calculations of data from the State Statistics Service of Ukraine (2023) https://www.ukrstat.gov.ua/)

Ukrainian wheat due to problems with irrigation in the Zaporizhia, Dnipropetrovsk, and Kherson regions.

According to preliminary estimates, the losses from damage to irrigation systems will amount to up to $10 billion. The damage from the destruction of fish resources is also estimated at $10.5 billion. As a result of the explosion, water supply to 31 field irrigation systems in the Dnipropetrovsk, Kherson, and Zaporizhia regions was stopped. In 2021, these systems provided irrigation for 584 thousand hectares, from which farmers harvested about four million tons of grain and oilseed crops worth about $1.5 billion.

According to estimates by Ukrainian scientists, in the post-war period, provided that the occupied territories are liberated, it will be necessary to irrigate at least 18 million hectares of agricultural land to combat the effects of climate change, which is increasingly reducing the availability of soil moisture.

7 Climate Change, Water, and Indigenous Food Systems

While Indigenous Peoples often experience significant costs related to climate change, they also actively contribute to creating sustainable food and water systems that help mitigate impacts from climate change (IPCC 2022b). For a closer look at the risks from climate change to Indigenous Peoples water and food systems and water and food security, see Case Study 4.2.

Case Study 4.2 Climate change and its impact on water and food security: Considerations for Indigenous Peoples

Author: Dr. Donald K. Warne

Indigenous Peoples and cultures, including the American Indian and Alaska Native (AI/AN) population in the USA, are distinctively influenced by climate change. Across a multitude of generations, AI/AN cultures and peoples have used a diverse variety of plants and animals for food and medicine, as well as for cultural practices, including ceremonies and other community-based traditions. Traditional foods, medicines, and practices that are key components of Indigenous and other AI/AN cultures are threatened by the changes to ecosystems caused by climate change (Lynn et al. 2013).

Access to water and to diverse types of traditional foods through hunting, gathering, fishing, and cultivating has decreased significantly in the face of changes to traditional ecosystems. These changes have occurred as a result of climate change as well as diversion of resources for other purposes. For example, damming rivers in Arizona (DeJong 2024) and the Dakotas (Johnson et al. 2015) has increased access to natural resources for the general population, but at the expense of traditional AI/AN ways of being. A substantial loss has occurred in terms of access to water, native fish, farmlands, and animals that can no longer depend on historic sources of water.

Climate change has a complex effect on ecosystems and natural processes. In Alaska and the Pacific Northwest, climate change has resulted in a severe reduction in the salmon population (Schoen et al. 2017; Siegel & Crozier 2019) both an affront to access to food, but also resulting in loss of Indigenous cultural practices and community identity (Atlas et al. 2021). The timing and quantity of salmon runs has been altered by climate change, and traditional fishing methods and timing of ceremonies associated with the salmon runs have changed in a negative manner that lies outside traditional cultural context (Jacob et al. 2010).

In addition to the impact of climate change on Indigenous food sovereignty and food security, AI/AN communities have seen access to safe drinking water impacted as well. In considering the water ecosystem, numerous Tribes in the USA have unique rights in terms of historical access to waterways (Anderson 2016). The pressures posed by climate change and population growth in states like Arizona (Shamir et al. 2015) place even greater risk of water access to the regional tribes (Chief 2018).

Climate change has led to the loss of access to traditional food sources and safe drinking water. This has had a direct impact on the health of AI/AN and other Indigenous peoples (Tanana 2023). The broader impacts of climate change and its effect on water and food security for Indigenous Peoples need to consider the role for promotion of traditional foods in addition to climate change mitigation (Milburn 2004). The complexity of climate change considerations for Indigenous populations requires a multifaceted and interdisciplinary approach to develop solutions in collaboration with the broader society and non-Indigenous populations.

8 Progress on Sustainable Food Systems, and Water, Food, and Nutrition Security

The Food Systems Countdown Initiative (2025) is tracking the progress of 50 food system indicators important for reaching the SDGs. Access to safe drinking water, availability of fresh fruits and vegetables, and prevalence of undernourishment have improved since 2000; however, cost of a healthy diet and share of population experiencing moderate or severe food insecurity have worsened across this same time span. Greenhouse gas emissions per kilogram produced from beef and milk, nitrogen use efficiency, and food product yield have improved, while pesticide use has worsened and water withdrawal for irrigation, overall greenhouse gas emissions, and greenhouse gas emissions per kilogram produced from cereal and rice has remained the same. Food safety capacity has remained unchanged, while food supply variability has improved, and food prod price volatility has worsened.

9 Solutions for Improving Sustainable Food Systems, and Water, Food, and Nutrition Security

Swinburne et al. (2019) recommend four actions to address the global syndemic of undernutrition, obesity, and climate change. These include strengthening governance at all levels to apply the concept of a right to wellbeing, countering commercial power at odds with planetary health, including redirecting $5 trillion in fossil fuel and food industry subsidies to more sustainable practices, creating strong accountability systems, and creating platforms for stakeholder funding and interaction.

Several near-term climate mitigation and adaptation options related to land, water, and food exist that have co-benefits for the climate, the environment, and human health, including food and water security (IPCC 2023). Mitigation options include reducing conversion of natural ecosystems, agriculture carbon sequestration, ecosystem restoration, afforestation and reforestation, shifting to healthy diets, improving sustainable forest management, reducing methane and nitrous oxides in agriculture, and reducing food loss and food waste. Climate responses and adaptation options include forest-based adaptation, improved cropland management, water use efficiency and water resource management, biodiversity management and ecosystem connectivity, agroforestry, sustainable aquaculture and fisheries, integrated coastal zone management, coastal defense, and efficient livestock systems. The IPCC also recognizes that several adaption options have synergies with the mitigation options, showing potential for scaling up climate action to improve food and nutrition security but notes that minimizing trade-offs from competing demands and cooperation with local communities and Indigenous People is imperative for successful adaption. In addition to shifting to sustainable healthy diets and reducing food loss and waste, other public health mitigation and adaption strategies with co-benefits include developing public health policies to improve nutrition, making health insurance accessible, providing social protections and safety nets, delivering public health educational and promotion campaigns, shifting to active mobility and public transport, shifting to clean energy sources, implementing early warning and response systems for high heat, water, food, and vector-borne diseases, improving access to clean water, reducing flooding impacts to water and sanitation systems, and increasing access to mental health care.

Like the IPCC, the FAO (2024a) provides similar recommendations to mitigate climate change that have co-benefits for human health, including reducing food losses and shifting diets. Specifically, shifting diets requires increasing intake of nuts and seeds, fruits, vegetables, and legumes, and lowering intake of sugars, sugar sweetened beverages, salt, processed and red meat, and processed foods. Shifting diets can be supported by regulations, financial incentives, promotion and education campaigns, and healthy and sustainable institutional procurement. The EAT-Lancet Commission on Food, Planet, Health (2019) also recommends shifting to a plant-based diet, increasing fruits, vegetable, whole grains, and legumes, and reducing red meat and sugar, to improve human health while staying within the planetary boundaries.

The IPCC (Mbow et al. 2019) proposes several examples of mitigation and adaptation strategies for climate change mapped to the four pillars of food security (Table 4.7). For example, the availability of food could be improved by reducing food waste, modifying diets to increase plant-based foods, adopting new technology, increasing productivity, adopting previously neglected crop varieties, and undertaking risk management practices, such as insurance. Access to food could be improved by reducing loss and waste, shortening supply chains, and utilizing integrated agricultural practices to build livelihoods that are more resilient. Utilization of food could be improved by increasing biodiversity, improving sanitation across food systems, and increasing consumption of healthier diets. Stability could be improved by diversifying local agriculture, reducing food loss and waste, investing in infrastructure, providing farmers with crop insurance, and building capacity to develop system resilience. Systematic impacts from all four pillars include reducing food loss and waste, shifting diets to include more plant-based foods and reduced energy intensive foods, increasing food system productivity, increasing production of healthy foods, reducing GHG, building system resilience, and including gender and equity in institutional and governance responses.

10 Conclusions

Global water and food systems currently experience significant impacts from climate change, which contributes to water, food, and nutrition insecurity. Implementing climate change mitigation and adaptation strategies can improve the water and food security status of millions of people around the globe. Adopting sustainable food systems that limit fossil fuel inputs, decrease food waste, and shift to plant-based diets will improve both human and planetary health while helping mitigate climate change.

11 Discussion Questions

1. Define water, food, and nutrition security.
2. How does the modern food system differ from sustainable food systems?
3. What are the impacts of climate change on food and water security?
4. What are practical solutions linked to climate change to improve water and food security?

References

Anderson RT (2016) Indian water rights and the Federal trust responsibility. Nat Resourc J 46(2):399–437. https://digitalrepository.unm.edu/nrj/vol46/iss2/5

Atlas WI, Ban NC, Moore JW, Tuohy AM, Greening S, Reid AR, Morven N, White E, Housty WG, Housty JA, Service CN, Greba L, Harrison S, Sharpe C, Butts KIR, Shepert WM, Sweeney-Bergen E, Macintyre D, Sloat MR, Connors K (2021) Indigenous systems of management for culturally and ecologically resilient Pacific salmon (*Oncorhynchus* spp.) fisheries. Bioscience 71(2):186–204. https://academic.oup.com/bioscience/article/71/2/186/6028542

Blankenship J, Blancato RB (2022) Nutrition security at the intersection of health equity and quality care. J Acad Nutr Diet 122(10S):S12–S19. https://www.sciencedirect.com/science/article/pii/S2212267222003896?via%3Dihub

Bolster CH, Mitchell R, Kitts A, Campbell A, Cosh M, Farrigan TL, Franzluebbers AJ, Hoover DL, Jin VL, Peck DE, Schmer MR, Smith MD (2023) Ch. 11. Agriculture, food systems, and rural communities. In: Crimmins AR, Avery CW, Easterling DR, Kunkel KE, Stewart BC, Maycock TK (eds) Fifth National Climate Assessment. U.S. Global Change Research Program, Washington, DC. https://repository.library.noaa.gov/view/noaa/61592

Borlaug N (1970) Norman Borlaug, Nobel lecture. https://www.nobelprize.org/prizes/peace/1970/borlaug/lecture/

Chief K (2018) Emerging voices of tribal perspectives in water resources. J Contemp Water Res Educ 163:1–5. https://doi.org/10.1111/j.1936-704X.2018.03266.x

Clark M, Springmann M, Rayner M, Scarborough P, Hill J, Tilman D, Macdiarmid JI, Fanzo J, Bandy L, Harrington RA (2022) Estimating the environmental impacts of 57,000 food products. PNAS 119(33):e2120584119. https://doi.org/10.1073/pnas.2120584119

Davis B, Mane E, Yonca Gurbuzer L, Caivano G, Piedrahita N, Azhar N, Benali M, Chaudhary N, Rivera R, Schneider K, Ambikapathi R, Winters P (2023) Estimating global and country-level employment in agrifood systems, FAO Statistics Working Paper Series 23–34. Food and Agriculture Organization of the United Nations. https://openknowledge.fao.org/server/api/core/bitstreams/08f5f6df-2ef2-45a2-85aa-b29c2ce01422/content

DeJong DH (2024) Damming the Gila. The Gila River Indian community and the San Carlos irrigation project, 1900–1942. University of Arizona Press. https://wrrc.cals.arizona.edu/sites/wrrc.arizona.edu/files/documents/DeJong_Revised_Damming_Gila.pdf

Department of Economic and Social Affairs (n.d.) Sustainable development goals. United Nations. https://sdgs.un.org/goals

EAT-Lancet Commission on Food, Planet, Health (2019) Healthy diets from sustainable food systems: Summary report of the EAT-Lancet Commission. https://eatforum.org/eat-lancet-commission/eat-lancet-commission-summary-report/

Economic Research Service (2017) Definitions of food security. United States Department of Agriculture. https://www.ers.usda.gov/topics/food-nutrition-assistance/food-security-in-the-us/definitions-of-food-security

Food and Agriculture Organization of the United Nations (1996) World Food Summit plan of action. United Nations https://www.fao.org/4/w3613e/w3613e00.htm

Food and Agriculture Organization of the United Nations (2001) 2001 The state of food insecurity in the world. United Nations. https://www.fao.org/4/y1500e/y1500e00.htm

Food and Agriculture Organization of the United Nations (2018) Sustainable food systems: Concept and framework. http://www.fao.org/3/ca2079en/CA2079EN.pdf

Food and Agriculture Organization of the United Nations (2022) The future of food and agriculture – drivers and triggers for transformation. doi:10.4060/cc0959en

Food and Agriculture Organization of the United Nations (2024a) The state of food and agriculture 2024 – value-driven transformation of agrifood systems. doi:10.4060/cd2616en

Food and Agriculture Organization of the United Nations (2024b) World food and agriculture – statistical yearbook 2024. doi:https://doi.org/10.4060/cd2971en

Food and Agriculture Organization of the United Nations (n.d.) Voices of the hungry. United Nations. https://www.fao.org/in-action/voices-of-the-hungry/background/en/

Food and Agriculture Organization of the United Nations & UN-Water (2024) Progress on the level of water stress – Mid-term status of SDG Indicator 6.4.2 and acceleration needs, with special focus on food security – 2024. United Nations. https://doi.org/10.4060/cd2179en

Food and Agriculture Organization of the United Nations and World Health Organization (2019) Sustainable healthy diets-guiding principles. https://www.who.int/publications/i/item/9789241516648

Food and Agriculture Organization of the United Nations, International Fund for Agricultural Development, United Nations Children's Fund, World Food Programme, & World Health Organization (2024) The state of food security and nutrition in the world 2024. https://openknowledge.fao.org/server/api/core/bitstreams/d5be2ffc-f191-411c-9fee-bb737411576d/content

Food Systems Countdown Initiative (2025) The food systems countdown report 2024: Tracking progress and managing interactions. Columbia University; Cornell University; Food and Agriculture Organization of the United Nations; Global Alliance for Improved Nutrition. https://doi.org/10.36072/fsci2024

High Level Panel of Experts (2020) Food security and nutrition: Building a global narrative towards 2030. United Nations. https://openknowledge.fao.org/server/api/core/bitstreams/8357b6eb-8010-4254-814a-1493faaf4a93/content

High Level Panel of Experts (2024) Conflict-induced acute food crises: potential policy responses in light of current emergencies. United Nations. https://www.fao.org/docs/devhlpelibraries/default-document-library/hlpe-fsn-issues-papers_conflicts-and-fsn.pdf?sfvrsn=823378b6_4

Integrated Food Security Phase Classification (n.d.) Understanding the IPC scales. https://www.ipcinfo.org/fileadmin/user_upload/ipcinfo/docs/communication_tools/brochures/IPC_Brochure_Understanding_the_IPC_Scales.pdf

Intergovernmental Panel on Climate Change (2019): Climate change and land: An IPCC special report on climate change, desertification, land degradation, sustainable land management, food security, and greenhouse gas fluxes in terrestrial ecosystems. [PR Shukla, J Skea, E Calvo Buendia, V Masson-Delmotte, H-O Pörtner, DC Roberts, P Zhai, R Slade, S Connors, R van Diemen, M Ferrat, E Haughey, S Luz, S Neogi, M Pathak, J Petzold, J Portugal Pereira, P Vyas, E Huntley, K Kissick, M Belkacemi, J Malley, (eds.)]. doi:10.1017/9781009157988

Intergovernmental Panel on Climate Change (2022a) Climate change 2022: Impacts, adaptation and vulnerability. Contribution of working group ii to the sixth assessment report of the Intergovernmental Panel on Climate Change. [PR Shukla, J Skea (eds.)]. https://www.ipcc.ch/report/ar6/wg2/downloads/report/IPCC_AR6_WGII_FullReport.pdf

Intergovernmental Panel on Climate Change (2022b) Climate change 2022: Mitigation of climate change. Contribution of working group iii to the sixth assessment report of the Intergovernmental Panel on Climate Change. [H-O Pörtner, DC Roberts, M Tignor, ES Poloczanska, K Mintenbeck, A Alegría, M Craig, S Langsdorf, S Löschke, V Möller, A Okem, B Rama (eds.)]. https://www.ipcc.ch/report/ar6/wg3/downloads/report/IPCC_AR6_WGIII_FullReport.pdf

Intergovernmental Panel on Climate Change (2023) Climate change 2023: Synthesis report. Contribution of working groups I, II and III to the sixth assessment report of the Intergovernmental Panel on Climate Change. [Core Writing Team, H Lee, J Romero (eds.)]. IPCC. https://www.ipcc.ch/report/ar6/syr/downloads/report/IPCC_AR6_SYR_FullVolume.pdf

International Dietary Data Expansion Project (n.d.) Data 4Diets: Food security indicators. https://inddex.nutrition.tufts.edu/data4diets/indicators

Jacob C, McDaniels T, Hinch SG (2010) Indigenous culture and adaptation to climate change: sockeye salmon and the St'át'imc people. Mitig Adapt Strateg Glob Change 15(8):859–876. https://doi.org/10.1007/s11027-010-9244-z

Johnson WC, Volke MA, Scott ML, Dixon MD (2015) The dammed Missouri: Prospects for recovering Lewis and Clark's River. Ecohydrology 8(5):765–771. https://doi.org/10.1002/eco.1534

Joint World Health Organization/Food and Agriculture Organization Expert Consultation (2005) Diet, nutrition and the prevention of chronic diseases. WHO technical report series 916. https://www.who.int/publications/i/item/924120916X

Loladze I (2014) Hidden shift of the ionome of plants exposed to elevated CO_2 depletes minerals at the base of human nutrition. eLife 3:e02245. https://doi.org/10.7554/eLife.02245

Lynn K, Daigle J, Hoffman J, Lake F, Michelle N, Ranco D, Viles D, Voggesser G, Williams P (2013) The impacts of climate change on tribal traditional foods. Clim Chang 120:545–556. https://link.springer.com/article/10.1007/s10584-013-0736-1

Mbow C, Rosenzweig C, Barioni LG, Benton TG, Herrero M, Krishnapillai M, Liwenga E, Pradhan P, Rivera-Ferre MG, Sapkota T, Tubiello FN, Xu Y (2019) Food security. In: Climate change and land: an IPCC special report on climate change, desertification, land degradation, sustainable land management, food security, and greenhouse gas fluxes in terrestrial ecosystems. Intergovernmental Panel on Climate Change. [PR Shukla, J Skea, E Calvo Buendia, V Masson-Delmotte, H-O Pörtner, DC Roberts, P Zhai, R Slade, S Connors, R van Diemen, M Ferrat, E Haughey, S Luz, S Neogi, M Pathak, J Petzold, J Portugal Pereira, P Vyas, E Huntley, K Kissick, M Belkacemi, J Malley, (eds.)]. https://doi.org/10.1017/9781009157988.007

Mekonnen M, Gerbens-Leenes W (2020) The water footprint of global food production. Water 12(10):2696. https://doi.org/10.3390/w12102696

Milburn MP (2004) Indigenous nutrition: using traditional food knowledge to solve contemporary health problems. Am Indian Q 28(3 & 4):411–434. https://muse.jhu.edu/article/181503

Mozaffarian D, Fleischhacker S, Andres JR (2021) Prioritizing nutrition security in the US. JAMA 325(16):1605–1606. https://jamanetwork.com/journals/jama/fullarticle/2778232

National Institute of Food and Agriculture (2025, January 22). What is food and nutrition security? United States Department of Agriculture. https://www.nifa.usda.gov/topics/food-nutrition-security#:~:text=Building%20on%20and%20complementing%20our,equity%20lens%20to%20our%20efforts

Park CS, Vogel E, Larson LM, Myers SS, Daniel M, Biggs B-A (2019) The global effect of extreme weather events on nutrient supply: a superposed epoch analysis. Lancet Planet Health 3(10):e429–e438. https://www.thelancet.com/journals/lanplh/article/PIIS2542-5196(19)30193-7/fulltext

Poblacion A, Ettinger de Cuba S, Cook JT (2022) For 25 years, food security has included a nutrition domain: is a new measure of nutrition security needed? J Acad Nutr Diet 122(10):1837–1840. https://www.sciencedirect.com/science/article/pii/S2212267222002362?via%3Dihub

Richardson K, Steffen W, Lucht W, Bendtsen J, Cornell SE, Donges JF, Druke M, Fetzer I, Bala G, von Bloh W, Feuler G, Fiedler S, Gerten D, Gleeson T, Hofmann M, Huiskamp W, Kummu M, Mohan C, Nogues-Bravo D et al (2023) Earth beyond six of nine planetary boundaries. Sci Adv 9(37). https://doi.org/10.1126/sciadv.adh2458

Rockström J, Steffen W, Noone K, Persson Å, Chapin FS III, Lambin E, Lenton TM, Scheffer M, Folke C, Schellnhuber H, Nykvist B, De Wit CA, Hughes T, van der Leeuw S, Rodhe H, Sörlin S, Snyder PK, Costanza R, Svedin U et al (2009) Planetary boundaries: exploring the safe operating space for humanity. Ecol Soc 14(2):32. http://www.ecologyandsociety.org/vol14/iss2/art32/

Schneider KR, Remans R, Bekele TH, Aytekin D, Conforti P, Dasgupta S, DeClerck F, Dewi D, Fabl C, Gephart JA, Masuda YJ, McLaren R, Salsana M, Aburto N, Ambikapathi R, Arellano Rodriguez M, Barquera S, Battersby J, Beal T et al (2025) Governance and resilience as entry points for transforming food systems in the countdown to 2030. Nat Food 6:105–116. https://doi.org/10.1038/s43016-024-01109-4

Schoen ER, Wipfli MS, Trammell EJ, Rinella DJ, Floyd AL, Grunblatt J, McCarthy MD, Meyer BE, Morton JM, Powell JE, Prakash A, Reimer MN, Stuefer SL, Toniolo H, Wells BM, Witmer FDW (2017) Future of Pacific salmon in the face of environmental change: lessons from one of the world's remaining productive salmon regions. Aust Fish 42(10):538–553. https://doi.org/10.1080/03632415.2017.1374251

Shamir E, Megdal SB, Carillo C, Castro CL, Chang H-I, Chief K, Corkhill FE, Eden S, Georgakakos KP, Nelson KM, Prietto J (2015) Climate change and water resources management in the Upper Santa Cruz River, Arizona. J Hydrol 521:18–33. https://doi.org/10.1016/j.jhydrol.2014.11.062

Siegel J, Crozier L (2019, December) Impacts of climate change on salmon of the Pacific Northwest: A review of the scientific literature published in 2018. National Oceanic and Atmospheric Administration. https://www.webapps.nwfsc.noaa.gov/assets/11/9835_03132020_140127_BIOP-Lit-Rev-2018.pdf

State Statistics Service of Ukraine (2023) Statistical Yearbook of Ukraine for 2023. Statistical bulletin "Agricultural products in constant 2021 prices for 2023". https://www.ukrstat.gov.ua/

Sutton WR, Lotsch A, Prasann A 2024 Recipe for a livable planet: achieving net zero emissions in the agrifood system. Agriculture and Food Series. © Washington, DC: World Bank. http://hdl.handle.net/10986/41468

Swinburn BA, Kraak VI, Allender S, Atkins VJ, Baker PI, Bogard JR, Brinsden H, Calvillo A, De Schutter O, Devarajan R, Ezzati M, Friel S, Goenka S, Hammond RA, Hastings G, Hawkes C, Herrero M, Hovmand PS, Howden M et al (2019) The global syndemic of obesity, undernutrition, and climate change: The Lancet Commission report. Lancet 393(10173):791–846. https://www.thelancet.com/journals/lancet/article/PIIS0140-6736(18)32822-8/fulltext

Tanana H (2023) Protecting tribal health from climate change. Northeast Univ Law Rev 15(1):89–160. https://nulawreview.org/volume-15-issue-1-articles/tanana

Taub DR, Miller B, Allen H (2008) Effects of elevated CO_2 on the protein concentration of food crops: a meta-analysis. Glob Change Biol 14(3):565–575. https://doi.org/10.1111/j.1365-2486.2007.01511.x

Tirado MC, Crahay P, Mahy L, Zanev C, Neira M, Msangi S, Brown R, Scaramella C, Costa Coitinho D, Müller A (2013) Climate change and nutrition: creating a climate for nutrition security. Food Nutr Bull 34(4):533–547. https://journals.sagepub.com/doi/10.1177/156482651303400415

United Nations Children's Fund (2021) Reimaging WASH: Water security for all. United Nations. https://www.unicef.org/media/95241/file/water-security-for-all.pdf

United Nations Convention to Combat Desertification (2022) The global land outlook, second edition. United Nations. https://www.unccd.int/resources/global-land-outlook/global-land-outlook-2nd-edition

United Nations Educational, Scientific and Cultural Organization (2012) United Nations Educational, Scientific and Cultural Organization (UNESCO) International Hydrological Programme (IHP) eighth phase "Water security: responses to local, regional, and global challenges" strategic plan IHP-VIII (2014–2021). United Nations. https://unesdoc.unesco.org/ark:/48223/pf0000218061

United Nations Environment Program (1987) Report of the World Commission on Environment and Development: Our common future. United Nations. https://digitallibrary.un.org/record/133790?ln=en&v=pdf

United Nations Systems Standing Committee on Nutrition (2010) 6th report on the world nutrition situation: Progress in nutrition. United Nations. https://openknowledge.fao.org/server/api/core/bitstreams/7ff77377-801f-450b-ae64-a6a5f0421dca/content

United Nations-Water (2013) Water security & the global water agenda: A UN-Water analytical brief. United Nations. https://www.unwater.org/sites/default/files/app/uploads/2017/05/analytical_brief_oct2013_web.pdf

United Nations-Water (2025, January) Water facts. United Nations. https://www.unwater.org/sites/default/files/2025-01/UN-Water_Water_Facts_one_pager_January_2025.pdf

World Bank (2006) Repositioning nutrition as central to development: A strategy for large scale action. https://openknowledge.worldbank.org/server/api/core/bitstreams/2c424cec-2752-5c88-b697-b1f9a672f3a7/content

Chapter 5
Climate Change Impacts on Water Resources

Ahmed Sefelnasr, Mohsen Sherif, Dalal Alshamsi, and Abdel Azim Ebraheem

1 Learning Objectives

1. Develop a framework for assessing spatial variation in water security.
2. Compare and analyze different studies by a variety of subjective, temporal, and spatial indicators or elements.
3. Develop a comprehensive formula that considers all potential factors impacting water security issues and calculates a specific area's Water Stress (Scarcity) Index.
4. Identify water security vulnerabilities and implement appropriate measures in a specific region, considering the dynamic conditions to identify appropriate and timely adaptation measures and resilience.

A. Sefelnasr · A. A. Ebraheem (✉)
National Water and Energy Center, United Arab Emirates University,
Al Ain, United Arab Emirates
e-mail: abdelazim.aly@uaeu.ac.ae

M. Sherif
Civil and Environmental Engineering Department, United Arab Emirates University,
Al Ain, United Arab Emirates

D. Alshamsi
National Water and Energy Center, United Arab Emirates University,
Al Ain, United Arab Emirates

Department of Geosciences, College of Science, United Arab Emirates University,
Al Ain, United Arab Emirates

A. R. Molla et al. (eds.), *Climate Change, Drinking Water Security, and Public Health*, https://doi.org/10.1007/978-3-032-12074-8_5

2 Introduction

The importance of water security is increasingly evident in how nations manage their water resources (ESCAP 2013; GWP 2000; Waughray 2011). Water security is generally the ability of a community to preserve feasible access to a sufficient supply of sufficient quality water to ensure livelihoods, human well-being, and socio-economic development, protect against water-borne pollution and water-related catastrophes, and conserve biodiversity and the environment (ESCAP 2013).

Water security is a national security issue for the United Arab Emirates (UAE), forming part of the country's economic growth, ongoing innovation, and entrepreneurship efforts. The UAE (Fig. 5.1) is experiencing substantial economic growth, population growth, and development in all sectors, inducing pressure on natural resources, especially water. Local policies and strategies on water security have been adequate to a certain extent. However, ensuring sustainable water security in the long term requires extensive efforts at various levels. In 2017, the UAE announced its Water Security Strategy 2036 to ensure sustainable access to water during both routine and emergency conditions in line with local regulations,

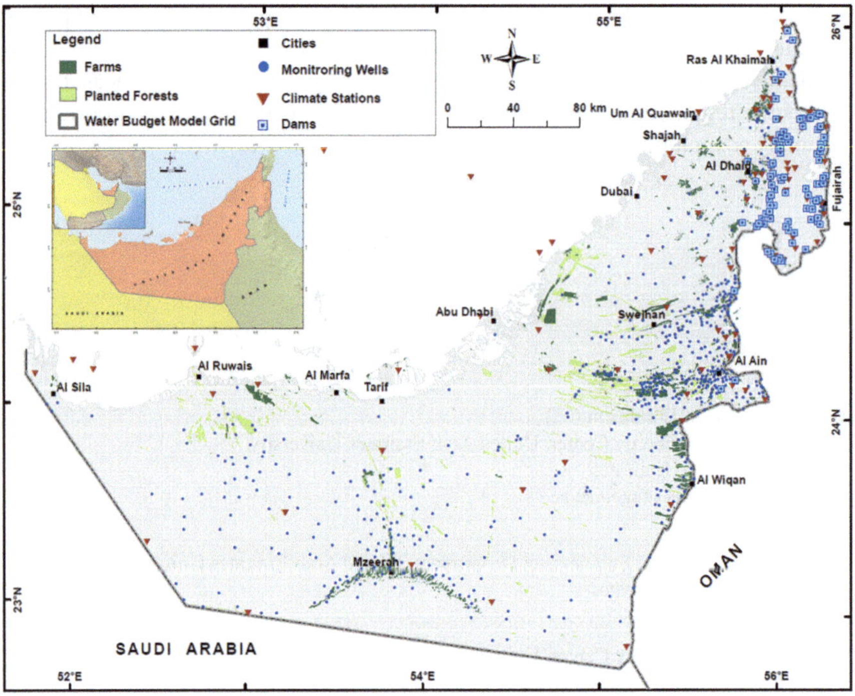

Fig. 5.1 Location map depicting climate stations, monitoring wells, farms, planted forests, dams, and the grid of the water budget model (adapted from Sherif et al. 2021) (Reproduced from Sherif et al. 2021, Figure 1. https://doi.org/10.3390/w13060864, licensed under the terms of the Creative Commons Attribution 4.0 International License (http://creativecommons.org/licenses/by/4.0/))

standards of the World Health Organization, and the country's vision to achieve prosperity and sustainability across all its territories. The UAE territories in the seven emirates are marked with various physiographic settings and land use (Sherif et al. 2021). Thus, the dependency on one average value for water scarcity/stress, as always has been done, is not proper for proposing the optimum alternatives of water security across the UAE.

Additionally, climate change impacts freshwater availability and, from a longer perspective, might endanger water security. The expected increase in climate variability will result in increased water variability. While some regions may experience reduced water availability due to increased temperatures and altered precipitation patterns, others might see changes in irrigation demands. Accurate predictions are essential for adapting to these challenges and ensuring sustainable water management across different regions. Freshwater use has increased rapidly during the past few decades in the UAE, where reliance on fossil groundwater resources is high. Prolonged droughts and the area's low precipitation rates significantly contribute to the nation's serious problems securing its agricultural water demand. Water storage requires the construction of many recharge dams.

Along with the growth in population, GDP, and irrigated regions throughout the desert region, fossil groundwater withdrawal for irrigation has been rising quickly in the UAE since the 1960s. Water security models should also include all elements of water availability, dependability, fair access, and environmental sustainability (ADB 2016). According to Hoekstra et al. (2018), today's definition of water security includes the four key components of economic prosperity, social equity, long-term resilience, and water-related hazards. Thus, water security aims to balance the many risks and hazards connected to water with the three pillars of economic, social, and environmental sustainability.

Researchers frequently employ global hydrological models (GHMs) to compute the water stress index at a spatial resolution of 30 × 30 arc minutes. Several studies utilize these models, including those by Falkenmark (1997), Falkenmark et al. (1989), Oki (2001), Arnell (2004), Hanasaki (2008a, b), Wada (2011), Gain et al. (2016), and Mekonnen and Hoekstra (2016), and Al Rashed et al. (2023), among others. In many of these global investigations, the Water Scarcity/Stress Index (WSI) is employed to quantify water scarcity. This index typically examines conventional water resources in terms of physical "water scarcity," which can result from either high demand (water stress) or limited availability (water deficit). These models lack effectiveness in assessing spatial disparities in water security across the UAE, making them unsuitable for informing decision-makers about enhancing water security indicators under climate change in the UAE due to the following:

1. The absence of non-renewable groundwater resources undermines the scale, accuracy, and reliability of current indicators when evaluating water scarcity in the agricultural regions of the UAE, where groundwater is the primary freshwater source.

2. Insufficient attention has been paid to the significance of non-traditional water supplies in meeting household water demands and achieving water security in urban areas.
3. Current water scarcity indices overlook the crucial role of food imports in ensuring food and water security in an arid country like the UAE.
4. These models do not account for the impacts of climate change on water resources.

The main objective of the present chapter is to provide a framework for assessing spatial variation in water security on a local scale (1 km^2) based on the physiographic setting and land use under climate change, while the specific objectives are listed below:

- Provide enhanced information on the status of water resources in UAE and promote the sustainable management of water resource and their associated ecosystems in UAE.
- Consider dynamic conditions to identify water security vulnerabilities and establish applicable water security measures. This will help determine appropriate and timely adaptation measures and resilience in the UAE, which is experiencing a continuously changing environment.
- Create calculation methods for the Water Stress (Scarcity) Index for the UAE that takes into account (1) only renewable water resources, (2) renewable and non-renewable water resources, and (3) conventional (renewable and non-renewable) and non-conventional water resources.
- Boost preparedness to respond to and address issues related to water scarcity without compromising the pace of economic growth or the standard of living.
- Support the development of adaptation strategies that align with other UAE strategies to achieve water security in climate change.

3 Materials and Methods

The UAE's surface area is mostly desert, and the Aeolian landform system dominates it. Some of the geomorphologic features include mountains, gravel plains, dunes, coastal regions, and drainage basins. The UAE's surrounding water bodies serve as potential sources of water vapor for this area. Convective cloud mechanisms, which produce short-lived storms, are the primary cause of precipitation (Sherif et al. 2021).

To create an average rainfall map for the UAE, historical rainfall records from different representative gauges for the years 1980 to 2021 are periodically employed because the study area experiences geographic variability in the rainfall patterns (Fig. 5.1). Particularly in metropolitan environments, fluctuations in rainfall intensity over time are crucial to the rainfall-runoff process. Figure 5.2 shows the yearly rainfall variance from 1934 to 2021 and the average annual rainfall deviations from the average yearly rainfall for the same period. The highlands and the East coast get

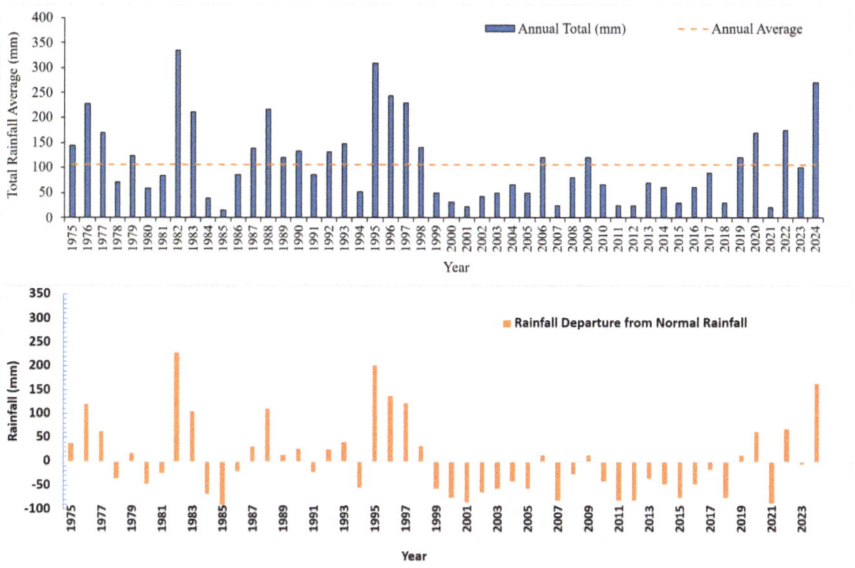

Fig. 5.2 The top graph illustrates the variation of annual rainfall from 1975 to 2024, while the bottom graph displays the average yearly rainfall departures from the usual rainfall during the same period (Modified from Sherif et al., 2021) (Adapted from Sherif et al. 2021, Figure 2. Some modifications were made. https://doi.org/10.3390/w13060864, licensed under the terms of the Creative Commons Attribution 4.0 International License (http://creativecommons.org/licenses/by/4.0/))

the most rain, while the desert foreland receives the least. The East Coast, Mountain, Gravel Plain, and Desert Foreland have mean rainfall volumes of 3, 16, 10, and 71%, respectively, of the total annual rainfall in UAE (Sherif et al. 2023).

Due to prolonged droughts and inherently low precipitation rates, the UAE is encountering significant challenges in safeguarding its agricultural water resources. Low precipitation levels induce the natural occurrence of drought. Droughts have a detrimental effect on society, the economy, agriculture, and the environment. Several academics calculated the frequency and length of droughts in the UAE (e.g., Sherif et al. 2014a, b; Yilmaz 2020). In the UAE, droughts can last anywhere from 1 to 7 years, lasting 2.8 years, and happen every 4.5 years on average. Based on projections from IPCC (2014), regions characterized by mid-latitudes and moist tropical climates are anticipated to undergo enhanced levels of intense rainfall incidents. This pattern has been consistently noticed in the UAE since 2015, except for the water year 2020–2021. Since 2016, there has been a general rise in the total amount of rainfall each year, accompanied by notable alterations in rainfall distribution. During periods of high precipitation, such as 2019–2020 and 2021–2022, the total annual rainfall in the northern and eastern regions of the UAE exceeds twice the average precipitation. In the water year 2019–2020, the cumulative rainfall in the east area of Al Ain city amounted to 500 mm, varying 300 mm from the average rainfall.

Annual water resources data for the entire UAE were obtained and gridded at 1 km intervals to demonstrate the suitability of the established Water Stress Index (WSI) calculation process at a precise local scale (see Fig. 5.1). Each grid cell's net water demand was calculated for the industrial, household, and agricultural sectors (including livestock) if containing a farm and planted forest irrigation. The groundwater budget model and farm inventory visits were used to compute the irrigation water demand for the irrigated regions in 2020 (Sherif et al. 2021). In the UAE, irrigation systems rely on extracting nonrenewable groundwater resources. The estimated replenishment percentage of the Quaternary aquifer in agricultural regions varied between less than 1% to 20% based on the current extraction rate and average wet year conditions, which is a concerning prospect.

As a result, the water supply security is threatened by declining water levels, the mining of nonrenewable groundwater, and saltwater intrusion. Along with the information on land use, land cover, and aquifer features, each grid cell was also given the available groundwater reserve, groundwater salinity, and recharge volume (Sherif et al. 2021). Arid nations (like the GCC countries, including the UAE) with a high domestic water demand heavily rely on unconventional water resources like desalination and water recycling. The UAE's near-total reliance inevitably impacts the ecosystem on desalination for domestic water supply.

3.1 *Data Acquisition*

Considerable efforts were dedicated to acquiring, processing, and validating the data to measure the WSI spatial variations in the UAE. This endeavor aimed to comprehensively evaluate the water security status of each region. The geographical data of the following characteristics were made accessible on a precise grid with a cell area of one km^2:

- Available conventional (surface and groundwater) and non-conventional (desalination and treated wastewater).
- Groundwater abstraction, depletion, recharge, and renewability ratio.
- Access to pure drinking water.
- Availability of a sanitary residential effluent drainage system.
- Drought periods return time.
- Food security.
- Water resources quality.
- Freshwater demand in agriculture, domestic, commercial, industry, and landscaping and planted forestry.

The availability and use of water resources were calculated for 2020 using the same authors' water balance model developed in 2015. The model and approaches are fully described in Sherif et al. 2021. Therefore, only the results of the above water scarcity parameters will be presented in the Results and Discussion section hereafter. The following section explains the newly developed frameworks for calculating the WSI and assessing water security in the UAE with and without considering food import.

3.2 WSI Calculation Methods

The quantitative water stress/scarcity index (WSI) is usually used to infer each nation's water security level. The calculation of WSI, which quantifies the proportion of available water consumed in response to demand, can be simplified by applying the Withdrawal/Renewability Ratio (Falkenmark et al. 1989; Falkenmark 1997). As stated by Wada (2011), the phrase "WSI" can denote the proportion of water availability that accounts for aquifer recharge about overall water withdrawal, which encompasses the utilization of renewable groundwater and surface water for agricultural and livestock purposes. The Water Stress Index for each country's environmental performance indicators was calculated using the methodology presented in Raskin et al. (1997) and Vorosmarty et al. (2000). The proportion of freshwater extraction from renewable water supply was determined using Eq. (5.1).

$$\text{Freshwater fraction } = \frac{\text{Freshwater withdrawal}}{\text{Renewable water available}} \tag{5.1}$$

The ideal outcome is 0.4%. This key fraction indicator uses the ratio between freshwater extraction and its renewability to assess water deficit. The target use fraction of 0.4 is subtracted from the freshwater part to determine water overuse. As it stands, this indicator only depicts water stress in terms of replenishable freshwater supplies. The total freshwater withdrawals are then divided by the total water withdrawals from all sources to determine the water overuse weight. The water stress index is then calculated by multiplying the weight by the water overuse (Eq. 5.2).

$$\text{WSI} = \frac{\text{Freshwater withdrawal x Water overuse}}{\text{Freshwater withdrawal} + \text{Desalinated water} + \text{treated wastewaster}} \tag{5.2}$$

This approach is only helpful for rapidly calculating a country's freshwater withdrawal replenishment rate. Various authors have proposed modifications to Eq. (5.2) to incorporate additional factors such as price, availability, quality, and safety (as expressed in Eq. 5.3), including Wada (2011), Gain et al. (2016), and Wada et al. (2014).

$$WSI_i = \left(\frac{W_{w,i} - RWi}{A_{w,i} - E_{w,i}} \right) \tag{5.3}$$

Water withdrawal is denoted as $W_{w,\, i}$, renewable water resources are represented as RW_i; water availability, which includes the clean non-renewable groundwater reserve, is denoted as $A_{w,\, i}$, and environmental flow restraints in the UAE are denoted as $E_{w,\, i}$. Nonrenewable groundwater and non-conventional water resources (including treated municipal water and desalinated water) were not considered in the calculation of the WSI index, highlighting one of the limitations of the index. Equation (5.4) can be derived by omitting the environmental flow criterion ($E_{w,i}$) from the denominator of Eq. (5.3), due to the absence of perennial surface water in the UAE.

$$WSI_i = \left(\frac{W_{w,i-RW}}{A_{w,i}} \right) \tag{5.4}$$

The equation above represents water withdrawal as W_w, I, renewable water resources such as RW_i, water accessibility including fresh non-renewable groundwater reserves such as A_w, I, and environmental flow constraints in the UAE such as $E_{w,\, i}$. One of the limitations of this index is that non-renewable groundwater and alternative water resources, such as treated municipal water and desalinated water, were not considered in the WSI calculation. To integrate treated effluent reuses, extraction of non-renewable groundwater, and desalination usage, the authors implemented the following adjustments to Eq. (5.4) (Al Rashed et al. 2023):

$$WSI_i = \left(\frac{DT_{net\,i} - \left(NRGW_i + DSW_i + RUWi \right)}{SFWA_i} \right) \tag{5.5}$$

NRGW represents the abstraction of non-renewable groundwater. $SFWA_i$ indicates the available surface clean water. DSW and RUW represent the utilization of desalinated water, and the reused treated wastewater, respectively. DT_{neti} refers to the total water use from all sectors in the UAE, including agricultural water supply, livestock, industrial, and domestic water usage. One key advantage of this indicator is its ability to quantify water stress based on water availability annually. The term "SFWAi" in the denominator of Eq. (5.5) can be replaced with "total recharge from all sources" due to the absence of perennial surface water in the UAE.

Utilizing the extensive water resources database available at a fine localized scale (grid of 68,955 cells of 1 km^2), Eq. (5.5) was applied three times for each grid cell: once considering only nonrenewable brackish water resources, a second time considering non-conventional water resources, and a third time considering both types of water resources. This approach aims to gauge how these resources can mitigate severe water stress in agricultural areas, particularly during drought.

In contrast to the water stress indicator suggested by Falkenmark (1997), Ohlsson (2000) introduced the Social Resource Water Stress Index, which considers the technological, economic, and societal factors that impact water stress. Ohlsson also acknowledged that institutions and technology are decisive factors in a nation's capacity to endure water shortages (2000). Alsaeed et al. (2024) developed a conceptual framework for water resources management based on four sustainability pillars: environment, economy, society, and infrastructure, incorporating a total of 24 indicators. In the interim, Gain et al. (2016) put forth an original approach for ascertaining the Water Stress Index specific to each country. The value function approach was employed, enabling a methodical, rational, and explicit examination of individuals' judgements and values. This methodology converted the performance of the indicators into a numeric score (resilience) ranging from zero (indicating the lowest performance) to one (indicating the highest performance). The Geographic Information System (GIS) was utilized to generate the value functions that specify the proximity or distance of each map piece to 1 for optimal security.

To date, the water scarcity and security indices have exclusively examined the physical dimensions of water availability, including demand and renewability, while neglecting to account for social aspects, including economy, governance, and human development (World Bank 2003, Vörösmarty et al. 2005, Varis et al. 2014, 2017; Al Rashed et al. 2023). To aid in guaranteeing food security, further consideration must be given to the accessibility, cost, quality, and safety of food and the proportion of net imports. The green-blue water (GBW) shortage indication technique (Sayyad et al. 2015) was utilized to measure water availability, considering green water (combination of transpiration from vegetation and evaporation from soil and water bodies and blue water (total surface water and deep aquifer recharge).

4 Vulnerability Assessment

Water availability in arid regions (such as the UAE) is often highly variable. The changing climate makes water availability even more unpredictable, making dependency on water for agricultural irrigation and domestic purposes even more vulnerable. This water security study aims to assess these communities' vulnerability using evidence-based decision-making methodologies in the context of climate change.

The approach is based on defining sustainable, flexible solutions through robust climate stress tests to identify the uncertainty of climate projections for water resources management. The proposed approach allows managers to develop appropriate and effective climate change adaptation measures by accounting for future climate scenarios in water resources planning.

4.1 Impact of Climate Change

Engineers and water stakeholders have always incorporated uncertainty in water resources operations, design, and planning using safety margins, redundancy, and, in some cases, risk-based planning. Many existing methods to address uncertainty in planning and design may not be sufficient to cope with the consequences of climate change. Strong scientific consensus exists globally that our planet has been warming, the anthropogenic emissions of greenhouse gases are the primary cause, and that warming will continue. Climate projections show global temperatures will rise by 1 to 2 °C over the next 20 to 60 years (IPCC 2014). Given the hot and arid summers, significant warming during the winter months would likely have the most profound effects of climate change in the UAE (Dougherty 2009). Figure 5.3a represents the surface area temperature obtained from the Climatic Research Unit (CRU), the University of East Anglia, over UAE from 1901 to 2021. Results depict an increase in temperature relative to 1950–1990, followed by a sharp increase from 1990 to 2021.

Statistical analysis revealed that after 1990, there was a temperature rise, indicating severe warming. Long-term mean analysis showed warming intensification in summer (Fig. 5.3b). From 1901 to 2021, there was an increase in temperature during the decadal period. In terms of temperature projections based on multi-model ensembles of 31 Coupled Inter-comparison Project Phase 6 (CMIP6) models under Shared Socioeconomic Pathways (SSPs), SSP 4.5 and SSP 8.5 were selected to calculate the approximate increase in temperature over UAE. SSP-4.5 is a middle scenario in which emissions stay around current levels before starting to fall around

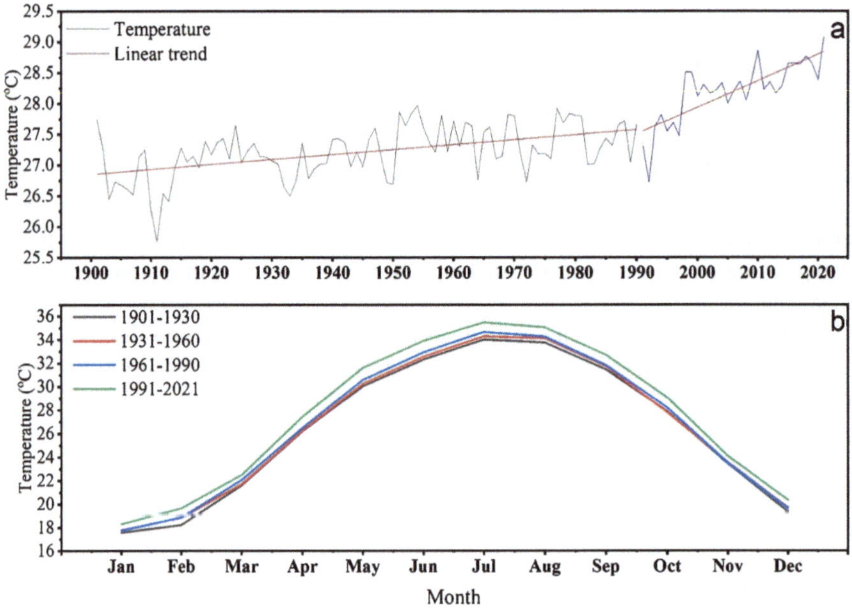

Fig. 5.3 (**a**) Temperature trend over UAE from 1901 to 2021 based on CRU datasets; (**b**) Seasonal, decadal temperatures (adapted from Sefelnasr et al. 2022) (Adapted from Sefelnasr et al. 2022, Figure 10. Some modifications were made. https://doi.org/10.3390/w14213448, licensed under the terms of the Creative Commons Attribution 4.0 International License (http://creativecommons.org/licenses/by/4.0/))

Table 5.1 Summary of UAE temperature projections based on CRU projection dataset (Sources: Environment Agency-Abu Dhabi, 2025; Shared Socioeconomic Pathways (SSPs), 2025)

Period	SSP-2–4.5	SSP5–8.5
2050	1.25 °C	2.02 °C
2100	1.40 °C	4.58 °C

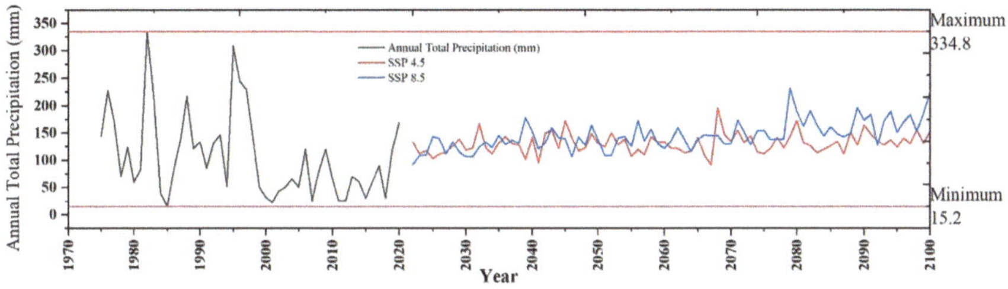

Fig. 5.4 Historical (1975–2020) and projected precipitation over UAE under CMIP6 SSPs (4.5 and 8.5)

mid-century but do not reach net zero by 2100. SSP-8.5 presents a future based on an intensified exploitation of fossil fuel resources.

The findings revealed that the temperature could potentially rise at a rate of 1.25 °C ~ 2.02 °C (SSP-4.5) and 1.40 °C ~ 4.58 °C (SSP5–8.5) up to 2100 (Table 5.1). Almazroui et al. (2020) also found increased projected warming over the Arabian Peninsula under SSPs. The study revealed that under SSP 8.5, the temperature range may rise from 4.1 to 5.8 °C by the end of 2100. The projections made by IPCC indicate that many mid-latitude land masses and wet tropical regions are likely to experience more intense high precipitation events.

We will conduct a preliminary assessment of rainfall variations from 1975 to 2020 using real-time observations from the National Centre for Meteorology, UAE. There is no discernible upward or downward trend in precipitation in the UAE, as shown in Fig. 5.4. Even after 2000, no significant rise has been comparable to the preceding years. However, climate changes or natural variations' impact on precipitation trends is difficult to quantify. In addition, we evaluated the precipitation projections of 31 CMIP6 models under SSP 4.5 and SSP 8.5 over the UAE for the short-term (2021–2050) and long-term (2050–2100) relative to 1975–2020 (Fig. 5.4). The mean annual precipitation over the UAE was found to be greater under both scenarios (SSP 4.5 (25%) and SSP 8.5 (38%)) compared to the reference period spanning 1975–2020. Precipitation forecasts for the UAE are, on average, more sporadic and volatile than temperature forecasts. As the impact of climate change on precipitation is more localized and not uniform throughout the region, the majority of studies indicate that recognizing precipitation patterns would not provide an accurate representation of changes (Almazroui et al. 2016, 2020; Basha et al. 2015; Dougherty 2009).

Groundwater in the UAE originates from deep geological formations, and the proximity of coastal areas to these sources already results in excessive salinity. Therefore, it is unlikely that the increasing sea levels along the Arabian Gulf will significantly affect groundwater quality in the next century. However, the main concern regarding climate change is the possibility of even more significant increases in water use, which might add to the already substantial strain on these limited groundwater resources.

Changes in winter weather patterns, including increased temperatures and erratic rainfall, could lead to more frequent periods of drought and, consequently, higher water consumption during the winter. Nevertheless, due to limited recharge and restricted rainfall, the supply of resources remains unaffected by climate conditions. As a result, Dougherty (2009) state that the fundamental adaptation measures lie in managing demand.

Since 2016, the total precipitation has generally increased, but precipitation patterns have undergone significant changes, except for the water year 2020–2021. In the northern and eastern regions of the UAE, the total annual rainfall during wet years, such as the water years (e.g., 2019–2020 and 2021–2022), exceeds twice the average precipitation. Specifically, in the area east of Al Ain City, the water year 2019–2020 received a total rainfall of 500 mm, which is 300 mm more than the average amount (as shown in Fig. 5.5a and b).

The results align with the findings of Sherif and Singh (1999), who projected that rising atmospheric greenhouse gas concentrations could increase global average temperatures by 1.5 to 4.5 °C by 2050. This anticipated rise, along with more extreme precipitation events, is expected to significantly impact evapotranspiration rates (ETo) and agricultural water demand in the UAE, with increases ranging from 2.7% to 12.4% compared to current rates (El-Rawy et al. 2023; Chowdhury et al. 2016). Specifically, they estimated that a 1 °C increase in temperature would lead to

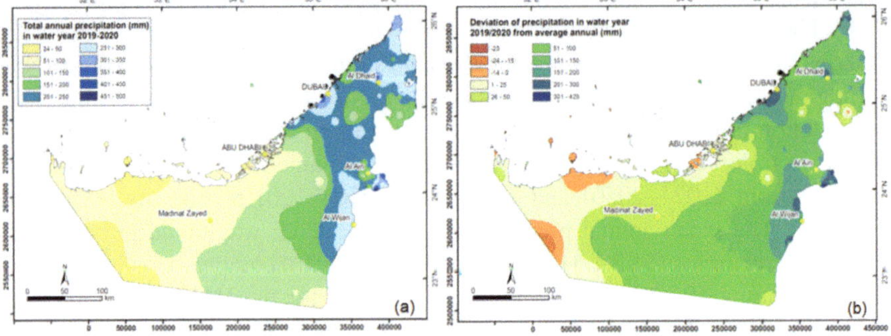

Fig. 5.5 (**a**) The yearly mean rainfall (in millimeters) for the water year 2019–2020 in the United Arab Emirates (UAE) and (**b**) the deviation of the average yearly precipitation (in millimeters) in the UAE for each year from the overall annual average precipitation (Sherif et al. 2023) (Reproduced from Sherif et al. 2023, Figure 2. https://doi.org/10.3390/w15040742, licensed under the terms of the Creative Commons Attribution 4.0 International License (http://creativecommons.org/licenses/by/4.0/))

a 2.9% rise in agricultural water demand. These findings underscore the need for adaptive water management strategies in response to climate change to ensure sustainable farming practices in the region.

It is anticipated that the UAE will encounter changes in its hydrological cycle, including variations in the volume and timing of runoff, because of the ongoing climate change and the warming of the Earth's surface. According to climate simulations, the UAE will see a rise in temperature and precipitation over the next few decades. The anticipated increase in precipitation is expected to lead to a surge in the volume of water available for runoff. The eastern region of the UAE, predominantly reliant on rainfall for its water supply, is anticipated to experience the most pronounced impacts of climate change on runoff. Prior studies based on different climate models showed increased precipitation projections from 50% to 100% in other parts of the country (Almazroui et al. 2020).

4.1.1 Progress in Combating Climate Change

UAE was among the first countries to ratify the Paris Agreement in September 2016, demonstrating its commitment to combating climate change. The UAE has implemented a variety of energy and water conservation measures to create national clean energy. According to the UAE State of Green Economy Report 2016, the UAE is attempting to regulate greenhouse gas emissions from important sectors, such as heavy industries, oil and gas, construction, and transportation through a national green growth framework termed Green Agenda 2015–2030. Several studies have been done by Abu Dhabi Global Environmental Data Initiative (AGEDI) to expand scientific understanding of the impact of climate change in the UAE to aid in better policymaking. According to the 2022 Environmental Performance Index (EPI), the UAE ranked 39th out of 180 countries in terms of its overall environmental performance. In terms of climate change, the UAE received a score of 34 out of 100 and was ranked 117th globally compared to 2020 (125th. EPI considers several key indicators, such as CO_2, CH_4, F-gas, NO_2, Black carbon, CO_2 from land cover, GHG intensity trend, and GHG per capita, to assess the environmental performance of UAE (Fig. 5.6).

The CO_2 growth rate is a key indicator of a country's contribution to global greenhouse gas emissions. The EPI indicator of the CO_2 growth rate over the UAE shows that the country's emissions have been growing steadily over the years, albeit at a relatively slow rate as of the latest EPI report in 2022. This score (42/100) indicates that the UAE's carbon footprint is decreasing because of its efforts to implement clean energy initiatives and reduce its reliance on fossil fuels. UAE's CH_4 growth rate EPI score in 2022 is 20.8, which is a relatively very low figure compared to some other countries and from the previous year (52.7 in 2020). However, the report also notes that the UAE's CH_4 emissions per capita are among the highest in the world, which underscores the need for the country to continue taking measures to reduce its overall greenhouse gas emissions. Some of the potential sources of CH_4 emissions in the UAE include oil and gas production, landfills, and livestock

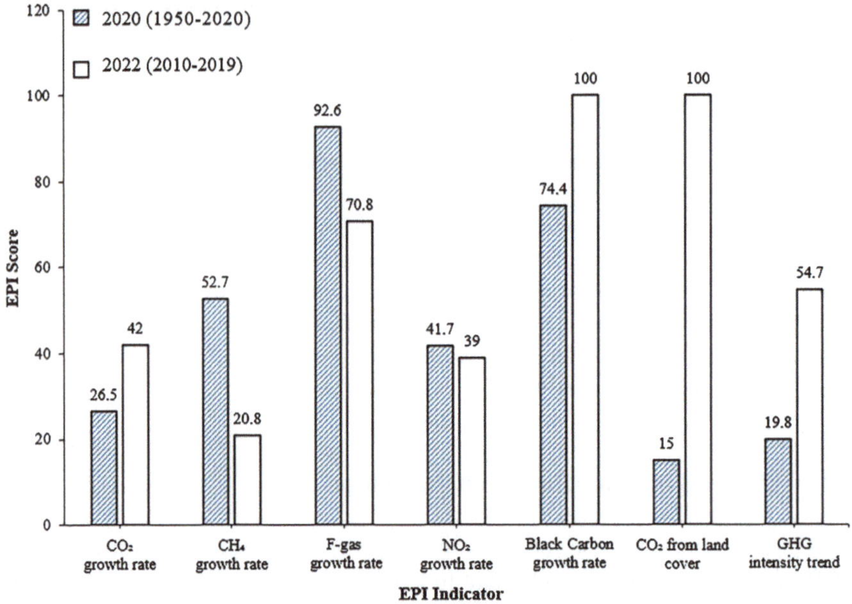

Fig. 5.6 EPI performance indicators over UAE for mitigating climate change (Source: YCELP and CIESIN, 2023)

farming. Similarly, Fluorinated gases, also known as F-gases, constitute a potent group of greenhouse gases extensively utilized in refrigeration and air conditioning systems, as well as in various industrial processes. The EPI also includes an indicator for F-gas and Nitrogen dioxide (NO_2) growth rate to assess a country's environmental performance concerning these emissions. The F-gas growth rate and NO_2 are relatively low compared to the previous report and UAE gets 70.8, and 39 scores out of 100, respectively. This suggests that the UAE's use of F-gases and NO_2 is increasing over time, which could have a significant impact on its overall greenhouse gas emissions (Fig. 5.6).

Black carbon is a type of fine particulate matter that is produced primarily by incomplete combustion of fossil fuels, biomass burning, and other sources. Black carbon contribution is null, based on the EPI score. Similarly, CO_2 emissions from land cover were also zero, with a score of 100 out of 100 (Fig. 5.6). This suggests that the country's black carbon emissions have been decreasing and the country is successful in preserving its natural areas, which can help mitigate climate change and provide other important ecosystem services. However, given that the UAE is a relatively arid country with limited natural vegetation areas, it may be important to focus on other land use changes that can affect carbon storage, such as soil degradation and desertification. The greenhouse gas (GHG) intensity trend is an indicator to assess a country's environmental performance to the decoupling of economic growth from GHG emissions. The GHG intensity trend has been decreasing in recent years and it is evident that the country has been successful in reducing its

GHG emissions relative to its economic growth. The EPI indicator also showed a similar picture as UAE has a 54.7 score out of 100 (Fig. 5.6).

In summary, the UAE's efforts to address climate change are heading in the right direction. As one of the world's most water-scarce nations, it grapples with significant challenges stemming from limited freshwater resources and heavy reliance on desalination and groundwater. Climate change compounds these difficulties by disrupting the region's hydrology and intensifying the frequency and severity of extreme weather events. To effectively tackle these challenges, the country must take proactive measures, such as investing in sustainable water management practices, reducing dependence on fossil fuels, and promoting initiatives for water conservation.

4.2 Climate Change and Water Availability

In UAE, conventional water sources include surface water (such as Aflaj water systems, springs, and accumulated water behind dams) and groundwater (both shallow and deep aquifers). However, fresh conventional sources are severely limited and are insufficient to meet the increasing water demands across all sectors. As a result, the UAE has been relying on non-conventional sources, including desalinated water, treated wastewater, and cloud seeding, since the early 1970s to bridge the widening gap between available water resources and demands (Alsharhan et al. 2001). Climate change has a multitude of serious consequences by changing the character (both quantity and quality) of the components of the hydrological cycle in space, time, and frequency domains. Therefore, the impacts of climate change on water availability as well as on sea water intrusion in coastal aquifers are briefly discussed below:

4.2.1 Surface Water

Despite the arid climate in the UAE, wet days can result in surface water runoff. The country currently has 158 dams, with a combined capacity of 130 million cubic meters (MCM). Over the past 40 years, the estimated annual runoff of surface water in the region fluctuated between 15 and 148 million cubic meters (MCM). However, in 2023, this figure surged to over 1050 MCM, attributed to significant changes in both the magnitude and pattern of precipitation (see Fig. 5.5). This dramatic increase highlights the potential for enhanced water resources, which could positively impact water availability and management strategies in the area. Such shifts also emphasize the importance of adapting to changing hydrological conditions as a result of climate variability. The primary purpose of these dams is to capture floodwater in reservoirs to enhance natural recharge (Sherif et al. 2014a, 2017). Only a small portion of the collected surface water is directly allocated for agricultural purposes. In the wet year of 2019–2020, the total water stored behind the dams was approximately 35 MCM, as reported by the Ministry of Energy and Infrastructure (MoEI

2021). For the hydrological year 2023–2024, the expected total volume of rainwater stored behind dams is projected to exceed previous levels, the several significant storm events that brought many dams to full capacity. However, much of the surface runoff from this unusually wet year flowed directly into the sea. This situation underscores the urgent need for improved rainwater harvesting strategies, not only to capture more water but also to mitigate flooding and optimize the use of surface runoff. By enhancing these practices, the region can better manage its water resources and reduce waste during extreme weather events.

4.2.2 Groundwater

Quaternary (upper) and deep Fragmented rock aquifers are the two main reservoirs that make up the UAE's groundwater system (Sherif et al. 2021). The first significant increase in the areas with fresh groundwater since 2005 is shown on the groundwater salinity map of the Quaternary aquifer in 2020 (Fig. 5.7a) due to the groundwater recharge (Fig. 5.7b). Comparing it to the groundwater salinity map for the year 2018 also reveals a minor increase in fresh and slightly brackish areas (Hydro-Atlas 2018). The groundwater reserve for each type of water (fresh, brackish, or salty) was calculated using Fig. 5.7a and the WBM, with the findings displayed in Table 5.2. The remaining lifespan, replenishment ratio, and groundwater abstraction for irrigation in 2020 were also estimated using the WBM. One advantage of this model is its utilization of an up-to-date Geographic Information System (GIS) water resources database, simplifying the process of inputting parameters into the water shortage model (Sherif et al. 2021).

The potential recharge approach (as illustrated in Fig. 5.7b) was utilized to estimate the recharge from rainfall and other sources to the Quaternary aquifer. This was accomplished using the analytical hierarchy process (AHP) submodules integrated into the same Water Balance Model (WBM) framework. With and without

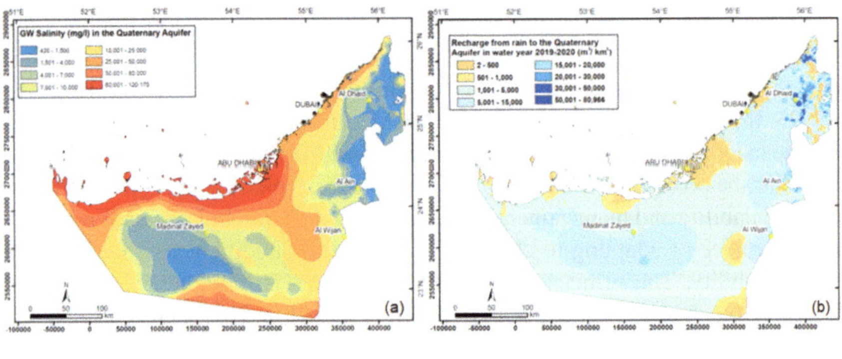

Fig. 5.7 (**a**) Groundwater salinity (mg/l) for 2020 and (**b**) Recharge to the Quaternary aquifer for the years 2019–2020 (m^3/km^2)

Table 5.2 Types of groundwater storage available in the Quaternary aquifer for the year 2020 (km^3), as computed from the Groundwater Balance Model (GWBM)

Water type	Fresh (km^3)	Brackish (km^3)	Saline (km^3)	Brine (km^3)	Total (km^3)
Salinity	(<1500)	(>1500–20,000)	(>20,000–65,000)	(>65,000)	–
2020	75	586	241	47	954

Table 5.3 Recharge to the Quaternary aquifer in the water year 2021–2022 (MCM)

Recharge type	Recharge volume (MCM)
Recharge from precipitation for a normal average rainy year with recharge augmentation	160
Recharge from precipitation for a normal average rainy year without recharge augmentation	145
Recharge from precipitation for a wet year with recharge augmentation (e.g., 2024)	686
Recharge from precipitation for a wet year without recharge augmentation (e.g., 2024)	626
Return flow from groundwater-fed irrigation	362
Return flow from desalination, treated wastewater, and network losses	427
Internal flux	52
Total recharge in the normal average rainy year	1001
Total recharge in the wet year	1527

augmentation, the Quaternary aquifer was recharged a total of 1197 and 1001 MCM, respectively (Table 5.3).

The impact of climate change on groundwater resources is long-term and has far-reaching consequences. A prominent issue is the increased inland migration of saltwater in coastal aquifers, as noted by Sherif and Singh (1999) and later in Sherif et al. (2014b). Excess rainfall or runoff that cannot be effectively stored ultimately ends up in groundwater basins or flows into the oceans. Climate change may lead to both increases and decreases in rainfall, complicating groundwater recharge dynamics. Thomsen (1989) highlighted that groundwater recharge would vary with regional rainfall distribution as predicted by global circulation models (GCMs). In Australia, Sharma (1989) found that a 20% increase in rainfall could result in a 30% rise in groundwater recharge under grasslands, while pine plantations might see an 80% increase. This indicates that significant changes in rainfall patterns could greatly influence aquifer recharge.

According to Table 5.3, recharge from precipitation during a wet year—with and without additional recharge measures—was projected at 686 and 626 million cubic meters (MCM), respectively. This is more than four times the recharge value observed in a normal year, emphasizing the importance of proactive water management strategies to maximize groundwater replenishment in changing climate conditions.

4.2.3 Climate Change and Seawater Intrusion

According to the Ghyben-Herzberg relationship, a one-meter rise in the free water table above mean sea level corresponds to 40 meters of freshwater below sea level. Consequently, a 1 m increase in sea level leads to a 40-meter decrease in the thickness of the freshwater layer. This principle typically holds true away from the coastal boundary, where water tables and piezometric heads are more influenced by pumping and recharge activities than by fluctuations in sea level. In the context of climate change, the rate of sea level rise is slow enough that groundwater levels along the coast will rise in tandem, rather than remaining static. However, this increase in groundwater levels near the coast will not be mirrored on land. As a result, the gradient of the water table and/or piezometric head will decrease, leading to greater saltwater intrusion. While the thickness of the freshwater layer near the coast may experience a slight reduction due to the rising water table, significant reductions are likely to occur further inland. Deep coastal aquifers with gentle hydraulic gradients are particularly vulnerable to the effects of climate change and rising sea levels (Sherif and Singh, 1999). Javadi et al. (2015) conducted a simulation study to evaluate three distinct management scenarios for controlling saline water intrusion (SWI). The scenarios included:

1. Recharge Well: This approach focuses on enhancing groundwater levels by injecting freshwater into the aquifer, aiming to dilute saline water and maintain water quality.
2. Abstraction Well: This method involves extracting groundwater from specific locations to lower the water table, thereby reducing the upward movement of saline water into freshwater aquifers.
3. Combination of Abstraction and Deep Recharge: This integrated strategy combines both abstraction and the injection of freshwater at deeper levels, creating a multi-layered defense against SWI.

The UAE has two major limestone aquifer systems: one is in Ras al Khaimah (northern limestone aquifer/wadi Bih aquifer) and the other is in the Al Ain region of Abu Dhabi emirate (Jabal Hafit limestone aquifer). Dolomite and fractured limestone make up the bulk of the wadi Bih aquifer. Gypsum, dolomite, and limestone make up the 1500 m thick Jabal Hafit aquifer, which evaporates Lower Eocene to Miocene-aged rocks. According to the hydrologic budget of the UAE, there is a total fresh groundwater reserve of 14,000 MCM, 9.57 MCM total annual recharge, 14 MCM total annual outflow, and 1.47 MCM total annual losses to the sea (Ebraheem et al. 2012; Ebraheem 2014; Mohamed et al. 2016; Sherif et al. 2019, 2021; Sefelnasr et al. 2022). As per Al Rashed et al. (2023), the water scarcity in the UAE falls below the true water scarcity benchmark of 500 cubic meters per capita per year. This is indicated by the country's annual per capita availability of green-blue water (GBW), which stood at 323 cubic meters per capita per year in 2020.

The restricted limestone aquifer in Abu Dhabi Emirate has a total groundwater reserve of about 360,000 MCM, with a total salinity that varies depending on the area from less than 10,000 to more than 250,000 mg/l. Due to its low potential, the Ophiolite aquifer will not be taken into account when calculating the water security index (IWACO 1986; NDC and USGS 1996; Sherif et al. 2021).

4.2.4 Desalination

The UAE relies extensively on desalination to meet the increasing water demand in the domestic sector. Currently, the UAE operates 45 large desalination facilities as well as more than 221 smaller ones, most of which are situated in coastal regions (Alsharhan and Rizk 2020). Now, 2.115 billion cubic meters (BCM) of desalinated water are produced annually. By the year 2026, 100% of residential and drinking water use in the UAE will be a result of desalination. It contributes 40% of the nation's total water budget (5.43 BCM in 2021). In the UAE, over 99% of people have access to clean drinking water (Wendling et al. 2018).

4.2.5 Reused Water

There are 140 wastewater treatment facilities in the United Arab Emirates that do not count the Sewage treatment plants (STP) operated by the private sector. Most of the major treatment facilities in the UAE use conventional tertiary treatment methods. The current rate of reuse is 85.8%, and it is rapidly increasing (Table 5.4). It contributes 12.16% of the nation's total water budget (5.43 BCM in 2021). A small quantity of the treated wastewater is being used for agriculture irrigation, with majority of treated wastewater in the UAE allocated for forest plantation, landscaping, and irrigation of green spaces.

Table 5.4 Generated and reused treated wastewater in the year 2022 (data obtained from MoEI Annual Statistical Book 2023)[a]

Wastewater type	Volume/percentage
Total generated wastewater (MCM) in UAE	977
Total received wastewater at the sewage treatment plants (MCM) in UAE	783
Total treated wastewater production (MCM) in UAE	769.5
Total reused treated wastewater (MCM) in UAE	660
Reuse Percentage for the entire UAE (660 MCM/769.5 MCM) in UAE	85.77%
Reuse Percentage for Abu Dhabi emirate (294/310)	94.77%
Reuse Percentage for Dubai emirate (277/293)	94.5%
Reuse Percentage for Sharjah emirate (58/97)	59.7%
Reuse Percentage for Ajman emirate (17/42)	40.5%
Reuse Percentage for Umm Al Quwain emirate (2.1/2.1)	100%
Reuse Percentage for Ras Al Khaimah emirate (11.2/16.1)	69.6%
Reuse Percentage for Fujairah emirate (0.7/9.3)	7.5%

[a]MoEI Annual Statistical Book 2023, licensed under the terms of the Creative Commons Attribution 4.0 International License (http://creativecommons.org/licenses/by/4.0/)

4.3 Accessibility to Water Services

Water access is measured by the percentage of individuals who have access to piped purified drinking water sources. Better drinking water sources boost water security as access grows on a scale from 0 (no access) to 1 (100% access). The fraction of the population that has access to better sanitation is known as access to sanitation. Water security is claimed further through improved access to sanitation (Hsu et al. 2014). These two indicators have values for the UAE that are more than 0.98 (Gain et al. 2016). According to EPI (2022), the UAE score on the water safety and quality index is only 0.517. This low score might be due to inaccurate or outdated data used in the evaluation and needs to be investigated by the UAE water authorities.

4.3.1 Groundwater Depletion

According to the Ministry of Energy and Infrastructure (MEI), just 6 MCM of groundwater were utilized for home purposes in 2020. To estimate the yearly groundwater abstraction for irrigation purposes, whether in farms or planted forests, agricultural statistics and the average annual crop water requirements are combined. The water budget model created by Sherif et al. (2021) was used to estimate the irrigation extraction based on data from the water resources GIS database on specific yield and yearly water table decline. The two approaches' calculated values of groundwater extraction fit each other quite well (Sherif et al. 2021).

An estimated 2.364 billion cubic meters (BCM) of groundwater are utilized for irrigation. In all agricultural areas, the replenishment percentage ranges from less than 1% to 20%, as indicated by the brown, red, and purple colors in Fig. 5.8. Despite significant recharge episodes occurring in December 2019 and from January to March 2020, 2022, 2023, and February 2024, these locations continue to experience severe groundwater depletion issues, except for the urban/settlement regions where water is sourced from nonconventional sources and the rural areas (depicted in yellow) where water extraction is limited. The areas marked in red in Fig. 5.8 denote regions with severe groundwater stress.

The water years 2019–2020, 2021–2022, and 2022–2023 were the wettest years since 2009. According to the results of the Water Balance Model (WBM), groundwater recharge exclusively from rainwater percolation amounted to approximately 356 million cubic meters (MCM)—almost twice the usual amount (refer to Table 5.3). There is also a tremendous decrease in the groundwater salinity, and the pattern of groundwater salinity in the Quaternary aquifer is remarkably similar to that reported by Sherif et al. (2021) for the year 2005. Field measurements revealed a range of 5 to more than 100 meters for the depth of groundwater (mbgl). Excessive groundwater withdrawal results in a local cone of depressions since the Quaternary aquifer system's annual replenishment is restricted. The Quaternary Aquifer's hydrologic budget showed that the current abstraction from this aquifer is 2.4 km^3, but the entire yearly recharge from rain in an average rainy year is only about

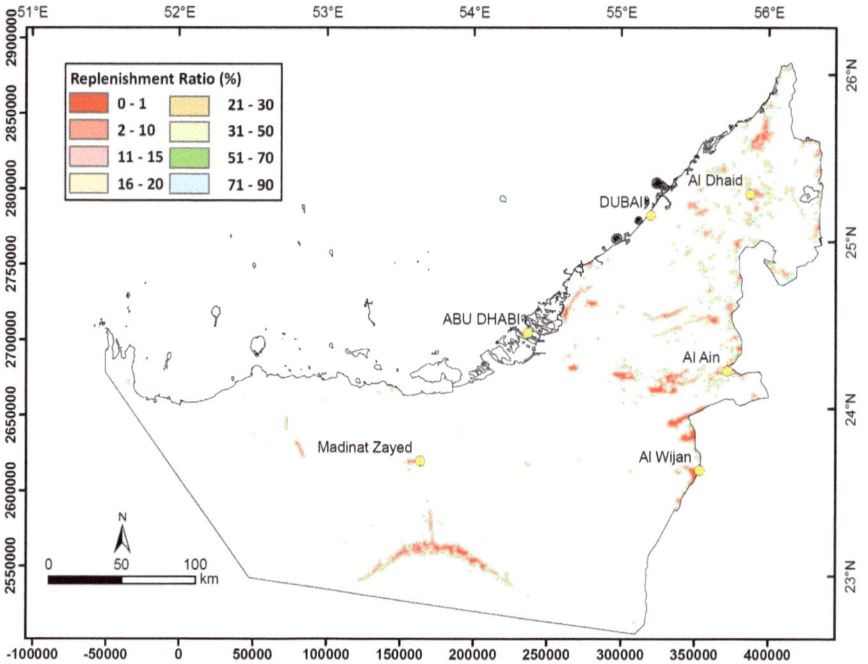

Fig. 5.8 Groundwater replenishment percentage (%) in the Quaternary aquifer. The white-colored areas represent a sandy desert where groundwater abstraction is nil and limited to natural discharge

0.2 km^3. Significant cones of depression and severe groundwater salinization were the results of this significant gap between recharge and abstraction. The estimated reserve of fresh groundwater in the UAE's Quaternary aquifer is projected to be 90 km^3 in 2020. However, the estimated storage of groundwater with slightly brackish to slightly saline quality is much larger, at 560 km^3, underscoring the importance of managing and utilizing brackish water. Overusing and mining groundwater resources can lead to several negative consequences, such as the rapid depletion and deterioration of groundwater supplies, as well as the loss of a reliable long-term water supply, and increased costs associated with the replacement of depleted groundwater resources.

4.3.2 Water Security

The United Nations has recognized water security as one of the 17 Sustainable Development Goals (SDGs). Although the physical availability of freshwater resources to meet water demand is crucial for water security, achieving this objective also depends significantly on social and economic factors (Gain et al. 2016). To generate the water scarcity map depicted in Fig. 5.9, the blue water scarcity index (WSI) calculation using Eq. 5.6 was combined with the WBM data.

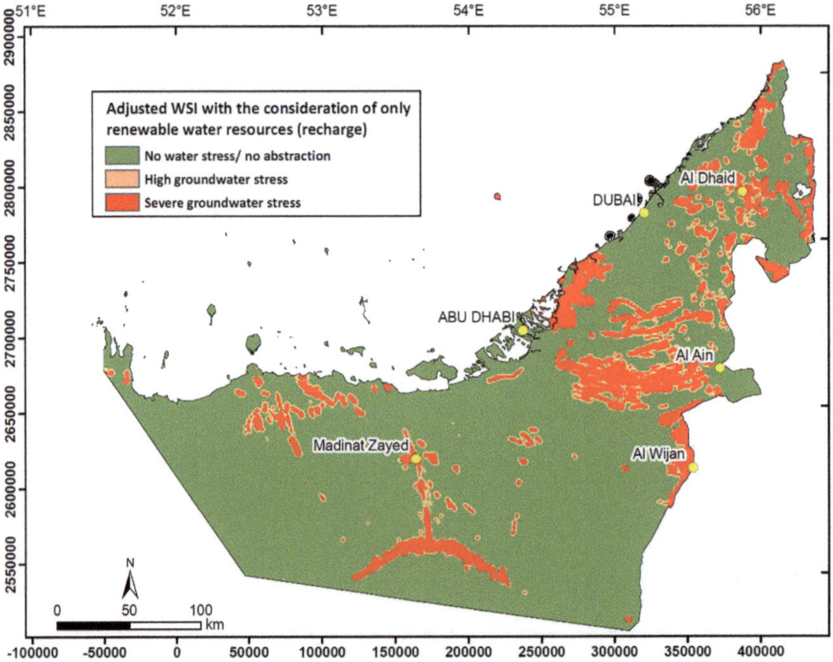

Fig. 5.9 Spatial variation of the adjusted blue water scarcity index (WSI) with the consideration of only renewable water resources

The use of irrigation in water-stressed regions relies heavily on fossil groundwater resources, which can have salinities as high as 5000 mg/l. The addition of fossil groundwater resources with a salinity of up to 5000 mg/l improves the WSI values slightly, as shown in Fig. 5.10. However, when considering unreplenished groundwater reserves with salinities up to 10,000 mg/l, WSI significantly improved throughout the UAE (Fig. 5.11). Figures 5.10 and 5.12 demonstrate significant water stress in agricultural areas, indicating a need for the UAE to utilize brackish water (up to 10,000 mg/l) for fruits and vegetables. However, carefully considering the technical, economic, and environmental implications is essential before desalination is used to treat this water. Additionally, non-traditional water resources (wherever they are accessible) are being used more and more to maintain the current crops. Small desalination facilities and zero-level discharge desalination technology investments are required for this. By the end of 2022, Abu Dhabi will reuse treated wastewater (TWW) on more than 4000 farms. TWW and desalinated water are already being used in some projects to combine with low to medium-brackish groundwater. Water stress on farms close to urban areas can be reduced by increasing the reuse percentage of treated wastewater (84%) and using tertiary treated wastewater for farm irrigation. Desalinated water output grew from 7.0 MCM in 1973 to 2105 MCM in 2021. Currently, desalinated water supplies more than 98% of home demand. Additionally, the use of treated wastewater for landscaping,

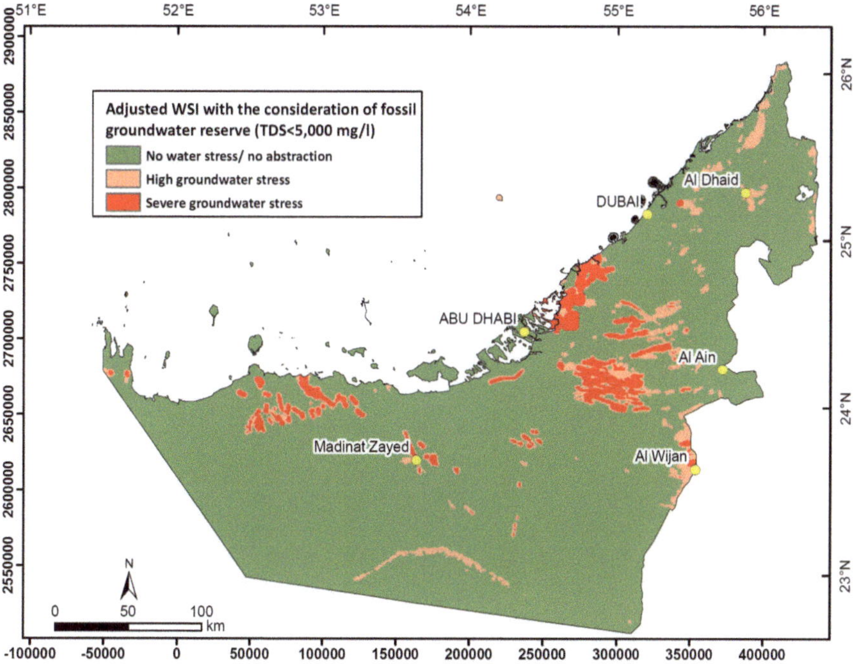

Fig. 5.10 The spatial variability of the adjusted blue water scarcity index (WSI) when fossil groundwater reserve (TDS up to 5000 mg/l) is considered

irrigation, and the establishment of new forests is constantly growing. It is anticipated that the use of treated wastewater would increase for agricultural irrigation and reduce the strain on dwindling freshwater aquifer systems. About 52% of the nation's overall water budget comes from non-conventional water sources (5.43 BCM in 2021). As a result, Fig. 5.12 depicts a blue water shortage index map that considers non-conventional water resources (Falkenmark 1997).

Comparison of Figs. 5.9 and 5.12, depicting the WSI with and without the incorporation of non-conventional water resources, respectively, indicates that the utilization of non-conventional water resources to fulfill water demands results in a significant reduction of water stress and increase of urban water security in urban and settlement areas. However, in rural and agricultural areas, where either there is no exploitation or water demand is fulfilled through fossil groundwater, the change in WSI values with the incorporation of non-conventional water resources is minimal, as these areas are already experiencing either low or high water stress. These findings also suggest that the current production of non-conventional water resources and the available fossil groundwater resources with salinities of up to 5000 mg/l are insufficient to ensure water and food security in the UAE. As a result, the UAE exploits its rapid economic growth, solid diplomatic and trade ties, and efficient trade ports, making it simple to obtain food supplies and attain food and water security.

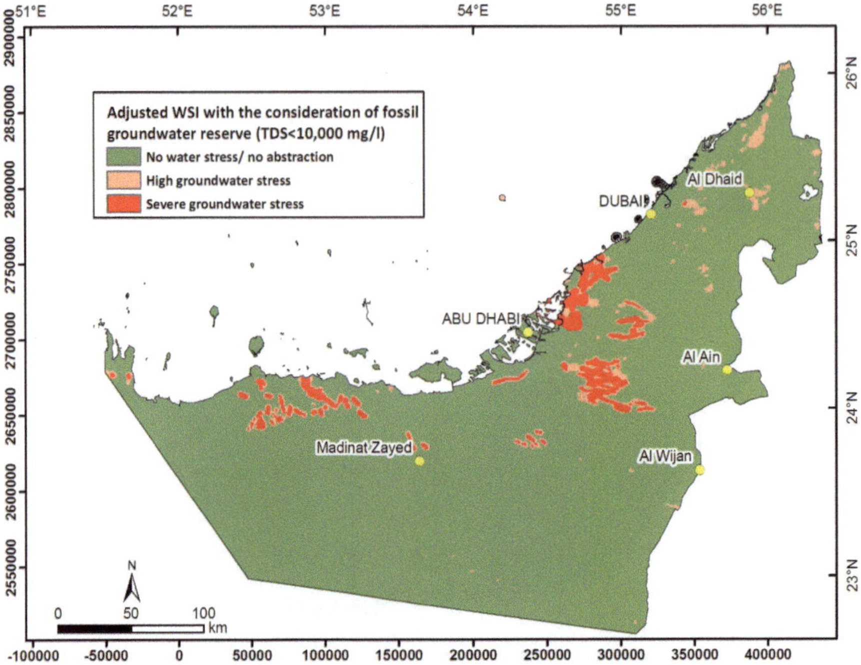

Fig. 5.11 The spatial variability of the adjusted blue water scarcity index (WSI) when fossil groundwater reserve (TDS up to 10,000 mg/l) is considered

This amount corresponds to around 697,484 hectares of arable land. In the United Arab Emirates, there are 286,860 hectares of genuine arable land throughout all three classifications (S1: Highly appropriate, S2: Moderately suitable, and S3: Marginally suitable) (113,260: already exploited and 173,600: unexploited). The volume of water needed for irrigation to use this undeveloped arable land is 3.147 BCM (Table 5.5).

This is impractical unless there is support for significant climate change that results in the desired change in precipitation patterns for the following reasons:

- The groundwater resources that are currently available (as of the year 2020) are only sufficient for brief periods of time between 5 and 13 years (Table 5.5).
- Serious issues with groundwater depletion and water availability.

Recent studies (Ebraheem et al. 2023, Sherif et al. 2014a, b, 2017, 2018, 2023) have revealed that the volume of water stored behind dams is only about 10% of the total runoff in an average wet year in the UAE. Additionally, it is important to highlight that only 40% of the water stored behind the country's 145 dams and breakers contributes to recharging the underlying aquifer. The remaining 60% is lost, with 15% lost to direct evaporation and 45% to percolation or evaporation as soil moisture levels rise in the unsaturated zone. The fate of this water is contingent upon air temperature and precipitation patterns. To improve the situation, there is a need to

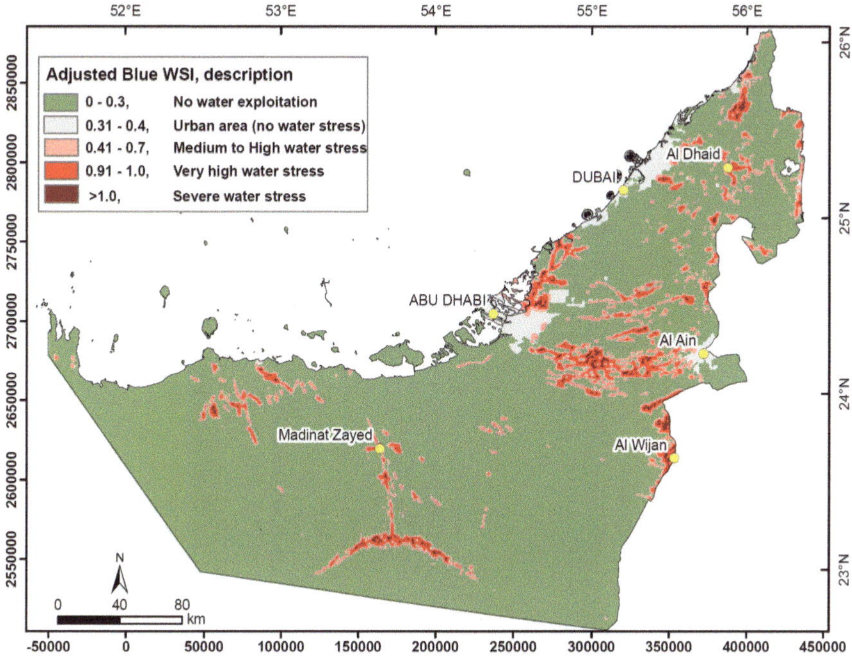

Fig. 5.12 The spatial variability of the amended blue water scarcity index (WSI) when fossil groundwater reserves and non-conventional water resources are considered

revise dam operations in the UAE and consider alternative methods, such as direct utilization of collected water and early recharge techniques.

Because of the UAE's high human development, the social water stress index (SWSI) values are comparatively close to the HWSI values. However, only the availability and renewability of conventional water resources are considered when computing the HWSI and SWSI indexes (Table 5.6).

Table 5.6 demonstrates that aside from the method solely considering conventional water resources, five additional methods were employed to compute the Water Security Index (WSI), considering various types of water resources. This factor indeed plays a pivotal role in assessing water security. When solely considering conventional water resources, the Water Stress Index (WSI) stands at 7.4, indicating severe water scarcity in the UAE. However, upon factoring in non-conventional water resources and non-renewable groundwater resources, the WSI value significantly decreases to 0.318, highlighting the importance of non-conventional water resources in ensuring water security in the UAE.

Table 5.5 Groundwater stcrage versus water demand

	GW TDS	Priority	Arable area	Groundwater storage (up to 5000 mg/l)	BAU irrigation volume	Imported water	Saved lands	Durability (years)	
	mg/l		Hectares	MCM	MCM/yr	MCM	Hectares	Actual	Virtual
S1	<2000	Not exist or already exploited							
S1	2000–3000								
S1	3000–4000								
S1	4000–5000								
S2	<2000	1	26,500	3777	470	1846	104,131	8	Renewable
S2	2000–3000	2	34,900	8120	619	2431	137,139	13	
S2	3000–4000	3	32,100	7354	569	2236	126,137	13	
S2	4000–5000	4	36,600	7375	718	2821	159,144	12	
S3	<2000	5	15,500	1635	275	1080	60,907	6	
S3	2000–3000	6	6800	643	121	474	26,721	5	
S3	3000–4000	7	11,000	1723	195	766	43,224	8	
S3	4000–5000	8	10,200	1672	181	711	40,081	9	
Total			173,600	32,300	3147	12,365	697,484	10	

Table 5.6 The water scarcity/stress index was computed utilizing seven distinct methodologies

Index	HWSI (1)	SWSI (2)	WSI (3)	WSI (4)
Current	18.9 (Scarcity)	10 (stress)	16 (dry year) 9.2 (wet year) 4.3 (dry year and return flow included) 3 (wet year and return flow included)	0.318
Target	5–10 (1000–1700 m^3 /capita/yr)	0–5 (Relative sufficiency)	0.4	0.4

(1) Index of hydrological water scarcity or stress = (obtainable renewable water (MCM) divided by the number of individuals, in the hundreds, who are annually required to share one million m3)
(2) The Social Water Scarcity Index (SWSI)
(3) The WSI was determined employing the methodology utilized to compute the EPI (2010) indicators (Raskin et al. 1997; Vorosmarty et al. 2000). The calculation excluded irrigation return flow
(4) WSI takes non-renewable groundwater resources and nonconventional water resources into account

5 Conclusions

The groundwater resources in Abu Dhabi are primarily sourced from inland Quaternary and deep aquifers, which are less susceptible to salinity issues than the northern emirates' coastal aquifers. As sea levels rise and climate change progresses, the risk to groundwater quality in Abu Dhabi is expected to remain low over the next century. In contrast, the northern emirates face significant challenges due to their reliance on coastal aquifers, which are at high risk for saltwater intrusion. As these aquifers are over-extracted, reducing the water table and piezometric head can exacerbate this intrusion, leading to a decline in freshwater availability. While there might be a minor decrease in the thickness of the freshwater lens near the coast, more significant reductions are anticipated inland, particularly in deep coastal aquifers that have gentle hydraulic gradients. The vulnerabilities of these systems highlight the need for careful management and monitoring to ensure sustainable groundwater use in the face of climate change and rising sea levels. Addressing these challenges will be crucial for maintaining water quality and availability in both regions.

Climate change profoundly affects groundwater resources and surface runoff, which influences the potential for rainwater harvesting. As rainfall patterns become increasingly erratic and intense, existing flood mitigation infrastructure may struggle to cope, heightening the risk of flooding and diminishing the efficiency of rainwater capture and storage. Additionally, climate change impacts the annual volume of irrigation water available. Fluctuations in precipitation, evaporation rates, and temperatures can alter crop water requirements and soil moisture levels, prompting the need for adjustments in irrigation strategies to sustain agricultural productivity. To address these challenges, it is essential to implement adaptive water management practices that bolster the resilience of both groundwater and surface water systems. This includes upgrading flood infrastructure, enhancing rainwater harvesting techniques, and optimizing irrigation methods to ensure sustainable water use in the face of changing climatic conditions.

Incorporating nonrenewable resources into a water security approach for the UAE involves recognizing the potential of mildly to moderately brackish groundwater (TDS between 1500 mg/l and 10,000 mg/l) as an asset. Given the abundance of this groundwater, it can be strategically utilized in agricultural areas facing significant water scarcity. By blending this brackish water with non-conventional water sources, such as treated wastewater or desalinated water, farmers can enhance water availability while mitigating the salinity levels. This approach helps sustain agricultural productivity and promotes more efficient water use across the region. Furthermore, assessing spatial differences in water quality and availability under changing climate conditions will be crucial. Tailoring management strategies to local contexts can optimize available resources, ensuring that the environment and agricultural needs are met sustainably. This integrated approach will be vital for enhancing water security in the UAE amid growing challenges related to climate variability and water scarcity.

Any changes in temperature and precipitation could severely strain the already perilously low groundwater and surface water supplies in the UAE. Precipitation volume and pattern may benefit from climate change as well. Both potential climate change scenarios, whether advantageous or disadvantageous, warrant serious concern, meticulous research, and prudent planning. Until 2015, the UAE's depletion rate (i.e., the ratio of conventional water resources withdrawal to available water resources) was alarmingly high. As a result, the calculated remaining lifespans of available conventional freshwater resources varied across the UAE, ranging from less than 5 years to over 100 years, depending on geographic location and land use. As a result, considering the onset of climate change and ongoing population growth, the UAE may encounter difficulties in enduring extended periods, especially if it continues to rely solely on conventional water resources for domestic food production (although this scenario is nearly impossible).

The UAE and the other nations in this region have no choice but to look for new water supplies. Water imports and the development of non-traditional water sources, such as desalination and water recycling, are beneficial for effective water management. Enhancing the utilization of non-conventional water resources, curbing population growth, augmenting water storage in reservoirs, and enhancing household and industrial water consumption efficiency, irrigation practices, and agricultural water productivity are imperative measures. Growing saline crops or using seawater for greenhouses are two methods, in addition to using wastewater more effectively, minimizing water use, and reducing food waste, to increase water and food security.

The storm events that occurred on July 27, 2022, in Fujairah (200 mm) and on February 12, 2024, in Al Ain City (152 mm), were unprecedented in the last 40 years. These intense storms overwhelmed the infrastructure of both cities, resulting in severe flooding. The daily precipitation rates during these events exceeded the total annual precipitation in a normal year. Apart from the water year 2020–2021, there have been significant changes in annual precipitation and precipitation patterns since 2016, indicating potential impacts of climate change. There is an urgent need to develop and implement strategies to reduce the high rate of evaporation from water accumulated behind dams, which currently stands at about 50%. Given the observed changes in precipitation size and pattern, regulated aquifer recharge from these dams can be utilized to enhance rainwater harvesting and increase recharge, thereby improving water security by augmenting volumes of renewable water resources.

A considerable area of concern pertains to the presence of leaky distribution networks, with estimates indicating that these leaks result in the loss of 10–15% of the water distributed. Furthermore, it is noteworthy that a significant volume of 110 million cubic meters (MCM) of treated wastewater continues to be discharged into the Gulf. This situation highlights the potential for enhancing water resources management practices and optimizing the utilization of these resources. Instead of permitting the discharge of treated wastewater into the Gulf, exploring recycling and reutilization methods can significantly contribute to water conservation and meeting various water needs in the region. The implementation of improved water resources management is crucial in this regard.

WSI calculations in the UAE example show the necessity of adopting a comprehensive strategy, such as a trade-WFL (water-food-land) nexus in dry regions. Addressing water security should involve a dual approach that incorporates both a developmental perspective (focusing on water demand and supply management) and a risk-based one (encompassing emergency production and distribution programs) to establish a comprehensive water resource management framework.

6 Discussion Questions

1. What is meant by conventional and non-conventional and renewable and non-renewable water resources?
2. Discuss the elements beyond the conventional metrics of availability and renewability that need to be considered for assessing water security in a specific region.
3. Explain the possible impacts of climate change on water security and discuss the possible means to mitigate such effects, specifically concerning arid and semi-arid regions.
4. Discuss possible strategies to achieve water security in arid regions.

References

ADB (2016) Asian water development outlook 2016: strengthening water security in Asia and the Pacific. Asian Development Bank

Al Rashed M, Sefelnasr A, Sherif M et al (2023) Novel concept for water scarcity quantification considering nonconventional and virtual water resources in arid countries: application in Gulf Cooperation Council Countries. Sci Total Environ 882:163473. https://doi.org/10.1016/j.scitotenv.2023.163473

Almazroui M, Saeed F, Islam MN, Alkhalaf A (2016) Assessing the robustness and uncertainties of projected changes in temperature and precipitation in AR4 Global Climate Models over the Arabian Peninsula. Atmos Res 182:163–175

Almazroui M, Islam MN, Saeed S et al (2020) Future changes in climate over the Arabian Peninsula based on CMIP6 multimodel simulations. Earth Syst Environ 4:611–630

Alsaeed B, Hunt D, Sharifi S (2024) A Sustainable Water Resources Management Assessment Framework (SWRM-AF) for arid and semi-arid regions—Part 1: Developing the conceptual framework. Sustainability 16:2634. https://doi.org/10.3390/su16072634

Alsharhan AS, Rizk ZE (2020) Water resources and integrated management of the United Arab Emirates, vol 3. Springer Nature

Alsharhan AS, Rizk ZA, Nairn AEM et al (2001) Hydrogeology of an arid region: the Arabian Gulf and adjoining areas. Elsevier B.V, Amsterdam

Arnell NW (2004) Climate change and global water resources: SRES emissions and socio-economic scenarios. Glob Environ Chang 14:31–52. https://doi.org/10.1016/j.gloenvcha.2003.10.006

Basha G, Ouarda TB, Marpu PR (2015) Long-term projections of temperature, precipitation and soil moisture using non-stationary oscillation processes over the UAE region. Int J Climatol 35:4606–4618

Chowdhury S, Al-Zahrani M, Abbas A (2016) Implications of climate change on crop water requirements in arid region: an example of Al-Jouf, Saudi Arabia. J King Saud Univ Eng Sci 28(1):21–31. https://doi.org/10.1016/j.jksues.2013.11.001

Dougherty W (2009) Climate change impacts, vulnerability, and adaptation. Environment Agency, Abu Dhabi, Abu Dhabi

Ebraheem A (2014) Mapping groundwater conditions in different geological environments in the northern area of UAE using 2D earth resistivity imaging survey. Environ Earth Sci 72:1599–1614

Ebraheem A, Sherif M, Al Mulla M et al (2012) A geoelectrical and hydrogeological study for the assessment of groundwater resources in Wadi Al Bih, UAE. Environ Earth Sci 67:845–857

Ebraheem AA, Alshamsi D, Sherif M et al (2023) Mapping groundwater potential zones in UAE using GIS-based hydrogeological modeling. In: Sherif M, Singh VP, Sefelnasr A, Abrar M (eds) Water resources management and sustainability. Water Science and Technology Library. Springer Nature, Cham, pp 255–268

El-Rawy MO, Batelaan N, Al-Arifi A, Alotaibi F, Abdalla MG (2023) Climate change impacts on water resources in arid and semi-arid regions: a case study in Saudi Arabia. Water 15:x

Environment Agency-Abu Dhabi (2025) Abu Dhabi Climate Change Adaptation Plan – Environment Sector 2025–2050. Internal Report, Environment Agency-Abu Dhabi, Abu Dhabi, UAE. Table 1, page 13. Also available online: https://www.ead.gov.ae/-/media/Project/EAD/EAD/Documents/EN_Abu-Dhabi-Climate-Change-Adaptation-Plan-for-the-Environment-v4.pdf

ESCAPU (2013) Water security & the global water agenda: a UN-Water analytical brief

Falkenmark M (1997) Meeting water requirements of an expanding world population. Philos Trans R Soc Lond B Biol Sci 352:929–936

Falkenmark M, Lundqvist J, Widstrand C (1989) Macro-scale water scarcity requires micro-scale approaches: aspects of vulnerability in semi-arid development. Nat Resour Forum Wiley Online Library 13:258–267

Gain AK, Giupponi C, Wada Y (2016) Measuring global water security towards sustainable development goals. Environ Res Lett 11:124015

GWP (2000) Towards water security a framework for action. United Nations, New York

Hanasaki N (2008a) An integrated model for the assessment of global water resources – Part 2: Applications and assessments. Hydrol Earth Syst Sci 12:1027–1037. https://doi.org/10.5194/hess-12-1027-2008
Hanasaki N (2008b) An integrated model for the assessment of global water resources – Part 1: Model description and input meteorological forcing. Hydrol Earth Syst Sci 12:1007–1025. https://doi.org/10.5194/hess-12-1007-2008
Hoekstra AY, Buurman J, Ginkel KC (2018) Urban water security: a review. Environ Res Lett 13:053002
Hsu A, Emerson J, Levy M et al (2014) The 2014 environmental performance index. Yale Center for Environmental Law & Policy, New Haven, CT
Hydro-Atlas (2018) Hydro-Atlas of Abu Dhabi Emirate. Internal Report. Environmental Agency of Abu Dhabi, Abu Dhabi, UAE
IPCC Climate Change (2014) Synthesis report. Contribution of Working Groups I, II and III to the Fifth Assessment Report of the Intergovernmental Panel on Climate Change
IWACO (1986) Ground water study, drilling of deep-water wells at various locations in the UAE. In: Groundwater Development in the Northern Agricultural Region. p 7
Javadi A, Hussain M, Sherif MM, Farmani R (2015) Multi-objective optimization of different management scenarios to control seawater intrusion in coastal aquifers. Water Resour Manag. https://doi.org/10.1007/s11269-015-0914-1. Published online 29 January 2015
Mekonnen MM, Hoekstra AY (2016) Four billion people facing severe water scarcity. Sci Adv 2:1500323. https://doi.org/10.1126/sciadv.1500323
MoEI (2021) Annual statistical book. Internal Report, Ministry of Energy and Infrastructure, Dubai, UAE
MoEI (2023) Annual statistical book. Internal Report, Ministry of Energy and Infrastructure, Dubai, UAE
Mohamed MM, Al-Suwaidi N, Ebraheem AM, Al Mulla MM (2016) Groundwater modeling as a precursor tool for water resources sustainability in Khatt area, UAE. Environ Earth Sci 75:1–18. https://doi.org/10.1007/s12665-016-5261-x
NDC; USGS. (1996). Groundwater Research Project; National Drilling Company and USGS: Al Ain/Abu Dhabi, United ArabEmirates.
Ohlsson, L., (2000). Water conflicts and social resource scarcity. Phys Chem Earth B Hydrol Ocean. Atmos 25, 213–220. https://doi.org/10.1016/S1464-1909(00)00006-X
Oki T (2001) Global assessment of current water resources using total runoff integrating pathways. Hydrol Sci J 46:983–995. https://doi.org/10.1080/02626660109492890
Raskin P, Gleick P, Kirshen P, Pontius G, Strzepek K (1997) Comprehensive assessment of the freshwater resources of the world. Stockholm Environ Inst. https://www.ircwash.org/sites/default/files/210-97CO-14037.pdf
Sayyad G, Vasel L, Besalatpour AA, et al (2015) Modeling blue and green water resources availability in an Iranian data scarce watershed using SWAT. JWMM. https://doi.org/10.14796/JWMM.C391
Sefelnasr A, Ebraheem AA, Faiz MA et al (2022) Enhancement of groundwater recharge from Wadi Al Bih Dam, UAE. Water 14:3448. https://doi.org/10.3390/w14213448
Shared Socioeconomic Pathways (SSPs) (2025). https://climatedata.ca/resource/understanding-shared-socio-economic-pathways-ssps/
Sharma ML (1989) Impact of climate change on groundwater recharge. In Conference on Climate and Water. Publication of the Academy of Finland. Vol. 1, pp. 511–520
Sherif MM, Singh VP (1999) Effect of climate changes on Seawater Intrusion in Coastal Aquifers. Hydrol Process 13(8):1277–1287. https://doi.org/10.1002/(SICI)1099-1085(19990615)13:8<1277::AID-HYP765>3.0.CO;2-W
Sherif M, Almulla M, Shetty A, Chowdhury RK (2014a) Analysis of rainfall, PMP and drought in the United Arab Emirates. Int J Climatol 34:1318–1328

Sherif MM, Sefelnasr A, Ebraheem A, Javadi A (2014b) Quantitative and Qualitative Assessment of Seawater Intrusion in Wadi Ham under Different Pumping Scenarios. J Hydrol Eng 19(5):855–866

Sherif M, Ebraheem A, Shetty A (2017) Groundwater Recharge from Dams in the United Arab Emirates. In: World Environmental and Water Resources Congress 2017. Presented at the World Environmental and Water Resources Congress 2017. American Society of Civil Engineers, Sacramento, California

Sherif M, Ebraheem A, Al Mulla M, Shetty A (2018) New system for the assessment of annual groundwater recharge from rainfall in the United Arab Emirates. Environ Earth Sci 77:1–17

Sherif M, Ksiksi T, Neumann E, Abuelgasim (2019) A review of the current status of climate change modeling in UAE and the Region; Ministry of Climate Change and Environment: Dubai, United Arab Emirates

Sherif M, Sefelnasr A, Ebraheem AM, Al Mulla M, Alzaabi M, Alghafli K (2021) Spatial and temporal changes of groundwater storage in the Quaternary Aquifer, UAE. Water 13:864. https://doi.org/10.3390/w13060864

Sherif M, Sefelnasr A, Al Rashed M et al (2023) A review of managed aquifer recharge potential in the Middle East and North Africa Region with Examples from the Kingdom of Saudi Arabia and the United Arab Emirates. Water 15:742. https://doi.org/10.3390/w15040742

Thomsen R (1989) The effect of climate variability and change on groundwater in Europe. In: Conference on Climate and Water, vol 1. Publication of the Academy of Finland, Helsinki, Finland, pp 486–500

Varis O, Kummu M, Lehr C, Shen D (2014) China's stressed waters: societal and environmental vulnerability in China's internal and transboundary river systems. Appl Geogr 53:105–116. https://doi.org/10.1016/j.apgeog.2014.05.012

Varis O, Keskinen M, Kummu M (2017) Four dimensions of water security with a case of the indirect role of water in global food security. Water Secure 1:36–45. https://doi.org/10.1016/j.wasec.2017.06.002

Vorosmarty CJ, Green P, Salisbury J, Lammers RB (2000) Global water resources: vulnerability from climate change and population growth. Science 289:284–288

Vörösmarty CJ, Douglas EM, Green PA, Revenga C (2005) Geospatial indicators of emerging water stress: an application to Africa. AMBIO A J Human Environ 34:230–236

Wada Y (2011) Global monthly water stress: 2. Water demand and severity of water stress. Water Resourc Res 47

Wada Y, Gleeson T, Esnault L (2014) Wedge approach to water stress. Nat Geosci 7:615–617

Waughray D (2011) Water security, the water-food-energy-climate Nexus: the world economic forum water initiative. Island Press, Washington, DC

Wendling Z, Emerson J, Esty D et al (2018) Environmental performance index. Yale Center for Environmental Law & Policy, New Haven, CT

World Bank (2003) World Bank water resources sector strategy: strategic directions for World Bank engagement. World Bank, Washington, DC

Yale Center For Environmental Law And Policy-YCELP-Yale University, & Center For International Earth Science Information Network-CIESIN-Columbia University (2023) 2022 Environmental Performance Index (EPI) (Version 2022.00) [Data set]. Palisades, NY: NASA Socioeconomic Data and Applications Center (SEDAC). https://doi.org/10.7927/DWT2-9K25

Yilmaz (2020) Climate change effects on drought in Sharjah, UAE. Int J Environ Sci Dev 11:116–122

Part II
Drinking Water Quality, Quantity, Scarcity, and Safety

Chapter 6
Drinking Water Quantity and Climate Change

Fayzul K. Pasha

1 Learning Objectives

1. Discuss the health consequences of restricted drinking water quantity.
2. Review the mechanisms through which climate change is expected to affect water supply.
3. Identify the regions that are expected to be affected the most by climate changes.

2 Introduction

Two terms—safe and healthy—are interlinked with drinking water. There may be an abundance of water but if the water is not clean, safe, healthy, and readily accessible there is no use of that water. Drinking water can be defined as the contaminant free water that is healthy and safe for human consumption. Drinking water is also known as potable water. While drinkable water can be used for domestic, commercial, agricultural, and industrial purposes, it strictly needs to be risk free of any health concerns. Contaminated water with fecal and high doses of undesirable chemicals can be a source of many waterborne diseases such as typhoid, diarrhea, cholera, hepatitis A, etc. (WHO 2024a).

F. K. Pasha (✉)
Department of Civil and Geomatics Engineering, Lyles College of Engineering, California State University, Fresno, CA, USA
e-mail: mpasha@csufresno.edu

A. R. Molla et al. (eds.), *Climate Change, Drinking Water Security, and Public Health*, https://doi.org/10.1007/978-3-032-12074-8_6

3 Health Consequences of Restricted Drinking Water Quantity

3.1 Drinking Water

One of the important requirements to ensure safe and healthy drinking water is to have an improved water source with a water treatment facility on the premises. Waterborne diseases can affect water quality during water transmission. Therefore, clean and safe water distribution infrastructure to deliver water when it is needed can be just as important as having an improved water source. Overall, water management can play an important role in ensuring safe and healthy drinking water.

The degree of water treatment can vary, depending on the source of water and the economic conditions of the jurisdiction. As a result, the water quality standard for drinking water can vary from agency to agency and country to country. The USEPA (United States Environmental Protection Agency) categorized and published the contaminants, treatment byproducts, and the presence of chemicals with their limiting values. Broad categories of the contaminant and byproducts include few broad groups; microorganisms, disinfectants, disinfection byproducts, inorganic chemicals, organic chemicals, etc. (USEPA 2024a). Table 6.1 includes some key National Primary Drinking Water Regulations (NPDWR) standards developed by the USEPA. The standards apply to treated water quality in public water systems in the USA.

3.2 Drinking Water Requirement Per Person

Water consumption per person varies significantly as it is a function of many parameters ranging from human physiology to weather and climatic conditions of a region. The United States Centers for Disease Control and Prevention (CDC) found that water intakes per person per day for their physical fitness varies by age, gender, body weight, race, education, etc., in a study conducted in 2015–2018 (CDC 2024). While the average consumption for all children and adolescents is 23 ounces/day, the consumption can vary by age and gender even for children and adolescents. For example, children between 12 and 19 years of age consume average 31 ounces/day and children between 6 and 11 years old consume 20 ounces/day. Similarly, the water consumption per person also varies by race.

The average water intakes per person per day by an adult is 44 ounces per day. However, the intake is affected by many parameters. The average intake per person per day decreases as the adults get older. For an adult between 20 and 39 years the intake is 51 ounces per day; intake for an adult between 40 and 59 is 43 ounces per day; intake for an adult age greater than 60 years is 36 ounces per day. Similar to intake by children, the intake rate varies by other parameters, including gender, race, education, etc.

Table 6.1 Key National Primary Drinking Water Regulations (NPDWR) standards developed by the USEPA (2024a)

Contaminant type	Contaminant	MCLG (mg/L)	Potential health risk
Microorganisms	Cryptosporidium	0	Diarrhea, vomiting, cramps, etc.
	Giardia lamblia	0	Diarrhea, vomiting, cramps, etc.
	Legionella	0	A type of pneumonia
	Total Coliforms	0	Indication of potentially harmful bacteria
	Viruses (enteric)	0	Diarrhea, vomiting, cramps, etc.
Disinfectants	Chlorine (Cl_2)	MRDL: 4.0	Eye and nose irritation and stomach discomfort
	Chlorine dioxide (ClO_2)	MRDL: 0.8	Anemia and nervous system effects for infants and young children
Disinfection byproducts	Bromate	0	Increased risk of cancer
	Chlorite	0.8	Anemia and nervous system effects for infants and young children
	Haloacetic acids (HAA5)	–	Increased risk of cancer
	Total Trihalomethanes (TTHMs)	–	Liver, kidney, or central nervous system problems; increased risk of cancer
Organic chemicals	Acrylamide	0	Nervous system and increased risk of cancer
	Benzene	0	Anemia, decrease blood platelets and increased risk of cancer
	Chlorobenzene	0.1	Liver and kidney problems
	Ethylene dibromide	0	Liver, kidney, stomach, and reproductive system problems
	Glyphosate	0.7	Kidney problems and reproductive issues
	Heptachlor	0	Liver damage and increased risk of cancer
	Hexachlorobenzene	0	Liver, kidney, and reproductive problems
	Toluene	1	Kidney, liver, and nervous system problems
	Trichloroethylene	0	Liver problems and increased risk of cancer
	Vinyl chloride	0	Increased risk of cancer
	Xylenes (total)	10	Nervous system damage

(continued)

Table 6.1 (continued)

Contaminant type	Contaminant	MCLG (mg/L)	Potential health risk
Inorganic chemicals	Arsenic	0	Skin damage and increased risk of cancer
	Barium	2	High blood pressure
	Cadmium	0.005	Kidney damage
	Cyanide	0.2	Nerve damage or thyroid problems
	Lead	0	Infants and children: Delays in physical or mental development in children and kidney problems and high blood pressure in adults
	Mercury	0.002	Kidney damage
	Nitrate (measured as nitrogen)	10	Shortness of breath and blue-baby syndrome including infant death

MCLG Maximum Contaminant Level Goal, *MRDL* Maximum Residual Disinfectant Level

In a 2010 general assembly, the United Nations (UN) defined access to 50 to 100 liters of safe drinking water per person per day for personal and domestic use as a right (UN 2024a). The personal and domestic use can include many water needs such as drinking, cooking, washing, growing foods, etc. The World Health Organization (WHO 2013) has distributed the minimum water consumption number as follows; for short-time survival including drinking and cooking: 20 liters per person per day; for medium-term maintaining including sanitation, growing food, cleaning home, washing clothes, personal washing, and all other short-term survival: 70 liters per person per day.

However, socio-economic and access to improved water sources affect the water consumption for domestic and personal use. For example, according to the U.S. Environmental Protection Agency (EPA) the average water consumption per person per day in the USA is 82 gallons (USEPA 2024b) (310L), which is more than 3 times the recommended water consumption number by the U.N.

3.3 *Water Availability and Access to Safe Drinking Water*

The water consumption per person per day, which eventually is used to estimate the total water demand for a community, or a city, or a region is used to plan, design, and construct water supply infrastructure. The water consumption per person can vary significantly as it is influenced by a multitude of factors such as lifestyle, climate/weather conditions, socioeconomic conditions, etc. While access to safe drinking water is readily available in many developed countries due to its improved source and developed water supply infrastructures, the developing country on the other hand suffers due to their lack of infrastructure even though the water can be available in the region. The access to safe and healthy drinking water is thus critical. As a result, the water availability and thus the water consumption per person per day may fall much below the recommended level in the rural and poorer community. Understanding the availability and access to clean and safe drinking water is therefore important for water management to maintain a sustainable and healthy quality of life. This section provides an overview of the access to safe drinking water worldwide.

The percent of population with access to safe drinking water data for 190 countries has been collected from Food and Agricultural Organization (FAO 2024a). Data has been analyzed further, which has been shown in Table 6.2 and Fig. 6.1. Generally, the population in rural areas, which lacks improved infrastructure, has much less access to safe drinking area compared to urban areas. Combining access to safe drinking water in rural and urban area, the percentage of access to safe drinking water for the total population ranges from 31.70% to 100%. While 61 countries out of 190 have 99% of their population have access to safe drinking water, the percent of population access to safe drinking water varies for the rest of the countries. For example, the numbers of countries that have 90% of their total population with safe drinking water is 126; 75% of their total population with safe drinking water is 159; 50% of their total population with safe drinking water is 186, etc. As seen in Table 6.2, 19 countries do not have access to safe drinking water for even 50% of their rural population.

As access to safe drinking water and infrastructure for improved source of water and water supply system are related, an attempt has been made to observe the correlation between gross domestic product (GDP) per capita and percent of population with access to safe drinking water for each country (Fig. 6.2). World Bank 2022 GDP data (World Bank 2024) has been used for this analysis. As seen, the percent of total population access to safe drinking water increases as GDP increases approximately up to US $10,000. Table 6.2 summarizes the findings. While the average

Table 6.2 Statistics of number of countries with greater than or below than a certain percentage of population with access to safe drinking water (data sources: FAO 2024a[a]; World Bank 2024)

Percent of population with access to safe drinking water	Number of countries with population access to safe drinking water greater than certain threshold			
	Rural	Urban	Total	Average GDP (US $/capita)
= 100%	38	47	36	$ 51,394
> = 99%	53	78	61	$ 38,949
> = 90%	97	158	126	$ 24,730
> = 75%	134	179	159	$ 20,117
> = 50%	168	190	186	$ 17,520
Below 50%	19	0	4	$ 3,673
Total counties	187	190	190	187

[a]Source: Food and Agriculture Organization of the United Nations. AQUASTAT. Rural population with access to safe drinking-water (JMP), Urban population with access to safe drinking-water (JMP), Total population with access to safe drinking-water (JMP). Latest update: 2022. Accessed: September 8, 2024. https://data.apps.fao.org/aquastat/?lang=en&share=f-6b6ec785-1e2a-4add-906e-5bfde8018cca

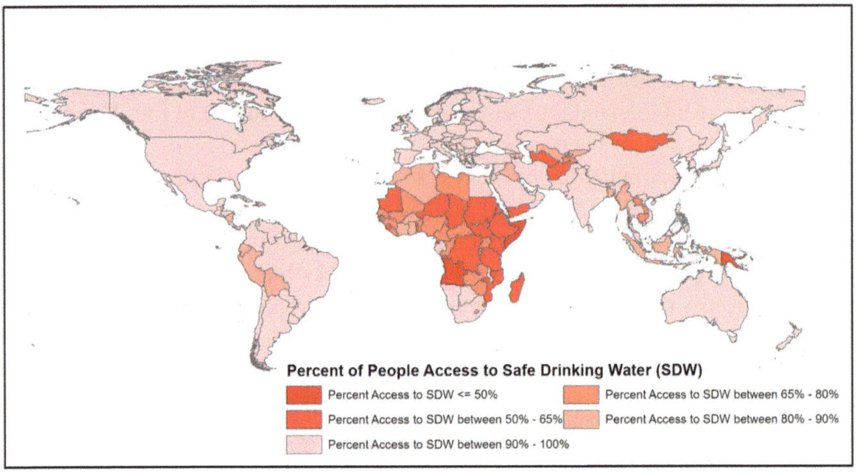

Fig. 6.1 Percent of people in a country which have access to safe drinking water (SDW) (data source: FAO 2024a) (Source: Food and Agriculture Organization of the United Nations. AQUASTAT. Total population with access to safe drinking-water (JMP). Latest update: 2022. Accessed: September 8, 2024. https://data.apps.fao.org/aquastat/?lang=en&share=f-6b6ec785-1e2a-4add-906e-5bfde8018cca)

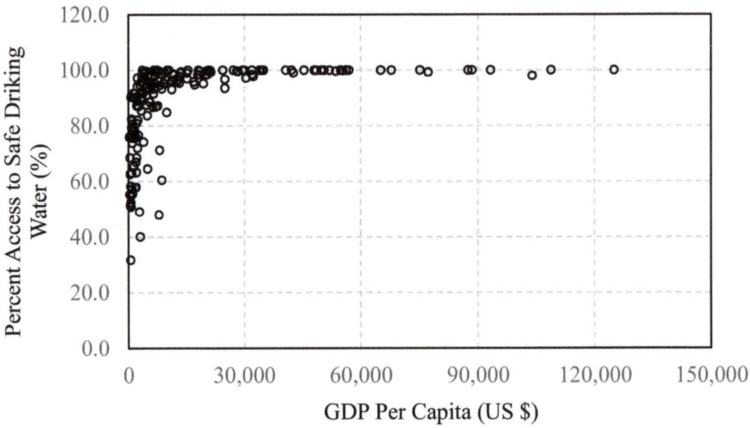

Fig. 6.2 Percent access to safe drinking water as a function of GDP per capita (data sources: FAO 2024a; World Bank 2024) (Source: Food and Agriculture Organization of the United Nations. AQUASTAT. Total population with access to safe drinking-water (JMP). Latest update: 2022. Accessed: September 8, 2024. https://data.apps.fao.org/aquastat/?lang=en&share=f-6b6ec785-1e2a-4add-906e-5bfde8018cca)

GDP per capita for the countries with 100% of their population with access to safe drinking water is US $51,394, the same for the countries with below 50% of their population with access to safe drinking water is US $3,673.

3.4 Diseases Due to Lack of Safe Drinking Water (SDW)

Access to safe drinking water is a function of country's economic status, which often plays an important role to prevent and mitigate public health issues. Access to safe drinking water is critical for public health safety and it can be considered as a fundamental human right. The lack of sufficient but safe drinking water and sanitation facilities is still the serious and important issue in many low- and middle-income countries (LMICs), which has led to severe public health concerns and psychological health consequences (Bisung and Elliott 2017). Poor water resources management and inadequate safety measures can lead to the same health consequences even in the high-income countries (Lee et al. 2023).

Contaminated water can spread waterborne diseases, which include diarrhea, cholera, amebiasis, typhoid, hepatitis, etc. An estimated 1.1 billion people worldwide do not have access to safe drinking water. Another 2.4 billion people do not have adequate sanitation facilities. Many of these people suffer from water related diseases, including 3.4 million deaths per year (APEC Water 2024). Per UNICEF (2023) about 4000 people die, including 1000 children, every day due to inadequate water, sanitation, and hygiene (WASH). An anticipated 3.0 billion people are expected to have water stress due to limited access to safe water (Shayo et al. 2023) which will result in a higher number of waterborne related deaths.

One of the most common types of waterborne diseases is diarrhea caused by organisms, including bacteria, virus, and some parasitic organism Cryptosporidium, Salmonella, etc. Globally, approximately 1.7 billion children suffer from diarrhea each year, causing 443,832 deaths among children under 5 years of age and 50,851 deaths in children 5 to 9 years of age. Diarrhea is the third leading cause of death in children aged 1 to 59 months (WHO 2024b). The water contaminated specifically with feces can be a reason for this infection, which leads to severe dehydration resulting in malnutrition and eventually death of young children. The key to preventing this disease is to ensure access to safe drinking water by providing improved water sources with water distribution systems. Similarly, ensuring developed sanitation systems can prevent the spread of diarrhea.

Cholera, a type of very acute diarrhea, can be fatal if not treated within hours. The bacteria, *Vibrio cholerae*, is responsible for this infectious disease, which spreads through the contaminated water. Globally, an estimated number of 1.3 to 4.0 million people each year are affected by cholera and approximately 21,000 to 143,000 people die from this disease. Similar to diarrhea, access to safe drinking water and ensuring improved sanitation can be impactful to the prevention of this disease.

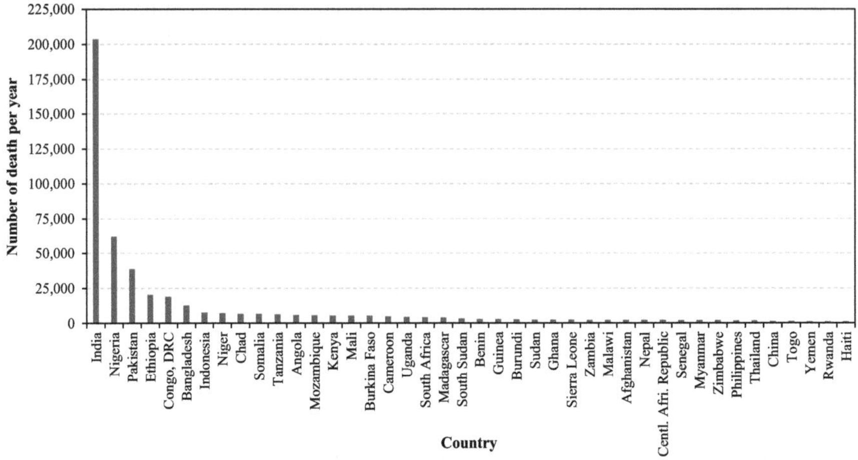

Fig. 6.3 Countries with at least 1000 diarrheal death per year (data source: WHO 2024c) (World Health Organization (2024c) data.who.int, {Number of diarrhoea deaths from inadequate water} [{numeric}]. {https://www.who.int/data/gho/data/indicators/indicator-details/GHO/number-of-diarrhoea-deaths-from-inadequate-water} (Accessed on September 8, 2024))

Considering any amount of death is significant and devastating, no attempt has been taken to normalize the data when compared to diarrheal death per country. The following diagram (Fig. 6.3) shows the total number of diarrheal deaths in 2019 for the countries that have at least 1000 diarrheal deaths per year (WHO 2024c). While Haiti has at least 1000 deaths per year, India has more than 20,000 deaths per year resulting in it being the top diarrheal death country in the world. Second in the rank is Nigeria followed by Pakistan, Ethiopia. As seen from the graph, most of the countries in the rank are from Africa and Asia, where many countries suffer from the lack of improved water source and inadequate water supply systems.

4 Climate Change and Water Supply

With only 0.5% usable freshwater on earth for the entirety of the world's population (8.2 billion people), water supply becomes a water crisis under climate change conditions, as climate change makes water resources prediction, which inherently is stochastic, more unpredictable, and scarcer. The obvious natural phenomena of climate change (temperature rise and hence drought conditions, increase in snowmelt, and sea level rise) have significant impacts on the world's fresh water supply. Temperature affects precipitation which directly impacts watershed runoff. Similarly, untimely snowmelt can create uncertainty in the availability of water resources. Among many other unwanted hydrological events, the rise in sea level causes intrusion of sea water in the ground, causing more salinity in the coastal groundwater, resulting in less water supply for drinking and other freshwater use purposes.

Drought, which has been found to be a more pronounced effect of climate change due to the rise in temperature, results in more water demands ranging from agricultural to domestic use. Soil moisture depletion increases water demand for agriculture, including with more irrigation water requirements. Abrupt reduction of soil moisture can affect natural habit for flora and fauna, causing more water demand for ecological need which may affect the percentage of water dedicated for water supply for drinking water and domestic purposes. In water resources management, even without the climate change conditions, it has been a constant struggle to optimize water allocations for different needs. Climate change just adds a new but significant and difficult dimension by making the availability of this resource more unpredictable.

Untimely snowmelt due to temperature rise additionally adds another level of uncertainty for the availability of water resources. Generally, higher temperature reduces snowfall and this reduction in fresh water in the mountainous area where pumping fresh water from valley can be a challenge. Additionally, in some months of the year, snowmelt can add more water when it's not required, while in other months of the year when water is needed, more reduction of snowmelt causes water scarcity.

The untimely snowmelt and uncertainty in precipitation due to temperature change can make streamflow more unpredictable. In some cases, for certain months of the year, higher storage facilities will be required to store additional inflow due to basin runoff and snowmelt. The current infrastructure, including dams and dam-lakes, may not be sufficient. The current water resources management strategies and operation policies may not be adequate and resilient enough to meet the water resources needs for many purposes, resulting in higher uncertainty in water supply for domestic use.

4.1 Source of Drinking Water

Less than 2.5% of total water on earth meets different freshwater demands, including industrial, agricultural, commercial, and domestic demands of 8.2 billion people. Domestic water use, which is also known as water use for municipal purposes, is for human needs, including drinking, cleaning, sanitation, cooking, etc. The distribution of the fresh water varies by countries as the freshwater comes from different sources, including glaciers and ice, groundwater, freshwater lakes and reservoirs, rivers, etc.

The freshwater open to the atmosphere known as surface water includes water from rivers, lakes, and reservoirs. Generally, about 80% of total daily water used for different purposes is the surface water (NGWA 2024). Surface water makes up a significant portion for public water supply and irrigation. For domestic use, including drinking purpose surface water, this usually comes from lakes, reservoirs, and rivers, and this generally needs more treatment than its counterpart groundwater. As a result, countries with lack of improved water sources and distribution systems rely more on groundwater. Groundwater, which can be defined as the stored water in earth's aquifers, can be made readily available by pumping a well. While in the USA about 70% of drinking water is surface water and 30% is groundwater (DWRF 2024), the number can vary significantly worldwide especially in the countries where water infrastructure is not developed. Especially for countries with lower precipitation and higher uncertainty for economic development and food production, groundwater can be the main source of water withdrawal (Wada et al. 2014).

Agriculture water use constitutes the majority (about 70%) of the freshwater withdrawal (DWRF 2024). Although about 35% of total fresh water withdrawn globally is groundwater (Döll et al. 2012), the number varies significantly in countries in arid and semi-arid areas, such as the Middle East, the northern portion of the Indian sub-continent, China, and northern Africa where the availability of surface water is limited. In those countries, groundwater is the main source of agricultural production and domestic use. While in Europe the groundwater withdrawal is generally constant at 30% of total freshwater withdrawal, the number varies in other parts of the world. In West Asia, the groundwater withdrawal tripled in the last few decades, making it 70% of water freshwater withdrawal. Similarly, in South and East Asia the groundwater withdrawal doubled in the last few decades (DWRF 2024).

The climate condition of a region and the hydrologic cycle can affect the amount of available renewable water resources, which can be defined as the rechargeable water that includes both surface and groundwater. Available per capita renewable water resources data from 47 countries has been divided into three categories and listed in Table 6.3. Category 1 includes the countries that have renewable water resources of more than 10,000 m^3/person/year. This category includes countries that generally have more snowpacks and glaciers. Examples include Iceland, Canada, Russia, etc. On the other hand, countries located generally in arid and semi-arid areas have less than 2,500 m^3/person/year. Examples include Israel, India, China, etc.

4.2 *Effect of Climate Change on Water Supply*

A simpler definition for climate change may be the shifting of weather patterns, including temperature and precipitation (UN 2024b). The change in climate affects weather properties including changing averages and/or the variabilities of the said property (WMO 2024). The change can affect the region for a longer amounts of times, such as for decades. The variabilities in the weather parameters and patterns make the climate change effect unpredictable and the impact on every aspect of life including hydrology and water supply more pronounced.

Climate change impact is very complex. It can affect every aspect of life from daily routine to food production, as all of the weather parameters and variables are affected due to the change. Some of the changes include changes in not only average temperature but also the variability of the temperature causing substantial extreme

Table 6.3 Worldwide available renewable water resources per capita (data source: OECD 2025)

Category 1		Category 2		Category 3	
Countries with per capita renewable water resources greater than 10,000 m³/yr		Countries with per capita renewable water resources in between 10,000 and 2,500 m³/yr		Countries with per capita renewable water resources less than 2,500 m³/yr	
Country	m³/person/year	Country	m³/person/year	Country	m³/person/year
Ireland	10,545	Armenia	2,500	Israel	291
Hungary	12,441	Denmark	2,748	South Africa	519
Georgia	14,125	Azerbaijan	2,991	India	1,308
Slovak Republic	14,498	France	3,100	Czechia	1,381
Bulgaria	14,510	Japan	3,394	Korea	1,474
Slovenia	14,869	Mexico	3,439	Poland	1,501
Costa Rica	17,024	Moldova	4,218	Belgium	1,889
Sweden	17,649	Netherlands	4,917	China	2,007
Latvia	17,803	Switzerland	5,914	Germany	2,040
Australia	18,519	Belarus	6,331	Türkiye	2,087
Estonia	18,610	Greece	6,927	Romania	2,102
Finland	19,607	Portugal	7,082	Spain	2,107
Russia	29,429	Lithuania	7,847	Luxembourg	2,457
Croatia	30,777	Austria	9,442	United Kingdom	2,474
Brazil	38,958				
New Zealand	63,426				
Norway	67,569				
Canada	88,216				
Iceland	443,907				

climatic events (Olonscheck et al. 2021), rising sea levels due to increases in global temperature, melting glaciers, change in precipitation patterns causing changes in average annual rainfall and higher variabilities, droughts, and flooding, etc. Specifically, the impact on the availability of water resources can affect human life in a multitude of ways. Flooding can spread waterborne diseases and droughts can impact food production resulting in malnutrition in human health (NOAA 2021).

While the average global temperature is expected to rise due to climate change, the regional variabilities in temperature and other weather phenomena can affect water resources availabilities. According to NOAA, the global temperature rose 1.1 °C (1.98 °F) from 1901 to 2020 (Olonscheck et al. 2021). WMO reported 1.45 °C increase in mean global temperature in 2023 compared to the mean global temperature in 1850–1900 (UN 2024b). Due to the increase in global temperature, the changes in sea level rise, snowmelt, and reduction in glacier thickness are visible. The rate of sea level rise has increased from 1.7 mm/year in twentieth century to 3.2 mm/year since 1993. The glacier thickness reduced more than 60 ft. since 1980. In the arctic, the area covered by the sea ice end of summer reduced 40% since 1979 (Olonscheck et al. 2021).

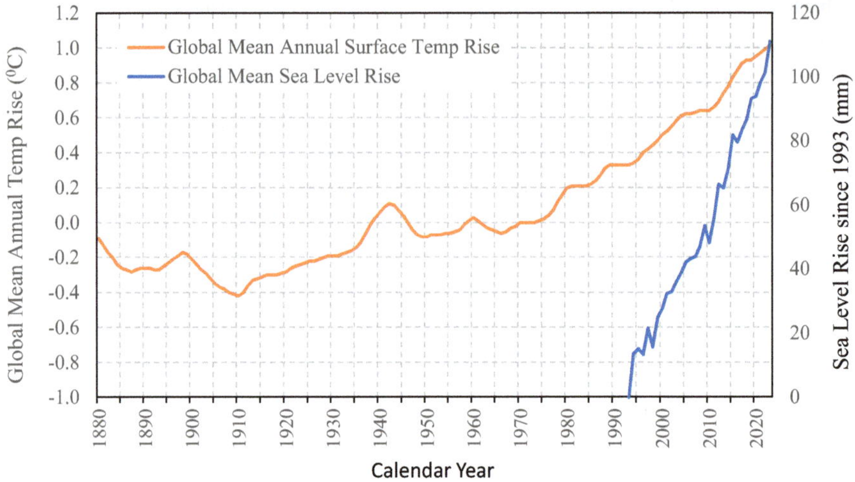

Fig. 6.4 Global mean annual surface temperature rise compared to average temperature in 1951–1980 and global mean sea level rise since 1993 (data source: NASA 2024)

Figure 6.4 shows the global mean annual surface temperature change compared to the long-term average temperature in 1951–1980 and global mean sea level rise since 1993. While the annual temperature anomaly is less than or close to zero until 1975, the anomaly has a distinct positive upward trend after 1975, which indicates that global mean annual temperature has been rising since 1975. The global mean sea level rise is due to many factors resulting from the increase in mean annual temperature. As seen in Fig. 6.4, the sea level has been constantly rising since 1993.

The rate of change of mean annual temperature in recent years is much more pronounced than ever before. The 10 warmest years ever from 1850 to 2023 happened to be in the last decade (2014–2023) with annual anomaly ranging from 0.77 °C (1.39 °F) to 1.18 °C (2.12 °F). The warmest year in the record is 2023. Compared to the average 1951–1980, the year 2023 was 1.8 °C warmer, which has exceeded the Paris Agreement's limit of 1.5 °C. While Europe had the highest temperature increase with 2.4 °C, the Oceania had the lowest increase with 0.9 °C. Asia had the second largest increase in temperature with 1.8 °C. Within Europe and Asia, the warmest area is found to be Western and Eastern Europe and Central Asia with an increase of temperature of 2.5 °C. Out of 121 countries that experienced increase in temperature by 1.5 °C, 68 countries had temperatures greater than 2.0 °C. With 3.6 °C increase in temperature, the Svalbard Islands in Norway recorded the highest warming. Top 10 countries with increases in temperature above 2.5 °C are in Europe (FAO 2024b). The countries with mean annual temperature increase of 1.5 °C in 2023 have combined population of more than 4 billion and the total cropland area in these countries cover more than the half of the world's total cropland area (NASA 2024).

Climate change will affect the availability and quality of water resources severely which can threaten the access to freshwater and result in poor quality of life for billions of people worldwide (Pltonykova et al. 2020). Climate change will impact hydro-climatology such that it will alter the water resources availability and pattern by increasing the variabilities and frequencies in extreme events including droughts and floods (Hoang et al. 2024). Prolonged droughts can be visible after heavy flooding; decrease in snowpacks after untimely snowmelt due to increases in temperature will cause changes in surface runoff and streamflow. Additionally, rises in sea level and tidal waves affect coastal area and groundwater quality and quantity.

4.2.1 Case Study: Effect of Climate Change on California's Water Supply

Baseline and Projected Timeline

A collaborative study was conducted by the University of Massachusetts, Hydrosystems Research group, and California Department of Water Resources (DWR) to observe the effects of climate change, specifically the risk of climate change on the California State Water Project (SWP) (DWR 2019). This case study section reviews the study conducted by DWR (2019) and provides additional analysis of the data obtained from that study, as presented in the subsequent sections. The baseline is defined as the period from 1971 to 2000, while the future refers to the period of 2035–2065, which is known as the mid-century conditions. Recent changes in the weather pattern in California indicate that the change has begun (Pierce et al. 2018). With mean temperature increasing from 0.6 to 1.11 °C since 1900, the northern part of California has become wetter and southern California has become drier since 1970. Overall, 1976–2004 was found to be significantly wetter in the western USA (DWR 2015; Higgins et al. 2007; Killam et al. 2014). As outlined in the report, the changing weather pattern will worsen both drought and flood risks. Projections include overall higher increases of temperature in summer than in the winter with more severe extreme events.

Since temperature and precipitation vary significantly in California both spatially and temporally, separate analyses have been conducted to observe the historical change in temperature and precipitation. Daily climate data from 1950 to 2015 has been analyzed to observe the warming patterns in four different seasons including winter, spring, summer, and fall. Each season consists of 3 months with December to February under winter and June to August in summer. As shown in Fig. 6.5, the mean temperature increases for all seasons. While spring has the highest mean temperature change, the fall has the lowest. The area considered under the study will likely experience the temperature change of 0 °C to 4 °C and precipitation change from −20% to 30%. Thus, different combinations of temperature and precipitation have been run to observe the effect.

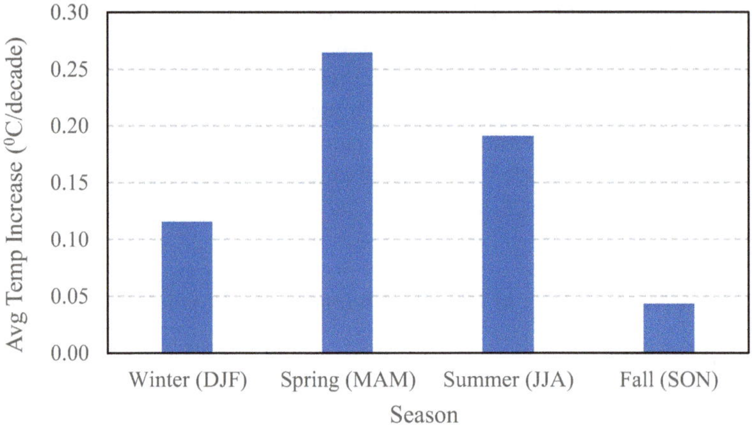

Fig. 6.5 Mean season warming pattern in California (data source: DWR 2019)

Modeling Approach

An advanced hydrologic model, Sacramento Soil Moisture Accounting (SAC-SMA) distributed hydrologic model (SAC-SMA-DS), was used to generate streamflow, which was the input to a water system model in the study (DWR 2019). The SAC-SMA-DS is a derived model from its original version SAC-SMA, which is a lumped hydrologic model developed by the National Weather Service of the NOAA (Burnash 1973). The SAC-SMA was coupled with a river routing model, hence known as SAC-SMA-DS to include the large, distributed watershed system in the Central Valley of California (Lohmann et al. 1998). SAC-SMA-DS used historical temperature and precipitation data as the input to generate streamflow at 32 locations in the Central Valley Watershed System. The streamflow was then used as the inputs to the water system model known as CalLite 3.0 that modeled the complex water supply systems in California considering different operation rules and regulatory compliance.

CalLite 3.0 is a simplified and hence faster water resources system planning tool derived from its original complex version CalSim-II (Draper et al. 2004). It simulates complex coordinated operations of the California Central Valley Water System (CVS) and the man-made constructed facilities in the Central Valley Project (CVP). CalLite 3.0 uses mixed integer linear program to optimize monthly water supplies, while considering system constraints including regulatory compliance and operating rules.

Study Area

Twelve major river systems, including the American River, Merced River, Stanislaus River, San Joaquin River, Mokelumne River, Calaveras River, Feather River, Tuolumne River, Sacramento River, Trinity River, Yuba River, and Clear Creek interconnected with the man-made facilities under the Central Valley Project (CVP) and State Water project (SWP) were defined as the Central Valley Water System (CVS) and were considered in the study (DWR 2019). To give a perspective—the watersheds of the San Joaquin and the Sacramento River supply a significant portion of water for two-thirds of the population of California. Most of the farmland in California depends on the water from these two river systems and half of the average annual streamflow in California flows in these river systems to the California Delta (Lund et al. 2010).

The Central Valley Project (CVP) supplies water for irrigation to about three million acres of farmland in California from its 20 reservoirs that have a total of 13 million ac-ft of storage capacity. The CVP also supplies water for municipal and wetland needs to meet different regulatory compliances and standards. Additionally, the State Water Project (SWP) that manages more than 30 lakes and reservoirs supplies water for more than 25 million people of California and for irrigation of about 750,000 acres of farmland (DWR 2019).

Effects of Climate Change on Water System

In the study (DWR 2019), the water system was modeled by CalLite 3.0 by taking the streamflow obtained from SAC-SMA-DS as input to this model. The performance of each model, i.e., the performance of hydrologic model SAC-SMA-DS and the performance of CalLite 3.0 were tested separately. Genetic algorithm (GA) was used in the SAC-SMA-DS to maximize NSE (Nash–Sutcliffe efficiency) coefficient. A model can be considered performing satisfactorily if the NSE coefficient value is found greater than 0.5 (Moriasi et al. 2007). As seen in Fig. 6.6, the NSE coefficient for both calibration (from 1951 to 1980) and validation period (from 1981 to 2002) gave NSE coefficient values higher than 0.5. For most of the subbasins the NSE values for both calibration and validation periods were greater than 0.90, with as high as 0.97 (refer to Fig. 6.6).

CalLite 3.0, which simulates complex water supply interactions, was validated with different historical scenarios at different locations to test whether the model can reproduce the same output as historical actual water supplies. The historical model outputs and the actual water supplies were plotted on a scatter plot to compare them for three different scenarios, including North Delta Storage, Delta Outflow, and State Water Project (SWP) deliveries. As mentioned, in all three cases the scatter plots produced R^2 value greater than 0.95, showing satisfactory performance of the water system model.

Using the calibrated and validated hydrologic and water system models, the responses of different climate scenarios were observed at different key locations and

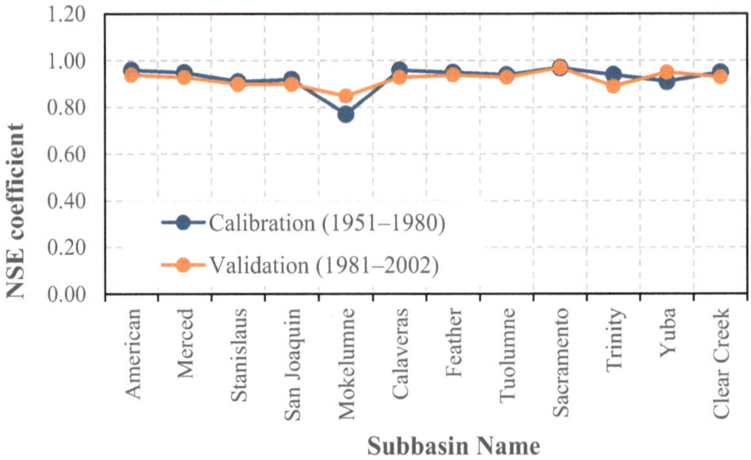

Fig. 6.6 Hydrologic model NSE (Nash–Sutcliffe efficiency) coefficients values for different subbasins (data source: DWR 2019)

for key scenarios, including State Water Project (SWP) Deliveries, Net Delta Outflows, Oroville Storage Levels, and System Shortages in the study (DWR 2019). The system response was observed by changing two meteorological variables: projected change in precipitation and projected change in temperature. To present system response considering different combinations of these two variables, a 3D surface plot was used. As seen from these surface plots in the study, the percent change in water supply in general decreases with the increase of temperature. Similarly, the water supply can increase with the increase of precipitation. However, since it's projected that the temperature will increase in the coming decades due to climate change, a certain amount of percent precipitation increase is required to offset the water supply shortage due to higher temperature. Also, the sensitivity of the water supply response to temperature and precipitation at some locations can be higher than other locations. For example, the storage in the Oroville dam at the end of April is moderately sensitive to temperature and precipitation, resulting in only a 10% increase in precipitation offsetting the impact of 2 °C increase in temperature. On the other hand, the Oroville Carryover Storage is found to be highly sensitive to the temperature and precipitation. To offset the impact due to 2 °C increase in temperature, it requires 20% increase in precipitation. The SWP water deliveries are more sensitive to lower precipitation than to higher precipitation (DWR 2019).

The effect of temperature change can be observed by considering no change in precipitation. This can provide an insight of how temperature itself can affect the water supply systems under projected climate change. The data has been extracted (with 5% rounding) from the report (DWR 2019) to observe this impact. As seen in Fig. 6.7, the percent change in system response when no precipitation change is considered can decrease up to 40% at 4 °C temperature increase for the Oroville Carryover storage. For the SWP deliveries, the system response can decrease about

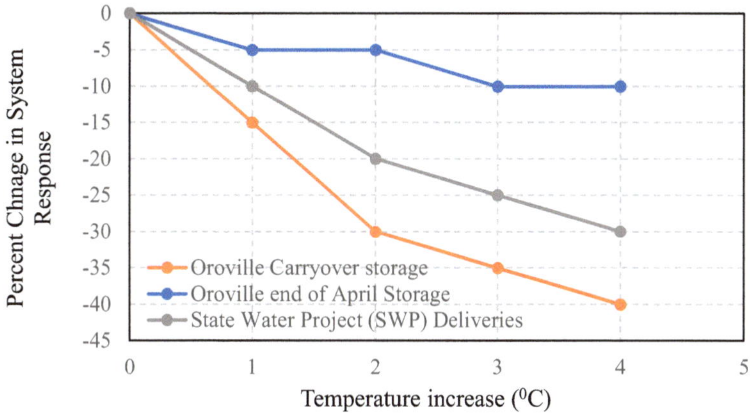

Fig. 6.7 Percent change in water system response in California as a function of temperature change (data source: DWR 2019)

30% at 40 °C temperature increase. As seen, the Oroville end of April storage system response is less sensitive to temperature change with a maximum change of up to 10% reduction in storage.

The average long-term system performance for projected conditions has been compared with current conditions to calculate the probability of system response to be inferior to the current conditions by calculating the cumulative density function (cdf) and probability density function (pdf) of the system response. To calculate cdf and pdf yearly, system performance has been calculated for all of the combinations of precipitation and temperature changes for the projected 2050 conditions and current conditions. The cdf calculates the probability of occurrence at or below a certain value of system response. Figure 6.8 summarizes the probability of inferior performance of the mid-century 2050 conditions to the current condition performance. As seen in Fig. 6.8, while there is a 93% likelihood of inferior performance of SWP, the percent of likelihood of inferior performance for the Summer Net Delta Outflow is 21.

4.2.2 Other Watershed Hydrology Analysis for Water Supply

Bekele and Knapp (2010) used Soil and Water Assessment Tool (SWAT) to observe the impact of climate change on water supply of the Fox River watershed which spans from Wisconsin to Illinois, USA. Their study found significant variability in monthly flows, which ranged from 66% increase in flow to 51% decrease in monthly flow. As the temperature increases, the annual water yield decreases. At 1.7 °C temperature increase, the average annual water reduces by 7% and at 3.3 °C of temperature increase, the average annual water yield reduction is 13%. The monthly

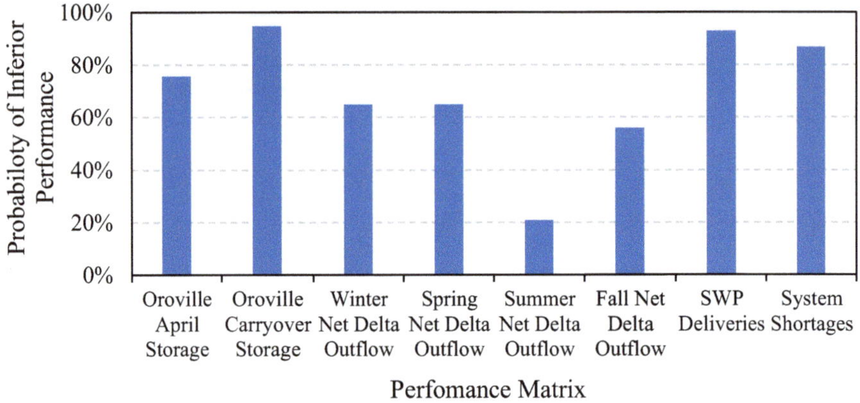

Fig. 6.8 Probability of inferior performance of different water system matrix in California under mid-century climate conditions (data source: DWR 2019)

streamflow reduces more (up to 39%) in late spring to summer compared to the winter months when the increase in temperature results in higher snowmelt causing an increase in monthly streamflow.

With an objective to understand climate trend and its hydrologic impacts, Yaning et al. (2009) conducted a study on the Tarim River basin located in the northern arid region of China. The Tarim River is one of the most important basin systems in China to supply freshwater water for different purposes, including domestic, agriculture, and ecological needs. With a catchment area of 1.02x106 km2 the Tarim River basin integrates 114 rivers to supply freshwater. Historical weather data has been analyzed to observe trends. Studies found that both temperature and precipitation had increased significantly since the 1980s with the 1990s found to be the warmest decade in the last 50 years. In the 1990s, the average annual temperature increased more than 100C compared to the average temperature in 1958–1989. While the streamflow in some rivers has a tendency to increase, some rivers have significant reduction in streamflow including in the Hotan River.

Maharajan et al. (2021) employed different climate models to observe the changes in temperature and precipitation in Thuli Bheri River basin located in Karnali province in Nepal. Different climate scenarios were then used in the Soil and Water Assessment Tool (SWAT) model to generate streamflow to observe the impacts of climate change water management for various purposes, including water use for hydropower, agriculture, ecosystem, etc. The study found significant variability in both temperature and precipitation. Temperature variability included change in temperature from −7 °C to +7 °C. Precipitation variability included an increase in precipitation between 12% and 50%. The high variabilities in these two variables under projected climate conditions resulted in higher streamflow variability ranging from −40% to +85%. High correlations between these three variables; temperature, precipitation, and streamflow were established through this study.

A study conducted by Tarekegn et al. (2021) observed the impact of climate change on the watershed hydrology of the Andasa watershed in Ethiopia. The study observed the impacts on watershed potential evapotranspiration (PET), soil moisture, and streamflow. Located in the upper Blue Nile basin, the Andasa watershed covers around 576 km^2. The Soil and Water Assessment Tool (SWAT) tool has been used to simulate the climate impact for the period of 1993 to 2099, with 1993–2012 used to calibrate the model. Study showed significant impacts on meteorological variabilities which affected the watershed hydrology. For the projected conditions, the mean monthly temperature increased in a range from 0.40 °C to 8.50 °C and the precipitation decreases in a range from 9.2% to 97.1% depending on various scenarios. Decrease in both temperature and precipitation resulted in a significant decrease in streamflow and soil moisture and increase in potential evapotranspiration. Model scenarios showed the decrease in streamflow and soil moisture ranged from 48.8% to 95.6% and 12.7% to 76.8%, respectively. The increase in potential evapotranspiration ranged from 4.4% to 17.3%.

5 Most Affected Regions

Although there are a lot of uncertainties associated with the availability of water resources under climate change conditions, it is expected that the number of countries that will have water stress will rise. When there were only seven (7) countries under water stress conditions in 1955, the number grew to 20 in 1990 and it is expected that the number will be 30–35 by 2025 (Gosain et al., 2006). Under projected climate change conditions, it is estimated that about 75% of the Himalayan glaciers will disappear by 2035, and a 10% reduction of precipitation in Sub-Saharan Africa by 2050 will reduce streamflow by 17% (Misra 2014). It is predicted that almost two-thirds of world's population will be under water stress (Gosain et al., 2006).

Lack of well calibrated hydrologic regional and (or) local models to include accurate climate change variabilities into models and associated uncertainties in the model parameters results in uncertainties and, in many cases, lack of water use data under projected climate conditions. The World Resources Institute (WRI) used a hydrologic model, which is global in nature to estimate the worldwide renewable water under projected climate conditions. Decreasing precipitation and increasing temperature trend will make Mediterranean countries hotspots of climate change (Cos et al. 2022). A recent analysis completed by the WRI shows that there are 25 countries that face extreme water stress every year and about 4 billion of the world's population will be under water stress for at least 1 month in a year. The top five most water stressed countries include Bahrain, Cyprus, Kuwait, Lebanon, Oman, and Qatar (Kuzma et al. 2023a).

If global temperature rises to 1.3 °C to 2.4 °C by 2050, an additional 1 billion people will live under extreme water stress. It is projected that by 2050 the water demand will increase by 20% to 25% with the number of watersheds that are

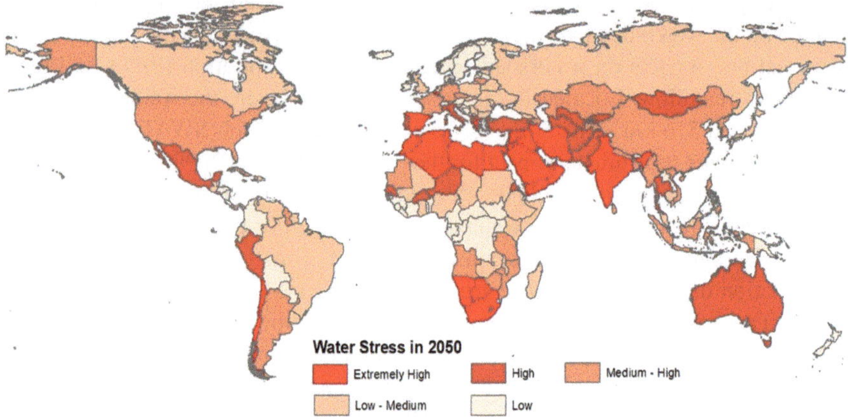

Fig. 6.9 2050 water stress regions in the world (data source: Kuzma et al. 2023b) (Kuzma et al. 2023, Aqueduct40_rankings_download_Y2023M07D05.xlsx. Copyright 2023 World Resources Institute. https://doi.org/10.46830/writn.23.00061, licensed under the terms of the Creative Commons Attribution 4.0 International License (http://creativecommons.org/licenses/by/4.0/))

difficult to predict due to their high variabilities that will increase by 19%. The report concluded that 100% of the population in North Africa and Middle East will face extreme high water stress by 2050 (Kuzma et al. 2023a). Although currently the Sub-Saharan countries in Africa are not in extreme water stress, they will face the highest change in water demand by 2050 to meet the increased water demand of 163% by 2050. The rate of increase of water demand in the Sub-Saharan countries is four times faster than the rate of increase in the Latin American countries. With a rate of 43% increase of water demand by 2050 in Latin America, it will become the second highest water demand region in the world (Kuzma et al. 2023a).

Figure 6.9 shows the water stress by country in the world with 2050 climatic change model conditions. Out of 164 countries worldwide, 29 countries are expected to have extremely high water stress conditions in 2050. With additional 22 countries worldwide having high water stress, the total number of countries that are expected to have at least high-water stress is 51 under 2050 climate change conditions. Except for a few, most of these countries are in the Middle East and North Africa. Most of the countries that are second in rank for high water stress are in Europe and Central Asia. Out of 24 countries in Latin America and Caribbean, one country is expected (Chile) to have extremely high-water stress; two countries, Mexico and Peru, are expected to have high water stress; the rest are expected to have moderate to low water stress for 2050 climate change conditions in Latin America and Caribbean.

Table 6.4 lists the top 25 countries that will be under extremely high-water stress under different time periods of climate change conditions, including the current conditions considered as baseline. For all four periods including the current time period, 2030, 2050, and 2080, most of the extremely high water stress countries will be in the Middle East and North Africa. Out of the top 25 countries, about 16 countries are in the Middle East and North Africa for all four different time periods of climate change conditions, making this area a hotspot of climate change. While

Table 6.4 Top 25 countries under extreme high-water stress under different time period of climate change (data source: Kuzma et al. 2023b)[a]

	Baseline		2030		2050		2080	
Rank	Country name	Region code*	Country name	Region code*	Country name	Region code*	Country name	Region code*
1	Bahrain	1	Bahrain	1	Bahrain	1	Bahrain	1
2	Cyprus	2	Cyprus	2	Cyprus	2	Cyprus	2
3	Kuwait	1	Qatar	1	Lebanon	1	Israel	1
4	Lebanon	1	United Arab Emirates	1	Qatar	1	Jordan	1
5	Oman	1	Lebanon	1	United Arab Emirates	1	Lebanon	1
6	Qatar	1	Israel	1	Oman	1	Qatar	1
7	United Arab Emirates	1	Oman	1	Israel	1	San Marino	2
8	Saudi Arabia	1	Saudi Arabia	1	Tunisia	1	United Arab Emirates	1
9	Israel	1	Tunisia	1	Libya	1	Oman	1
10	Egypt	1	Jordan	1	Jordan	1	Tunisia	1
11	Libya	1	Libya	1	Saudi Arabia	1	Iraq	1
12	Yemen	1	San Marino	2	Algeria	1	Libya	1
13	Botswana	3	Iran	1	Iraq	1	Algeria	1
14	Iran	1	Belgium	2	San Marino	2	Syria	1
15	Jordan	1	Chile	4	Iran	1	Saudi Arabia	1
16	Chile	4	Greece	2	Andorra	2	Iran	1
17	San Marino	2	Egypt	1	Greece	2	Chile	4
18	Belgium	2	South Africa	3	Chile	4	Greece	2
19	Greece	2	Botswana	3	Yemen	1	Andorra	2
20	Tunisia	1	Yemen	1	Spain	2	Spain	2
21	Namibia	3	Algeria	1	Belgium	2	Botswana	3
22	South Africa	3	Spain	2	Egypt	1	Egypt	1
23	Iraq	1	Namibia	3	Botswana	3	Belgium	2

(continued)

Table 6.4 (continued)

	Baseline		2030		2050		2080	
24	India	5	Iraq	1	South Africa	3	South Africa	3
25	Syria	1	Andorra	2	Kuwait	1	Yemen	1

*Region Code:
1 = Middle East & North Africa; 2 = Europe & Central Asia; 3 = Sub-Saharan Africa; 4 = Latin America & Caribbean; 5 = South Asia
[a]Kuzma et al. 2023, Aqueduct40_rankings_download_Y2023M07D05.xlsx. Copyright 2023 World Resources Institute. https://doi.org/10.46830/writn.23.00061, licensed under the terms of the Creative Commons Attribution 4.0 International License (http://creativecommons.org/licenses/by/4.0/)

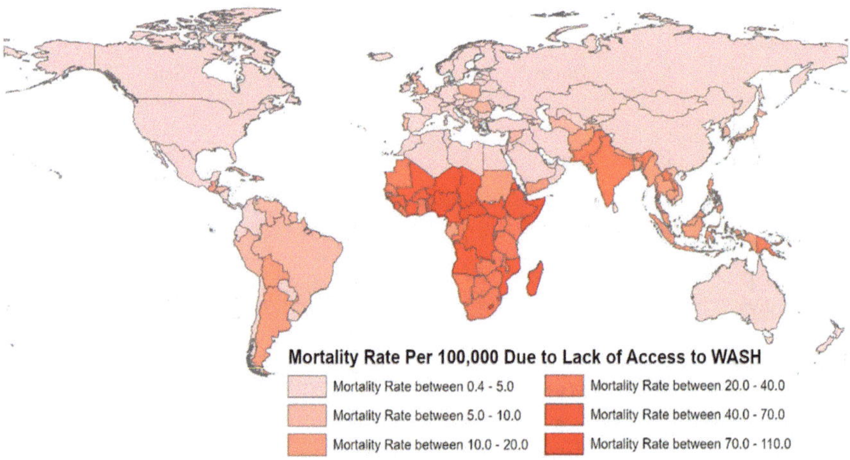

Fig. 6.10 Mortality rate per 100,000 due to lack of adequate access to safe water, sanitation, and hygiene (WASH) (data source: WHO 2024d) (World Health Organization (2024d) data.who.int, {Mortality rate attributed to exposure to unsafe WASH services (per 100,000 population) (SDG 3.9.2)} [{numeric}]. {https://www.who.int/data/gho/data/indicators/indicator-details/GHO/mortality-rate-attributed-to-exposure-to-unsafe-wash-services-(per-100-000-population)-(sdg-3-9-2)} (Accessed on September 15, 2024))

under current conditions only four countries are extremely high water stress in Europe and Central Asia, the number is expected to rise in future climate change conditions. Out of 25 top ranked extremely high-water stress countries, 20 countries are common under all climate change and current conditions, and 14 of these 20 common countries are in the Middle East and North Africa.

Access to safe clean water including safe drinking water is the key to prevent many waterborne diseases. Lack of access to safe clean water can contribute to lack of sanitation and hygiene resulting in higher waterborne diseases and higher subsequent death rates. Using the data collected by WHO (2024c), an analysis has been conducted which shows that the mortality rate per 100,000 can go as high as 108 due to lack of access to safe water and inadequate sanitation and hygiene (WASH). The minimum morality rate per 100,000 people is 0.4. The following map (Fig. 6.10)

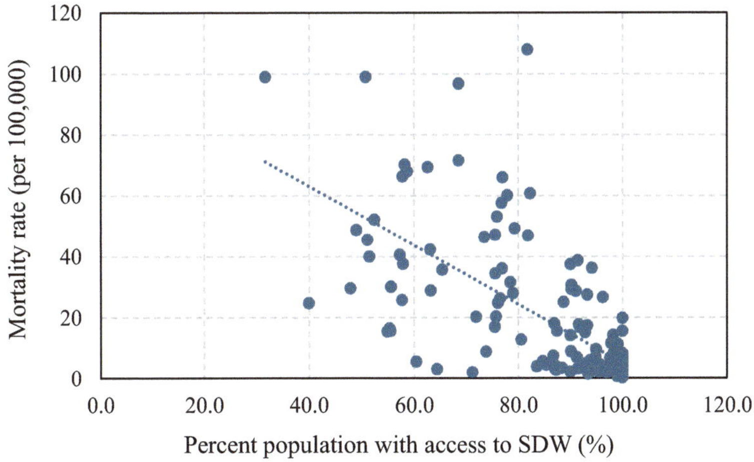

Fig. 6.11 Mortality rate as a function of percent of total population in a country with access to safe drinking water (SDW) (data sources: FAO 2024a; WHO 2024d) (Source: Food and Agriculture Organization of the United Nations. AQUASTAT. Total population with access to safe drinking-water (JMP). Latest update: 2022. Accessed: September 8, 2024. https://data.apps.fao.org/aquastat/?lang=en&share=f-6b6ec785-1e2a-4add-906e-5bfde8018cca)

shows the world mortality rate due to unsafe WASH. As seen, about 24 countries have the mortality rate above 40.0 and most of them are in Africa. Another 20 countries have the mortality rate between 20.0 and 40.0. Most of the countries that have mortality rate higher than 10.0 are either in Africa, Asia, or South America.

Since access to safe clean water is critical for human health, analysis has been conducted to observe the trend of WASH related mortality rate with percent of population with access to safe drinking water using the data obtained from WHO (WHO 2024d) and FAO (FAO 2024a). As seen in Fig. 6.11, the trend is very clear. As expected, the morality rate is inversely proportional to the percent of total population in a country with access to safe drinking water except for a few outliers. There might be several socio-economic and public health related issues and factors that contribute to the outliers. It is also important to keep in mind that access to safe drinking water is one out of many components of sanitation and hygiene. Many of the countries that are currently suffering from extremely high-water stress have higher mortality rates than other parts of the world. With climate change, these countries are expected to suffer even more with higher mortality rates due to inadequate WASH, if socio-economic conditions and public health awareness are not improved.

6 Conclusions

This chapter discusses three important broader topics related to drinking water; the health consequence of restricted safe drinking water, mechanism of climate change that affects water supply, and the regions that are affected to be the most due to climate change. Each of these topics is further discussed by subtopics with data, analysis, and case study.

The restricted drinking water including lack of improved water source or access to safe drinking water contributes significantly to the spread of waterborne diseases, like typhoid, diarrhea, cholera, hepatitis A. A country with lower GDP is found to have higher percent of people without access to safe drinking water resulting in higher mortality rate due to lack of WASH, including higher diarrheal death.

Climate change affects world water supply system as the temperature increases resulting in droughts, decrease in streamflow, and untimely snowmelt. Change in weather variables adds additional higher uncertainties to the already stochastic watershed hydrologic models causing the water supply more unpredictable both temporally and spatially. As a result, water supply system response can vary system to system. Under projected climate change conditions, it is expected that at least 51 countries will be facing at least high water stress in 2050 due to additional water demand up to 25%.

7 Discussion Questions

1. Explain why the drinking water consumption per person varies from country to country.
2. Explain how the mortality rate due to lack of adequate WASH including the diarrheal death can be reduced by ensuring higher percent of population with access to safe drinking water.
3. Explain why climate change affects the water supply system.
4. It is expected that the variabilities of both temperature and precipitation will increase under climate change conditions. If precipitation remains constant explain how temperature can affect water supply system response.
5. Explain with reasons which regions will be affected the most due to climate change.

Acknowledgments The author would like to acknowledge the support and inspiration that he had received from his wife, son, and daughter during the preparation of this manuscript. Thank you!

References

APEC Water (2024) Where do waterborne diseases rank in causing human health problems? https://www.freedrinkingwater.com/blogs/water-health/waterborne-disease-cause-human-health-problems. Accessed on September 8, 2024

Bekele EG, Knapp HV (2010) Watershed modeling to assessing impacts of potential climate change on water supply availability. Water Resour Manag 24(13):3299–3320. https://doi.org/10.1007/s11269-010-9607-y

Bisung E, Elliott SJ (2017) Psychosocial impacts of the lack of access to water and sanitation in low- and middle-income countries: a scoping review. J Water Health 15(1):17–30. https://doi.org/10.2166/wh.2016.158

Burnash RJ (1973) A generalized streamflow simulation system: Conceptual modeling for digital computers. US Department of Commerce, National Weather Service, and State of California, Department of Water Resources

California Department of Water Resources (DWR) (2015) California climate science and data for water resources management. California Climate Science and Data, June 2015, 1–28

Centers for Dieses Control and Prevention (CDC) (2024) Fast Facts: Data on Water Consumption National Health and Nutrition Examination Survey 2015–2018. https://www.cdc.gov/nutrition/php/data-research/fast-facts-water-consumption.html. Accessed September 7, 2024

Cos J, Doblas-Reyes F, Jury M, Marcos R, Bretonnière PA, Samsó M (2022) The Mediterranean climate change hotspot in the CMIP5 and CMIP6 projections. Earth Syst Dynam 13(1):321–340

Döll P, Hoffmann-Dobrev H, Portmann FT, Siebert S, Eicker A, Rodell M et al (2012) Impact of water withdrawals from groundwater and surface water on continental water storage variations. J Geodyn 59:143–156

Draper AJ, Munevar A, Arora SK, Reyes E, Parker NL, Chung FI, Peterson LE (2004) CalSim: generalized model for reservoir system analysis. J Water Resourc Plan Manag ASCE 130:480–489

Drinking Water Research Foundation (DWRF) (2024) Source comparison – facts about water. https://thefactsaboutwater.org/drinking-water-safety/source/ Accessed on Sep 9, 2024

DWR C (2019) Decision scaling evaluation of climate change driven hydrologic risk to the state water project. Technical Report, May 2019. https://water.ca.gov/-/media/DWR-Website/Web-Pages/Programs/All-Programs/Climate-Change-Program/Climate-Action-Plan/Files/CAP-III-Decision-Scaling-Vulnerability-Assessment.pdf

Food and Agricultural Organization (FAO) (2024a) AQUASTAT Dissemination System. https://data.apps.fao.org/aquastat/?lang=en&share=f-6b6ec785-1e2a-4add-906e-5bfde8018cca. Accessed on September 8, 2024

Food and Agriculture Organization of the United Nations (FAO) (2024b) Temperature change statistics 1961–2023. Global, regional and country trends. https://www.fao.org/statistics/highlights-archive/highlights-detail/temperature-change-statistics-1961-2023.-global%2D%2Dregional-and-country-trends/en. Accessed on Sep 10, 2024

Gosain AK, Rao S, Basuray D (2006) Climate change impact assessment on hydrology of Indian river basins. Curr Sci 90(3):346–353

Higgins RW, Silva VBS, Shi W, Larson J (2007) Relationships between climate variability and fluctuations in daily precipitation over the United States. J Clim 20:3561–3579

Hoang R, et al. (2024) The State Water Project Delivery Capability Report 2023. California Department of Water Resources, July 2024

Killam D, Bui A, LaDochy S, Ramirez P, Willis J, Patzert WC (2014) California getting wetter to the North, Drier to the South: natural variability or climate change? Climate 2:168180

Kuzma S, Saccoccia L, Chertock M (2023a) 25 Countries, housing one-quarter of the population, face extremely high water stress. World Resources Institute. https://www.wri.org

Kuzma S, Bierkens MFP, Lakshman S, Luo T, Saccoccia L, Sutanudjaja EH, Van Beek R (2023b) Aqueduct 4.0: Updated decision-relevant global water risk indicators. Technical Note. Washington, DC: World Resources Institute. https://doi.org/10.46830/writn.23.00061

Lee D, Gibson JM, Brown J, Habtewold J, Murphy HM (2023) Burden of disease from contaminated drinking water in countries with high access to safely managed water: a systematic review. Water Res 242:120244. https://doi.org/10.1016/j.watres.2023.120244
Lohmann D, Raschke R, Nijssen B, Lettenmaier DP (1998) Regional scale hydrology: I. Formulation of the VIC-2L model coupled to a routing model. Hydrol Sci J 43:131–141
Lund JR, Hanak E, Fleenor WE, Bennett WA, Howitt RE, Mount JF, Moyle PB (2010) Comparing futures for the Sacramento-San Joaquin Delta. Freshwater Ecology Series, vol. 3, 232 pp., Univ California Press, Berkeley; 2120 Berkeley Way, Berkeley, CA 94720 USA
Maharjan M, Aryal A, Talchabhadel R, Thapa BR (2021) Impact of climate change on the streamflow modulated by changes in precipitation and temperature in the North Latitude Watershed of Nepal. Hydrology 8(3):117. https://doi.org/10.3390/hydrology8030117
Misra AK (2014) Climate change and challenges of water and food security. Int J Sustain Built Environ 3(1):153–165
Moriasi DN, Arnold JG, Van Liew MW, Bingner RL, Harmel RD, Veith TL (2007) Model evaluation guidelines for systematic quantification of accuracy in watershed simulations. Trans ASABE 50:885–900
National Aeronautics and Space Administration (NASA) (2024) Global temperature and Sea Level. https://climate.nasa.gov/vital-signs/global-temperature/?intent=121. Accessed on Sep 10, 2024
National Ground Water Association (NGWA) (2024) Information on Earth's Water. https://www.ngwa.org/what-is-groundwater/About-groundwater/information-on-earths-water. Accessed on Sep 9, 2024
National Oceanic and Atmospheric Administration (NOAA) (2021) Climate change impacts. https://www.noaa.gov/education/resource-collections/climate/climate-change-impacts. Accessed on Sep 9, 2024
Olonscheck D, Schurer AP, Lücke L, Hegerl GC (2021) Large-scale emergence of regional changes in year-to-year temperature variability by the end of the 21st century. Nat Commun 12(1):7237
Organisation for Economic Co-operation and Development (OECD) (2025) Freshwater – resources, Measure: total renewable freshwater Period: Long-term annual average Time period: 2023. https://data-explorer.oecd.org/vis?fs[0]=Topic%2C1%7CEnvironment%20and%20climate%20change%23ENV%23%7CBiodiversity%252C%20water%20and%20ecosystems%23ENV_BIO%23&fs[1]=Topic%2C2%7CEnvironment%20and%20climate%20change%23ENV%23%7CBiodiversity%252C%20water%20and%20ecosystems%23ENV_BIO%23%7CWater%23ENV_BIO_WAT%23&pg=0&fc=Topic&snb=10&vw=tb&df[ds]=dsDisseminateFinalDMZ&df[id]=DSD_WATER_RESOURCES%40DF_WATER_RESOURCES&df[ag]=OECD.ENV.EPI&df[vs]=1.0&dq=.A.RENEW_RESOU.M3_PS.LTAA&pd=2023%2C&to[TIME_PERIOD]=false (Accessed on July 21, 2025)
Pierce DW, Kalansky JF, Cayan DR (Scripps Institution of Oceanography) (2018) Climate, drought, and sea level rise scenarios for the Fourth California Climate Assessment. California's Fourth Climate Change Assessment, California Energy Commission. Publication Number: CNRA-CEC-2018-006
Pltonykova H, Koeppel S, Bernardini F, Tiefenauer-Linardon S, de Strasser L (2020) The United Nations World Water Development Report 2020: Water and Climate Change
Shayo GM, Elimbinzi E, Shao GN, Fabian C (2023) Severity of waterborne diseases in developing countries and the effectiveness of ceramic filters for improving water quality. Bull Natl Res Cent 47(1):113
Tarekegn N, Abate B, Muluneh A, Dile Y (2021) Modeling the impact of climate change on the hydrology of Andasa watershed. Model Earth Syst Environ 8(1):103–119. https://doi.org/10.1007/s40808-020-01063-7
UNICEF (2023) Triple Threat How disease, climate risks, and unsafe water, sanitation and hygiene create a deadly combination for children. New York: United Nations Children's Fund (UNICEF), 2023. ISBN: 978-92-806-5438-7

United Nations (UN) (2024a) Global Issues – Water. https://www.un.org/en/global-issues/water. Accessed September 7, 2024

United Nations (UN) (2024b) What is climate change. https://www.un.org/en/climatechange/what-is-climate-change. Accessed on Sep 9, 2024

United States Environmental Protection Agency (USEPA) (2024a) National primary drinking water regulations. https://www.epa.gov/ground-water-and-drinking-water/national-primary-drinking-water-regulations. Accessed September 4, 2024

United States Environmental Protection Agency (USEPA) (2024b) Understanding your water bill. https://www.epa.gov/watersense/understanding-your-water-bill. Accessed September 7, 2024

Wada Y, Wisser D, Bierkens MF (2014) Global modeling of withdrawal, allocation and consumptive use of surface water and groundwater resources. Earth Syst Dynam 5(1):15–40

World Bank (2024) World Bank national accounts data, and OECD National Accounts data files. https://data.worldbank.org/indicator/NY.GDP.PCAP.CD. Accessed on September 8, 2024

World Health Organization (WHO) (2013) Technical notes on drinking-water, sanitation and hygiene in emergencies, July 2013. WEDC, World Health Organization

World Health Organization (WHO) (2024a) Drinking-water. https://www.who.int/news-room/fact-sheets/detail/drinking-water. Accessed September 4, 2024

World Health Organization (WHO) (2024b) Diarrhoeal disease – Key Facts. https://www.who.int/news-room/fact-sheets/detail/diarrhoeal-disease. Accessed on September 8, 2024

World Health Organization (WHO) (2024c) The Global Health Observatory – Explore a world of health data – Number of diarrhoea deaths from inadequate water, sanitation and hygiene. https://www.who.int/data/gho/data/indicators/indicator-details/GHO/number-of-diarrhoea-deaths-from-inadequate-water-sanitation-and-hygiene. Accessed on September 8, 2024

World Health Organization (WHO) (2024d) The Global Health Observatory, Explore a world of health data. https://www.who.int/data/gho/data/indicators/indicator-details/GHO/mortality-rate-attributed-to-exposure-to-unsafe-wash-services-(per-100-000-population)-(sdg-3-9-2). Accessed on Sep 15, 2024

World Meteorological Organization (WMO) (2024) Climate change. https://wmo.int/topics/climate-change. Accessed on Sep 9, 2024

Yaning C, Changchun X, Xingming H, Weihong L, Yapeng C, Chenggang Z, Zhaoxia Y (2009) Fifty-year climate change and its effect on annual runoff in the Tarim River Basin, China. Quat Int 208(1):53–61. https://doi.org/10.1016/j.quaint.2008.11.011

Chapter 7
Climate Change, Drinking Water Security, and Public Health

Kyana R. L. Young

1 Learning Objectives

1. Identify the climate change events which negatively impact drinking water quality.
2. Perform and understand the meaning of log inactivation as it pertains to waterborne pathogens.
3. Discuss current drinking water treatment technologies that will help alleviate the challenges related to climate change.

2 Introduction

Water quality is defined based on the intended use of water. For water labeled as wastewater/gray water, the term quality is not applied in the same manner as that of drinking water. Therefore, quality is objective based on circumstance. For this consideration, drinking water quality compliant standards are critical to ensuring that a community has a pathway to, at minimum, accessible safe drinking water. Therefore, the intention of monitoring quality is to assess whether the intended user can consume (drinking, cooking) the water without any adverse health impacts.

Drinking water quality is an agreed upon mandate that allows communities and regulating entities to determine the safety of water for consumption to prevent disease and sickness. The World Health Organization (WHO) and United Nations (UN) have regulated drinking water quality mandates that each governing member

K. R. L. Young (✉)
Department of Engineering, Wake Forest University, Winston-Salem, NC, USA
e-mail: youngk@wfu.edu

A. R. Molla et al. (eds.), *Climate Change, Drinking Water Security, and Public Health*, https://doi.org/10.1007/978-3-032-12074-8_7

state must comply (Hespanhol and Prost 1994). Additionally, each member state may also establish its own drinking water quality standard.

In the circumstances where safe drinking water is challenging to access, communities may begin to rely on commercially available drinking water if the quality of their relied source is compromised or if the trust of the source water is questionable. It is recommended that, to contribute to a healthy life, adult humans should consume 2 L of water every day. This water may come in the forms of foods and additional liquids. Regardless of the route of consumption, compliant drinking water quality must be maintained to prevent disease and death.

The United Nations Sustainable Development Goals recommends that each person has access to safe, clean drinking water (Browne 2012). Those adjectives are nebulous and require an assessment of meaning and application. Access—ability to obtain. Safe—without causing harm. Clean—no impurities. Many scenarios around the world drinking water may meet one or two about the specifics.

As of 2023, unsafe water sanitation and hygiene is a contributing factor for the death of 400,000 children worldwide (UNICEF 2023). While this is a grim reality in and of itself, there are also community impacts beyond the grieving the death of a child. The ability for a community to thrive and have members who contribute to the livelihood is a result of the available working bodies who can contribute to an economy. Water sources are continually introduced to contaminants.

To ensure that humans have access to safe drinking water before, during, and after extreme weather events, it is imperative that monitoring and protocol practices be implemented on a consistent and routine basis. Extreme weather impacts not just the community in immediate proximity to the events, but also those who may be experiencing the aftermath based on geography, providing assistance to remediate the challenge, and those who rely on a range of industries that directly impact one's livelihood.

3 Climate Change and its Impact on Water Systems

Climate change impacts the availability (quantity) of water for drinking, cooking, bathing, cleaning, and manufacturing purposes. Source waters, such as surface water (springs, lake, rivers, creeks, streams) and groundwater, are at risk for contamination and lack of availability depending on the climate-based challenge a region is experiencing. These challenges jeopardize livelihoods and health of communities who are relying on the water. Eutrophication of a water system is a prime example of this scenario.

When referring to the condition of surface water, trophy refers to the rate at which nutrients (phosphorus, nitrates) and microbes are introduced to the surface water system. The three levels of trophy, in this regard, are oligotrophic—low levels of nutrients and algae, eutrophic—high levels of nutrients and algae, and mesotrophic—balanced/intermediary levels of nutrients and algae. When surface water is deemed eutrophic, the probability for the development of harmful algal blooms

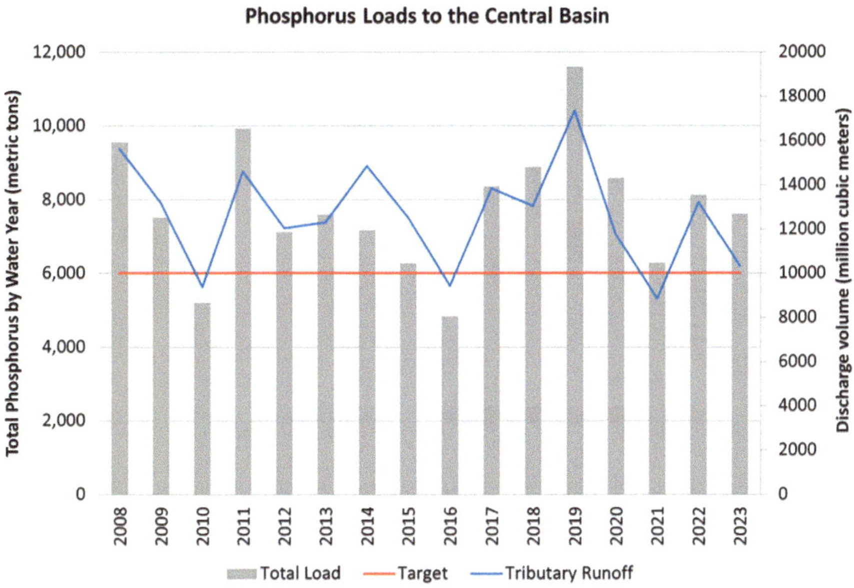

Fig. 7.1 The loading of phosphorus (nutrient) to the central basin of Lake Erie. (USEPA 2024)

(HAB) increases. Some of the more notable HAB's have occurred in the Asia [China (2017)], North America [United States (2015), Fig. 7.1], and Europe [(Ireland (2023)] (USEPA 2024).

When the surface water develops harmful algal blooms, the source water is now compromised by the availability of safe drinking water for the community. Communities must now rely on international aid, point of use water treatment systems, or use of water that is not of compliant water quality.

4 How Climate Change Affects Drinking Water Quality

4.1 Temperature Rise and Water Contamination

The term climate change has become a political argument for those who are debating whether observations are part of a natural trend or whether the results are anthropogenic. Essentially, data allows for an observance of how the climate is changing, which is factual, and the impacts of the change. Politicization focuses on the causes of the changes. Some think that the change is part of a natural fluctuation of temperature, rain events, snow events, droughts, and flooding. However, others think the combustion of fossil fuel, cloud management, and deforestation that the impacts climate are anthropogenic—caused by humans.

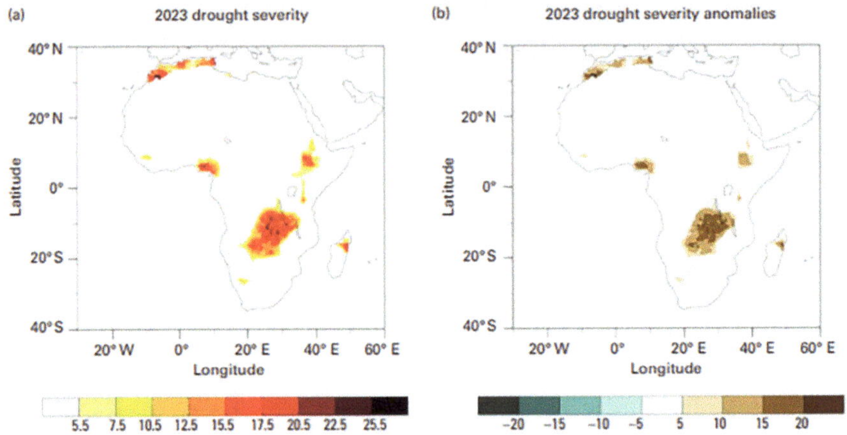

Fig. 7.2 The distribution of drought severities and anomalies on the continent of Africa, during the year 2023. (WMO 2024)

Documentation of weather has been available since 1659. However, when considering the life of the planet, this time frame is miniscule. Focusing on that "short" period allows for interpretation that is debatable. Even with the fluctuations, the trending of temperature is increasing. Thus, this section will refer to the impacts of climate as extreme weather events. These events of heavy rains, flooding, and droughts have increased significantly during the documented period of time.

Water, which has been frozen (less than 32 °F/0 °C), provides a stable and stagnant environment for pathogenic microbes to exist. These microbial pathogens emerge, from the warming of polar ice caps, to provide the opportunities for exposure to compromised source waters which inevitably negatively human health.

The earth's age is 4.543 billion years (Berkeley Earth 2024); the Berkeley Earth group has documented the land temperature of the earth since the year 1750 until now. The trend of the Earth's 12-month mean temperature has steadily increased every year since 1810.

As a continent, Africa experienced climate warming more than any other continent. Changing rainfall patterns throughout the continent of Africa has a variety of impacts based on the location (Fig. 7.2). Africa, with a total area of 11,730,000 mi^2, has a variety of climate and landscape including deserts. There are six major basins—Chad, Nile, Niger, Congo, Zambezi, and Orange—with all but one (Chad) leading to the discharge to the sea.

4.2 *Changes in Precipitation Patterns*

Weather patterns, on a global scale, have impacted precipitation patterns. Raining seasons, which have been defined by specific months and volumes of precipitation to be expected, have shifted. For example, the rain season in the Caribbean (North

America) were historically marked by a May–October timeframe, with an average of 5 in/month. Now the rainy seasons are shorter, with much higher burst of rains. The patterns change now requires the residents to adjust their lives to the volume of available water. Infrastructure and natural design not adjusted to the natural precipitation levels are not as sustainable; the integrity of these designs now compromised.

An increase in rainfall, over a shorter period, results in high runoff (volume/time), erosion, and the introduction of contaminants into source waters. The increase of rainfall has an impact of erosion for coastal regions and riverbeds. This impacts water quality as the contaminants in the sediments are now introduced to the source of water. The banks serve as a natural sorption mechanism to capture a variety of chemical and microbial in the stormwater. Release of these contaminants compromise the quality of water and may contribute to the larger ecological challenges such eutrophication. With eutrophic surface water classification, it is the combination of nutrients, temperature, and light availability which contributes to the development of harmful algae blooms (HAB). The HAB have historically rendered surface waters unusable for drinking, drinking water treatment, fishing, boating, and other recreational activities. Remediation of surface waters impacted by eutrophication and/or HAB development requires financial resources to assist with the reducing nutrient loading, as well as alternate sources of water which make accessibility a challenge.

4.3 Sea-Level Rise and Saltwater Intrusion

Saltwater intrusion is predicated on the increase in the high tides or surges of storms in coastal communities. High tides are caused by the melting of land-based glaciers, which is caused by the increase in temperatures. The melted glacial runoff discharges to receiving bodies of water, resulting in high sea levels and thus higher tides. The tides introduce saltwater to freshwater systems. At this current time, humans cannot properly process the concentration of salt in seawater (35 ppt—parts per trillion) within economical reasons; typically, freshwater has a concentration of 0.5 ppt. With ingestion of seawater salt concentrations, the kidney is not able to produce urine with less salt. This requires the body to increase urination and contributes to overall dehydration. When surface waters are compromised by the introduction of seaeater, the community who relies on the surface water for drinking, cooking, bathing, no longer have access to a drinking water source. Globally, ~40% of the humans live less than 100 km of a coast, who are relying on freshwater aquifers to serve as a potable water source. Thus, the saltwater infusion threatens the freshwater source for 40% of the world's population.

5 Extreme Weather Events

5.1 Case Studies

5.1.1 Algal Blooms in Freshwater Systems: United States

In the United States, the largest freshwater system in the world is known as The Great Lakes. Located in the Midwestern part of the United States, officially "belonging" to the state of Michigan, the lakes (Huron, Ontario, Michigan, Erie, and Superior) account for 94,250 mi^2. Lake Erie, accounts for ~9910 mi^2. The Lake serves as a natural border for the four states (Michigan, Ohio, Pennsylvania, and New York) as well as the Ontario province in Canada. Lake Erie is the freshwater source for ~11,000,000 people. Due to anthropogenic activity, high concentrations of nutrients in the watershed routinely discharge into the Lake. The introduction of high nutrient concentrations is not the sole contributing factor in the development of HAB (USEPA 2024) (Fig. 7.3). However, this is a component that can be immediately resolved with best management practices to reduce the use of nutrients. Another contributing factor to the development of the HABs is the increase in temperature in the region. The warmer temperatures allow the HAB to proliferate and grow in its ideal warm, nutrient rich environment. Unfortunately, the prevalence of HAB negatively impacts the accessibility of drinking water for people in the region.

Fig. 7.3 Lake Erie Harmful Algal Bloom (Doermann 2024) (NASA Earth Observatory image by Wanmei Liang, using Landsat data from the U.S. Geological Survey)

As a result, the nations of the United States and Canada have established the "Lake Erie Binational Nutrient Management Strategy," to reduce the high concentrations of phosphorus entering the lake. The Strategy details a variety of mitigation strategies, including, but not limited to, identifying sources that contribute the most phosphorus loading within the watershed and establishing policies that management the use and discharge of phosphorus. The Strategy does also note that warmer temperatures and shorter winters are also contributing factors to the development of HAB.

5.1.2 Droughts in Arid Regions: Zambia

In Zambia, a dam was built as a response to the growing need of electricity in the country. The use of hydroelectricity is slowly being abandoned based on the low availability of water from the Kariba Dam and low rainfall events. Rainy seasons have been occurring later in the year, based on recent (100 yr) trends. The delayed rainy seasons, accompanied by low rainfall events, have caused increase in dry events (UNOCH 2024). The dry events result in groundwater not being recharged. As a result, 3.5 million people are now at risk of having access to safe drinking water. Women and children, who are traditionally responsible for fetching water, are now required to find reliable water sources farther from their homes. When the water is identified, the quality is low; low volumes of water allow for the microbes to be concentrated. Furthermore, humans are now sharing these water sources with wildlife who are also searching for reliable water sources. The implications from the direct sharing are the higher risk of contracting communicable diseases. One of the recent reported diseases, with a direct connection to shared water sources, is cholera; cholera is a bacterial based microbe, which causes severe diarrhea and dehydration in humans. According to The World Health Organization estimates, there are 3–5 million cases and 100,000–200,000 deaths every year due to cholera. With drought, there are limited, if any, options to provide safe drinking water to communities. There are new technologies being implemented that would allow for rehabilitation of boreholes. Furthermore, as in the case in many communities on a global scale, many communities in Zambia rely on external assistance with the transport and distribution of water from other countries.

5.1.3 Saltwater Intrusion in Coastal Cities: Italy

Many coastal cities, such as Taglio di Po in Italy, have been experiencing saltwater intrusion which impacts their availability of freshwater and compromising the quality and quantity of the crops. Located 30 km from the Adriatic Sea, Taglio di Po has experiences a continual saltwater intrusion in the aquifers. The precipitation is recharging the aquifers at slower rates than the water being used by the community (Antonellini et al. 2008; Tarolli et al. 2023). This lack of freshwater in the aquifer allows for seawater to replace the freshwater. Due to extraction of methane from the subsoil, high incidences of subsidence—sinking of ground due to subsurface movement of material—has contributed to the main river (Po River) to exist below sea level. One of the temporary resolutions for the saltwater intrusion events includes the design and implementation of barriers to prevent the flow of saltwater to the aquifers (USGS 2019). Another adjustment that has been proposed by the community is the harvesting of rainwater. While this approach will not eliminate the continual allowing of saltwater to be intrusive in the aquifer, the community will have freshwater and not brackish—mixture of freshwater and seawater—to use for potable and agricultural purposes. A final recommendation proposed is the restoration of the soil structure. This will aid in the natural barrier between freshwater and seawater.

5.1.4 Flooding and Water Contamination: Thailand

While water is flowing through the cities and communities, it is also collecting a variety of contaminants from foreign to natural water. Flooding is a natural phenomenon, however, the increase in frequency, caused by anthropogenic activities, produces more floods in a short period of time. When the water is flowing over the land, it accumulates a variety of contaminants including microbes, pesticides, industrial waste sewage, and solid materials. The presence of these contaminants in the water makes the water treatment process challenging. Often after floods, communities rely on bottled water to replace the water they commonly obtain from a reliable source.

Flooding remediation efforts from Thailand include the development of the Climate Change Department, which aims to strategize about mitigation efforts to alleviate the community from the negative impacts of the changing climate (Mani and Pollitt 2024). The department is also responsible for managing water as it inundates communities. There are also new tools, such as satellite monitoring, which assist in detecting the flooding. These efforts are, however, not preventive but rather responsive. Another mitigation strategy is the building of 500 parks, which aims to provide more green space for the water to absorb in the ground. The water is then used as the source water for the drinking water treatment process.

6 Impacts on Public Health

6.1 Waterborne Diseases

The United States Surface Water Treatment Rule (SWTR), used to govern the public water systems using surface water as their source of drinking water treatment, allows for a specific requirement for treatment of water. Established in 1989, the SWTR required for all public water utilities, who provide water to ≥10,000 people with the United States, to inactivate or remove 3-log (99%) of Giardia lamblia and 4-log (99.99%) of viruses. The log reduction can be calculated using Eq. 7.1,

$$\text{log inactivation / reduction} = \log_{10} \frac{N_0}{N} \tag{7.1}$$

where N_0 is the initial concentration of microbes entering a system (influent) and N is the final concentration of microbes exiting of the system (effluent). Moreover, the WHO has drinking water guidelines which broadly encompass the full slate of microbial and chemical concentrations permissible in drinking water.

At times, the influent of water treatment systems can overwhelm the system causing failure of treatment systems. This is the exact condition which led to the Milwaukee 1993 *Cryptosporidium parvum.* After a series of heavy rains, over a three-day period, the "unnatural" rains inundated the drinking water treatment plant, with a flowrate 3x more than the normal flow. As a result, the treatment plant allowed *Cryptosporidium*—a waterborne pathogen, to be delivered from the public water distribution system to 800,000$^+$ people, where 400,000$^+$ people were infected. The event caused 69 deaths and 4000$^+$ hospitalizations. As a result, the United States with the enhancements of the SWTR, namely, the Long Term 1 and 2 Enhanced Surface Water Treatment Rules I and II (LT1ESWTR and LT2ESWTR) require an additional 2-log inactivation/removal of *Cryptosporidium.*

Furthermore, the increase in temperature provides conducive environments for waterborne pathogenic microorganisms to thrive or emerge from a dormant status. Exposure to these pathogens creates a pathway to disease development in humans, resulting in diarrhea. In the case of children under the age of 5, the UNICEF monitors diarrheal rates. This rate correlates to malnutrition, and sometimes death. Thus children, and other immunocompromised individuals who experience diarrhea, caused by pathogens, are at higher risks of death (UNICEF 2024).

7 Addressing the Challenges: Solutions and Adaptations

7.1 Improved Water Treatment Systems

The impact on drinking water quality from flooding, eutrophication, drought, and saltwater intrusion requires improved drinking water treatment systems that address the complexities of the water matrices. Point-of-use (POU) has long been a useful practice for humans who do not have a centralized drinking water treatment system. The POU treatment must be used and maintained by the individual, with the intention to treat the water to reduce contaminants (chemical and microbial) to the desired quality (Yale 2024).

Water treatment systems will need to add novel treatment technologies to address the complexities of water quality entering the system. In fact, the water treatment addendums may even require treatment separate from the infrastructure of the water treatment system. As an example with the harmful algal blooms, scientists and engineers are developing technologies that will address the capture of nutrients before they enter the source water used as the beginning stage of the drinking water treatment process. With the case of the saltwater intrusion systems, artificial recharge has become a popular option for coastal systems who are looking to preserve their freshwater source. The aquifers are recharged with high quality freshwater water to offset the intrusion of saltwater and to provide a higher water table which would not come into concerning contact with the seawater. Regardless of the improved water treatment system, it is imperative that the adaptations account for the changing climate.

8 International Agreements and Standards

8.1 World Health Organization Drinking Water Guidelines

The World Health Organization has established drinking water quality standards as a guide to ensure that, across the world, communities have access to drinking water with low risk of developing a disease. These standards were established with public health in mind. The United Nations has taken the standard to another level, declaring that communities should have "access" to "safe" drinking water (Hespanhol and Prost 1994; Browne 2012). Both of these words are subjective, as the UN has not defined what is considered safe—a clear distinction from the established WHO quality—and what is meant by access. In some communities, access is not a challenge. Rather, communities are concerned about the safety of the water due to historical challenges in municipalities providing safe drinking water. The standards are separated into Chemical, Microbial, Radiological impacts, and aesthetics.

While WHO and UN have standards and mandates to ensure that drinking water is safe, clean, and accessible, climate change prevents an additional challenge which comprises both the accessibility AND the safety of water. Climate change, and more broadly, extreme weather events continue to have a negative impact on the environment.

9 Conclusions

The extreme weather events taking place on a global scale have additional challenges beyond recovery and rebuilding. Community's access to safe drinking water becomes a subsequent result of compromised infrastructure, inundation of polluted source waters, and prevention of physical access to safe and clean drinking water sources. While governmental entities work to reverse the anthropogenic activities that contribute to extreme weather events, scientists, engineers, and community members must be prepared for the design, implementation, maintenance, and use of drinking water treatment systems to supplement those which have been compromised by extreme weather events.

10 Discussion Questions

1. Using the United Nations Sustainable Development Goals 6 and 13, how does your country fair with climate change actions and the status of accessibility of safe drinking water quality?
2. What do you recommend as a remedy for emergency drinking water treatment technologies post extreme weather events, where potable water is scarce?
3. Select an extreme weather event likely to impact your home/community. What are your strategies for securing access to safe, clean drinking water for at least 7 days after the event?
4. Discuss the trends of increased recorded temperature from 1760–current date. Are these trends part of a natural trajectory or are the trends indicative of a larger anthropogenic impact?

References

Antonellini M et al (2008) Salt water intrusion in the coastal aquifer of the southern Po plain, Italy. Hydrogeol J 16(8):1541–1556

Berkeley Earth (2024) Global land. https://berkeleyearth.org/temperature-region/global-land. Accessed Dec 14, 2024

Browne S (2012) United Nations development Programme and system (UNDP). Routledge

Doermann, L. (2024). Lake Erie Blooms. National Aeronautics and Space Administration https://earthobservatory.nasa.gov/images/153282/lake-erie-blooms. Accessed Jan 31, 2025

Hespanhol I, Prost AME (1994) WHO guidelines and national standards for reuse and water quality. Water Res 28(1):119–124

Mani M, Pollitt H (2024) Towards a green and resilient Thailand (No. 42397). The World Bank Group

Tarolli P et al (2023) Saltwater intrusion and climate change impact on coastal agriculture. PLOS Water 2(4):e0000121

United Nations Office for the Coordination of Humanitarian Affairs (UNOCH) (2024) In: Zambia: drought response appeal May 2024–December 2024. https://www.unocha.org/publications/report/zambia/zambia-drought-response-appeal-may-2024-december-2024-may-2024. Accessed Dec 14, 2024

United Nations Children's Fund (UNICEF) (2024) In: Diarrhoea. https://data.unicef.org/topic/child-health/diarrhoeal-disease/. Accessed Dec 14, 2024

United Nations Children's Fund (UNICEF) (2023) How disease, climate risks, and unsafe water, sanitation, and hygiene create a deadly combination in children. https://www.unicef.org/media/137206/file/triple-threat-wash-EN.pdf. Accessed Dec 14, 2024

United States Environmental Protection Agency (2024) In: Lake Erie water quality data. https://www.epa.gov/glwqa/lake-erie-water-quality-data. Accessed Oct 20, 2024

United States Geological Survey (USGS) (2019) Aquifers and Groundwater. https://www.usgs.gov/special-topics/water-science-school/science/aquifers-and-groundwater. Accessed Dec 14, 2024
World Meteorological Organization (WMO) (2024) State of the climate in Africa 2023. WMO-No 1360. https://library.wmo.int/idurl/4/69000. Accessed Dec 14, 2024
Yale Environment 360 (2024) In: As Drought Shrivels Hydro, This African Nation Pivots to Solar. https://e360.yale.edu/features/zambia-drought-hydro-solar-power. Accessed Dec 14, 2024

Chapter 8
A Model for the Release of Toxic Ions into Coastal Aquifers: The Release of Arsenic into Bangladesh's Drinking Well Water by Reduction and by the Salt Effect

Seth H. Frisbie

1 Learning Objectives

1. Explain the relationship between the annual flooding in Bangladesh from monsoons and cyclones, and the release of arsenic from sediments into Bangladesh's drinking well water by the process of reduction.
2. Explain the relationship between saltwater intrusion from the Indian Ocean, and the release of arsenic from sediments into Bangladesh's drinking well water by the salt effect.
3. Explain why the release of toxic ions from sediments into drinking well water by reduction and the salt effect is not limited to Bangladesh and arsenic.

2 Introduction

The saltwater intrusion during flooding caused by rising sea levels from climate change can increase the salinity of coastal aquifers and degrade drinking water quality. The mixing of sewage with floodwater can also degrade the quality of drinking

S. H. Frisbie (✉)
Department of Chemistry and Biochemistry, College of Arts and Sciences, Norwich University, Northfield, VT, USA
e-mail: sfrisbie@norwich.edu

A. R. Molla et al. (eds.), *Climate Change, Drinking Water Security, and Public Health*, https://doi.org/10.1007/978-3-032-12074-8_8

water. These two processes are well known (Frisbie et al. 2024; Schwartz et al. 2006; Sherif and Singh 1999; WHO 2022).

In contrast, this chapter will discuss two recently identified consequences of rising sea levels that can further degrade the quality of drinking water from coastal aquifers: the release of toxic ions from sediments into drinking water aquifers by reduction and the salt effect. Reviews of reduction and the salt effect are given. A case study of our recently published peer-reviewed research on sea level rise and the release of arsenic into Bangladesh's coastal drinking water aquifer by reduction and the salt effect is summarized. Finally, the global implications of these two recently identified consequences of rising sea levels are discussed (Frisbie et al. 2024).

3 Sea Level Rise and the Release of Inorganic Ions into Coastal Drinking Water Aquifers by Reduction

3.1 A Review of Oxidation and Reduction

An oxidation-reduction reaction is the transfer of one or more electrons from one atom or molecule to another atom or molecule. An oxidation reaction produces an electron (e^-) or electrons; therefore, an electron or electrons are written to the right of the arrow in an oxidation reaction as follows (Eq. 8.1).

$$H_3As(III)O_{3(aq)} + H_2O_{(l)} \rightleftarrows H_3As(V)O_{4(aq)} + 2H^+_{(aq)} + 2e^- \quad E^0_{ox} = -0.560\,V \quad (8.1)$$

The phase designations "aq" and "l" mean "aqueous" and "liquid," respectively. This reaction measures −0.560 volts (V) under standard conditions when all solutes are at 1.0 molar (M) concentration. The arsenic atom (As) in the reactant, arsenous acid ($H_3As(III)O_3$), is in the As(III) or "+3" oxidation state. The arsenic atom (As) in the product, arsenic acid ($H_3As(V)O_4$), is in the As(V) or "+5" or oxidation state. Therefore, the As(III) atom is reduced, and the As(V) atom is oxidized. More specifically, 2 electrons are taken from As(III) to make As(V); +3 = +5–2. A total of 2 electrons are transferred (Frisbie et al. 2024; Lide 1992).

A reduction reaction consumes an electron or electrons; therefore, an electron or electrons are written to the left of the arrow in a reduction reaction as follows (Eq. 8.2).

$$O_{2(g)} + 4H^+_{(aq)} + 4e^- \rightleftarrows 2H_2O_{(l)} \quad E^0_{red} = 1.229\,V \quad (8.2)$$

The phase designations "g," "aq," and "l" mean "gas," "aqueous," and "liquid," respectively. This reaction produces 1.229 volts under standard conditions when all gases are at 1.0 atmosphere (atm) pressure, and all solutes are at 1.0 molar concentration. The oxygen atoms (O) in the reactant, molecular oxygen ($O_{2(g)}$), are in the 0 oxidation state. The oxygen atom (O) in the product, water ($H_2O_{(l)}$), is in the −2 oxidation state. Therefore, the O atoms in $O_{2(g)}$ are oxidized, and the O atom in $H_2O_{(l)}$ is reduced. More specifically, 2 electrons are added to one of the atoms (O) in $O_{2(g)}$ to make one $H_2O_{(l)}$ molecule; 0–2 = −2. Another 2 electrons are added to the second atom (O) in $O_{2(g)}$ to make a second $H_2O_{(l)}$ molecule; 0–2 = −2. A total of 4 electrons are transferred (Frisbie et al. 2024; Lide 1992).

The combined oxidation-reduction reaction follows (Eq. 8.3).

$$2H_3As(III)O_{3(aq)} + O_{2(g)} \rightleftarrows 2H_3As(V)O_{4(aq)} \quad E^0_{cell} = 0.669\,V \tag{8.3}$$

The reactant, arsenous acid ($H_3As(III)O_3$), is the reducing agent; it causes the reduction of $O_{2(g)}$. The other reactant, molecular oxygen ($O_{2(g)}$), is the oxidizing agent; it causes the oxidation of $H_3As(III)O_3$. The standard electrode potential (E^0_{cell}) for this oxidation-reduction reaction is $E^0_{cell} = E^0_{ox} + E^0_{red} = -0.560\,V + 1.229\,V = 0.669\,V$. This positive E^0_{cell} means that the reaction is thermodynamically favored at standard conditions; therefore, the reactants are expected to spontaneously make products when all solutes are at 1.0 molar concentration, and all gases are at 1.0 atmosphere pressure (Frisbie et al. 2024; McMurry and Fay 2004).

This spontaneous oxidation of As(III) to As(V) by molecular oxygen in drinking water aquifers has enormous implications for public health (Eq. 8.3). First, the minimum voltage that is required for the conversion of liquid water ($H_2O_{(l)}$) to oxygen gas ($O_{2(g)}$) and hydrogen gas ($H_{2(g)}$) at standard conditions is 1.229 volts; at higher voltages, liquid water is expected to be converted to these gases and vaporize. Therefore, molecular oxygen is one of the most powerful oxidizing agents that can exist in liquid water (Eq. 8.2). Second, As(V) is relatively insoluble, and As(III) is relatively soluble. Therefore, arsenate ($H_{3-x}As(V)O_4^{-x}$) in the presence of molecular oxygen is expected to be bound with solid sediments and not dissolved in drinking aquifer water; this should decrease the risk of adverse health effects. In contrast, arsenite ($H_{3-x}As(III)O_3^{-x}$) in the absence of molecular oxygen is expected to be dissolved in drinking aquifer water and not bound with solid sediments; this should increase the risk of adverse health effects (Frisbie et al. 2024; Bhattacharya et al. 2002; Frisbie et al. 1999; Stumm and Morgan 1981; Zahid et al. 2008).

3.2 *Case Study: Sea Level Rise and the Release of Arsenic into Bangladesh's Coastal Drinking Water Aquifer by Reduction*

The People's Republic of Bangladesh, with its flat and low-lying coastal topography, familiar rice paddies and shrimp farms, annual monsoons, and frequent cyclones, is expected to be severely impacted by flooding as sea levels rise due to climate change. More specifically, Bangladesh is in one of the largest river deltas in the world. The Ganges, Brahmaputra, and Meghna rivers flow through Bangladesh into the Indian Ocean (Fig. 8.1). The average elevation is only 8 meters above sea

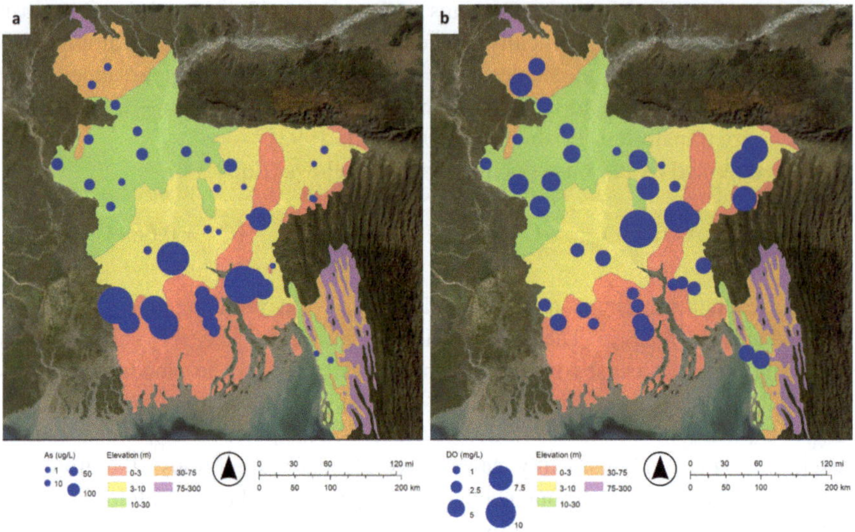

Fig. 8.1 Proportional symbol maps of (**a**) arsenic (As) in micrograms per liter (μg/L) and (**b**) dissolved oxygen (DO; $O_{2(aq)}$) in milligrams per liter (mg/L) in Bangladesh's drinking well water (1000 μg/L = 1 mg/L). (Frisbie et al. 2024) (Reproduced from Frisbie et al. 2024, Figures 2 and 3. https://doi.org/10.1371/journal.pone.0295172, licensed under the terms of the Creative Commons Attribution 4.0 International License (http://creativecommons.org/licenses/by/4.0/))

level. In a typical monsoon season, about 21% of Bangladesh's land is flooded with freshwater from its rivers in the north and saltwater from the Indian Ocean in the south. The largest recorded flood was during the 1998 monsoon season, covering 68% of Bangladesh's land. The monsoon season is from July to October; in contrast, cyclones are most likely to occur from April to May and from October to November and can rapidly yield over 9 meters of storm surge (Frisbie et al. 2024; Dewan et al. 2003; Monan 1995; Pethick and Orford 2013).

As sea levels continue to rise and floods and cyclones in Bangladesh increase in area and duration, the underlying aquifer can become more reducing. These floodwaters are a barrier to the diffusion of oxygen ($O_{2(g)}$), a powerful oxidizing agent, from the atmosphere into the aquifer. During flooding, heterotrophic organisms and abiotic processes can deplete the concentration of dissolved oxygen (DO; $O_{2(aq)}$) in the floodwater and in the drinking well water; this also helps to make a reducing environment. If the concentration of dissolved oxygen is relatively small, the reduction of insoluble arsenate ($H_{3-x}As(V)O_4^{-x}$) to soluble arsenite ($H_{3-x}As(III)O_3^{-x}$) is favored (Eq. 8.3); this will add arsenic to the drinking well water. The following data strongly suggest that this chemical change can increase the release of arsenic into Bangladesh's drinking well water by reduction (Frisbie et al. 2024; Alexander 1977; Bohn et al. 1979; Frisbie et al. 1999; Zahid et al. 2008).

Maps of the arsenic and dissolved oxygen concentrations in Bangladesh's drinking well water are shown in Fig. 8.1. A comparison of these maps suggests an inverse relationship; relatively low concentrations of dissolved oxygen are associated with relatively high concentrations of arsenic. In addition, Bangladesh's drinking well water has lower dissolved oxygen concentrations in the south near the Indian Ocean and higher dissolved oxygen concentrations in the north near the Himalayan mountains; this observation is consistent with floods and storm surges intruding from the south to the north (Fig. 8.1). The arsenic and dissolved oxygen concentrations of drinking well water in this study ranged from not detected to 448 micrograms per liter (μg/L) and from 0.16 to 12.70 milligrams per liter (mg/L; 1000 μg/L = 1 mg/L), respectively.

The linear regression equation for arsenic concentration versus dissolved oxygen concentration is shown in Eq. 8.4.

$$\mathrm{As}\left(\frac{\mu g}{\mathrm{L}}\right)=137-14.1\left(\mathrm{DO}\left(\frac{\mathrm{mg}}{\mathrm{L}}\right)\right) \tag{8.4}$$

This linear regression gives a statistically significant negative slope (−14.1 μg/mg; p-value = 0.0028); this suggests that arsenic is released from sediments into Bangladesh's drinking well water by reduction. More specifically, this inverse relationship between arsenic concentration and dissolved oxygen concentration suggests that in the absence of dissolved oxygen, insoluble arsenate ($H_{3-x}As(V)O_4^{-x}$) from sediments is reduced to soluble arsenite ($H_{3-x}As(III)O_3^{-x}$) and released into Bangladesh's drinking well water (Eq. 8.3; Frisbie et al. 2024; Bhattacharya et al. 2002; Frisbie et al. 1999).

3.3 Global Implications

Reduction is a global process; it is not limited to Bangladesh. If flooding allows heterotrophic organisms and abiotic processes to deplete dissolved oxygen (DO; $O_{2(aq)}$) in drinking well water faster than it is replaced by the atmosphere and other sources, then the aquifer can become more reducing. There are two requirements for releasing inorganic ions from sediments into drinking well water by reduction. First, the ion must be present in the aquifer. Second, the reduced ion must be more soluble than the oxidized ion.

The list of the relative abundances and average concentrations of elements in the earth's crustal rocks is given in Table 8.1. Since these 76 elements are in crustal rocks, they might also be in the drinking well water that contacts these rocks. Therefore, Table 8.1 helps us determine if an ion might be present in drinking well water (Greenwood and Earnshaw 1984).

Unfortunately, an exhaustive list of the possible oxidation states of these elements in drinking well water does not exist. However, a list of the known and common oxidation states of the first 92 elements is given in Table 8.2. This list is helpful but includes oxidation states that are impossible in natural waters. For example, Na^{-1} and Na^0 are violently oxidized by water, while Na^+ is stable in water. This list has no information about solubilities (Greenwood and Earnshaw 1984).

Table 8.1 The relative abundances and average concentrations of elements in the rocks of the earth's crust in milligrams per kilogram (mg/kg). (Greenwood and Earnshaw 1984)[a]

No.	Element		mg/kg	No.	Element		mg/kg
1	Oxygen	O	455,000	39	Thorium	Th	8.1
2	Silicon	Si	272,000	40	Samarium	Sm	7.0
3	Aluminum	Al	83,000	41	Gadolinium	Gd	6.1
4	Iron	Fe	62,000	42	Erbium	Er	3.5
5	Calcium	Ca	46,600	43	Ytterbium	Yb	3.1
6	Magnesium	Mg	27,640	44	Hafnium	Hf	2.8
7	Sodium	Na	22,700	45	Cesium	Cs	2.6
8	Potassium	K	18,400	46	Bromine	Br	2.5
9	Titanium	Ti	6320	47	Uranium	U	2.3
10	Hydrogen	H	1520	48a	Tin	Sn	2.1
11	Phosphorus	P	1120	48b	Europium	Eu	2.1
12	Manganese	Mn	1060	50	Beryllium	Be	2
13	Fluorine	F	544	51	Arsenic	As	1.8
14	Barium	Ba	390	52	Tantalum	Ta	1.7
15	Strontium	Sr	384	53	Germanium	Ge	1.5
16	Sulfur	S	340	54	Holmium	Ho	1.3
17	Carbon	C	180	55a	Molybdenum	Mo	1.2
18	Zirconium	Zr	162	55b	Tungsten	W	1.2
19	Vanadium	V	136	55c	Terbium	Tb	1.2
20	Chlorine	Cl	126	58	Thallium	Tl	0.7
21	Chromium	Cr	122	59	Thulium	Tm	0.5
22	Nickel	Ni	99	60	Iodine	I	0.46
23	Rubidium	Rb	78	61	Indium	In	0.24
24	Zinc	Zn	76	62	Antimony	Sb	0.2
25	Copper	Cu	68	63	Cadmium	Cd	0.16
26	Cerium	Ce	66	64a	Silver	Ag	0.08
27	Neodymium	Nd	40	64b	Mercury	Hg	0.08
28	Lanthanum	La	35	66	Selenium	Se	0.05
29	Yttrium	Y	31	67	Palladium	Pd	0.015
30	Cobalt	Co	29	68	Platinum	Pt	0.01
31	Scandium	Sc	25	69	Bismuth	Bi	0.008
32	Niobium	Nb	20	70	Osmium	Os	0.005
33a	Nitrogen	N	19	71	Gold	Au	0.004
33b	Gallium	Ga	19	72a	Iridium	Ir	0.001
35	Lithium	Li	18	72b	Tellurium	Te	0.001
36	Lead	Pb	13	74	Rhenium	Re	0.0007
37	Praseodymium	Pr	9.1	75a	Ruthenium	Ru	0.0001
38	Boron	B	9	75b	Rhodium	Rh	0.0001

[a]Used with permission of Elsevier, from *Chemistry of the elements*, Greenwood and Earnshaw 1984, Appendix 4; permission conveyed through Copyright Clearance Center, Inc

The World Health Organization (WHO) currently has recommendations for 28 inorganic drinking water contaminants, excluding disinfection agents and their byproducts: 16 are guideline values (GVs), 6 are health-based values (HBVs) without guideline values, and 6 are aesthetic values (AVs) without health-based or guideline values (Table 8.3). When calculating guideline values, the WHO reviews the toxicological literature, calculates a health-based value, and determines whether

Table 8.2 The known integer oxidation states of the elements. Nonintegral values, as in B_5H_9, C_3H_8, HN_3, S_8^{+2}, etc., are not included. The more common states are shown in bold. The symbol Z is the atomic number. (Greenwood and Earnshaw 1984)[a]

Z	Element		Oxidation states												
1	Hydrogen	H				**−1**	**0**	**+1**							
2	Helium	He					**0**								
3	Lithium	Li				−1	**0**	**+1**							
4	Beryllium	Be					**0**		**+2**						
5	Boron	B					**0**	+1	+2	**+3**					
6	Carbon	C	**−4**	−3	−2	−1	**0**	+1	+2	+3	**+4**				
7	Nitrogen	N		**−3**	−2	−1	**0**	+1	+2	**+3**	+4	**+5**			
8	Oxygen	O			**−2**	−1	**0**	+1	+2						
9	Fluorine	F				**−1**	**0**								
10	Neon	Ne					**0**								
11	Sodium	Na				−1	**0**	**+1**							
12	Magnesium	Mg					**0**		**+2**						
13	Aluminum	Al					**0**	+1		**+3**					
14	Silicon	Si	**−4**	−3	−2	−1	**0**	+1	+2	+3	**+4**				
15	Phosphorus	P		**−3**	−2	−1	**0**	+1	+2	**+3**	+4	**+5**			
16	Sulfur	S			**−2**	−1	**0**	+1	**+2**	+3	**+4**	+5	**+6**		
17	Chlorine	Cl				**−1**	**0**	**+1**	+2	**+3**	+4	**+5**	+6	**+7**	
18	Argon	Ar					**0**								
19	Potassium	K					**0**	**+1**							
20	Calcium	Ca					**0**		**+2**						
21	Scandium	Sc					**0**	+1	+2	**+3**					
22	Titanium	Ti				−1	**0**		+2	+3	**+4**				
23	Vanadium	V				−1	**0**	+1	+2	+3	+4	**+5**			
24	Chromium	Cr			−2	−1	**0**	+1	+2	**+3**	+4	+5	**+6**		

(continued)

Table 8.2 (continued)

Z	Element		Oxidation states												
25	Manganese	Mn		−3	−2	−1	**0**	+1	**+2**	+3	**+4**	+5	+6	**+7**	
26	Iron	Fe			−2	−1	**0**	+1	**+2**	**+3**	+4	+5	+6		
27	Cobalt	Co				−1	**0**	+1	**+2**	**+3**	+4	+5			
28	Nickel	Ni				−1	**0**	+1	**+2**	+3	+4				
29	Copper	Cu					**0**	+1	**+2**	+3	+4				
30	Zinc	Zn					**0**		**+2**						
31	Gallium	Ga					**0**	+1	+2	**+3**					
32	Germanium	Ge	**−4**				**0**	+1	**+2**	+3	**+4**				
33	Arsenic	As		**−3**			**0**		+2	**+3**		**+5**			
34	Selenium	Se			**−2**		**0**		**+2**		**+4**		**+6**		
35	Bromine	Br				**−1**	**0**	**+1**		**+3**	+4	**+5**		+7	
36	Krypton	Kr					**0**		**+2**						
37	Rubidium	Rb					**0**	**+1**							
38	Strontium	Sr					**0**		**+2**						
39	Yttrium	Y					**0**		+2	**+3**					
40	Zirconium	Zr					**0**	+1	+2	+3	**+4**				
41	Niobium	Nb				−1	**0**		+2	+3	+4	**+5**			
42	Molybdenum	Mo			−2	−1	**0**	+1	+2	+3	**+4**	+5	**+6**		
43	Technetium	Tc		−3		−1	**0**	+1	+2	+3	**+4**	+5	+6	**+7**	
44	Ruthenium	Ru			−2		**0**	+1	+2	**+3**	**+4**	+5	+6	+7	+8
45	Rhodium	Rh				−1	**0**	+1	+2	**+3**	+4	+5	+6		
46	Palladium	Pd					**0**		**+2**		**+4**				
47	Silver	Ag					**0**	**+1**	+2	+3					
48	Cadmium	Cd					**0**		**+2**						
49	Indium	In					**0**	+1	+2	**+3**					

50	Tin	Sn	**−4**				**0**		**+2**		**+4**				
51	Antimony	Sb		**−3**			**0**			**+3**		**+5**			
52	Tellurium	Te			**−2**		**0**		**+2**		**+4**	+5	**+6**		
53	Iodine	I				**−1**	**0**	**+1**		**+3**		**+5**		**+7**	
54	Xenon	Xe					**0**		**+2**		**+4**		**+6**		+8
55	Cesium	Cs					**0**	**+1**							
56	Barium	Ba					**0**		**+2**						
57	Lanthanum	La					**0**		+2	**+3**					
58	Cerium	Ce					**0**		+2	**+3**	**+4**				
59	Praseodymium	Pr					**0**		+2	**+3**	+4				
60	Neodymium	Nd					**0**		+2	**+3**					
61	Promethium	Pm					**0**			**+3**					
62	Samarium	Sm					**0**		+2	**+3**					
63	Europium	Eu					**0**		**+2**	**+3**					
64	Gadolinium	Gd					**0**	+1	+2	**+3**					
65	Terbium	Tb					**0**	+1		**+3**	+4				
66	Dysprosium	Dy					**0**		+2	**+3**					
67	Holmium	Ho					**0**			**+3**					
68	Erbium	Er					**0**			**+3**					
69	Thulium	Tm					**0**		+2	**+3**					
70	Ytterbium	Yb					**0**		+2	**+3**					
71	Lutetium	Lu					**0**			**+3**					
72	Hafnium	Hf					**0**		+2	+3	**+4**				
73	Tantalum	Ta				−1	**0**		+2	+3	+4	**+5**			
74	Tungsten	W			−2	−1	**0**	+1	+2	+3	**+4**	+5	**+6**		

(continued)

Table 8.2 (continued)

Z	Element		Oxidation states												
75	Rhenium	Re		−3		−1	**0**	+1	+2	+3	**+4**	+5	+6	+7	
76	Osmium	Os			−2		**0**	+1	+2	+3	**+4**	+5	+6	+7	+8
77	Iridium	Ir				−1	**0**	+1	+2	**+3**	**+4**	+5	+6		
78	Platinum	Pt					**0**		**+2**		**+4**	+5	+6		
79	Gold	Au				−1	**0**	+1	+2	**+3**		+5			
80	Mercury	Hg					**0**	**+1**	**+2**						
81	Thallium	Tl					**0**	**+1**		**+3**					
82	Lead	Pb	−4				**0**		**+2**		**+4**				
83	Bismuth	Bi		−3			**0**			**+3**		+5			
84	Polonium	Po			**−2**		**0**		**+2**		**+4**		+6		
85	Astatine	At				**−1**	**0**	**+1**		+3		+5		+7	
86	Radon	Rn					**0**								
87	Francium	Fr					**0**	**+1**							
88	Radium	Ra					**0**		**+2**						
89	Actinium	Ac					**0**			**+3**					
90	Thorium	Th					**0**		+2	+3	**+4**				
91	Protactinium	Pa					**0**			+3	+4	**+5**			
92	Uranium	U					**0**			+3	+4	+5	**+6**		

[a]Used with permission of Ǝlsevier, from Chemistry of the elements, Greenwood and Earnshaw 1984, Figure 2.5; permission conveyed through Copyright Clearance Center, Inc

Table 8.3 The current (2022) World Health Organization (WHO) drinking-water guideline values (GVs), health-based values (HBVs), and aesthetic values (AVs) for inorganic contaminants in milligrams per liter (mg/L). (Mitchell and Frisbie 2023; WHO 2022)

Inorganic		Value (mg/L)	Value type
Aluminum	Al	0.9	HBV
Ammonia	NH_3	1.5	AV
Antimony	Sb	0.02	GV
Arsenic	As	0.01	GV
Barium	Ba	1.3	GV
Beryllium	Be	0.12	HBV
Boron	B	2.4	GV
Cadmium	Cd	0.003	GV
Chloride	Cl^-	300	AV
Chromium	Cr	0.05	GV
Copper	Cu	2	GV
Cyanide	CN^-	0.5	HBV
Fluoride	F^-	1.5	GV
Hydrogen Sulfide	H_2S	1	AV
Iron	Fe	2	HBV
Lead	Pb	0.01	GV
Manganese	Mn	0.08	GV
Mercury	Hg	0.006	GV
Molybdenum	Mo	0.07	HBV
Nickel	Ni	0.07	GV
Nitrate	NO_3^-	50	GV
Nitrite	NO_2^-	3	GV
Selenium	Se	0.04	GV
Silver	Ag	0.1	HBV
Sodium	Na	200	AV
Sulfate	SO_4^{-2}	1000	AV
Uranium	U	0.03	GV
Zinc	Zn	5	AV

the health-based value should be adopted as a guideline value. The WHO also considers aesthetic aspects of drinking water quality, such as taste and the staining of laundry and plumbing fixtures, and adopts aesthetic values (AVs) for specific drinking water contaminants. It is essential to realize that the WHO has guideline or health-based values for only 21 (28%) of the 76 inorganic elements in drinking water; this is a tremendous limitation. For example, our team discovered thorium (Th), a known human carcinogen, in 31% of the drinking water wells in West Bengal village; however, the WHO does not have a guideline or health-based value for thorium, so we could not determine the safety of this drinking water (Bacquart et al. 2012; Mitchell and Frisbie 2023; WHO 2022).

Despite these limitations, the information from Tables 8.1, 8.2, and 8.3 helps us determine if the release of inorganic ions into drinking well water by reduction is a threat to global health. For example, arsenic, manganese, and iron are present in crustal rocks and, as a result, are also found in drinking well water (Table 8.1). These elements are more soluble when reduced (As(III), Fe(II), and Mn(II)) and less soluble when oxidized (As(V), Fe(III), and Mn(IV)) in drinking well water (Table 8.2). Finally, these elements have WHO drinking water guidelines or health-based values (Table 8.3). Therefore, increased exposure to arsenic, manganese, and iron in drinking well water by reduction from rising sea levels is a threat to global health.

In contrast, uranium is present in crustal rocks and, as a result, is also found in drinking well water (Table 8.1). This element has a WHO drinking water guideline (Table 8.3). However, reduced uranium (U(IV)) is less soluble than oxidized uranium (U(VI) in drinking well water (Table 8.2). Therefore, increased exposure to uranium in drinking well water by reduction from rising sea levels is not a likely threat to global health. Addressing the limitations of the information in these tables is an essential area of future research.

4 Sea Level Rise and the Release of Inorganic Ions into Coastal Drinking Water Aquifers by the Salt Effect

4.1 A Review of the Salt Effect

The salt effect, uncommon ion effect, or diverse ion effect causes an increase in the solubility of an ionic solid when the solid is in a solution of ions that are different than the ions in the solid. For example, the solubility of solid barium sulfate ($BaSO_{4(s)}$) increases as the concentration of dissolved potassium nitrate ($K^{+}_{(aq)} + NO_{3}^{-}{}_{(aq)}$) increases. More specifically, in this example, the uncommon ions dissolved in water ($K^{+}_{(aq)}$ and $NO_{3}^{-}{}_{(aq)}$) will make an ionic solid ($BaSO_{4(s)}$) release more of its ions ($Ba^{+2}_{(aq)} + SO_{4}^{-2}{}_{(aq)}$) into solution. Minerals are ionic solids, so the salt effect has significant implications for the chemistry of coastal drinking water aquifers. In contrast, the common ion effect, taught to most general chemistry students, causes a decrease in the solubility of an ionic solid (Fig. 8.2; Frisbie et al. 2024; Petrucci 1985; Kielland 1937).

Similarly, the salt effect releases arsenic from naturally occurring minerals into drinking well water. The salt effect is driven by the attraction of oppositely charged particles to each other. Therefore, knowing if a chemical is positively charged, negatively charged, or electrically neutral is essential. Positively charged ions (cations) are attracted to negatively charged ions (anions) and can take part in the salt effect. Electrically neutral chemicals cannot take part in the salt effect. Since arsenic in drinking well water commonly exists as weak acids, it is essential to know how their charges change with pH (Fig. 8.3; Frisbie et al. 2024; Spencer et al. 2008).

Figure 8.3 panel a on the left suggests that from the minimum observed pH in our study at 3.90 to pH = pK_{a2} at 6.77 most of the As(V) has a − 1 charge, and from pH = pK_{a2} at 6.77 to the maximum observed pH in our study at 7.96 most of the As(V) has a − 2 charge. Therefore, $H_2As(V)O_4^{-1}$ and $HAs(V)O_4^{-2}$ are the dominant forms of As(V) in the salt effect.

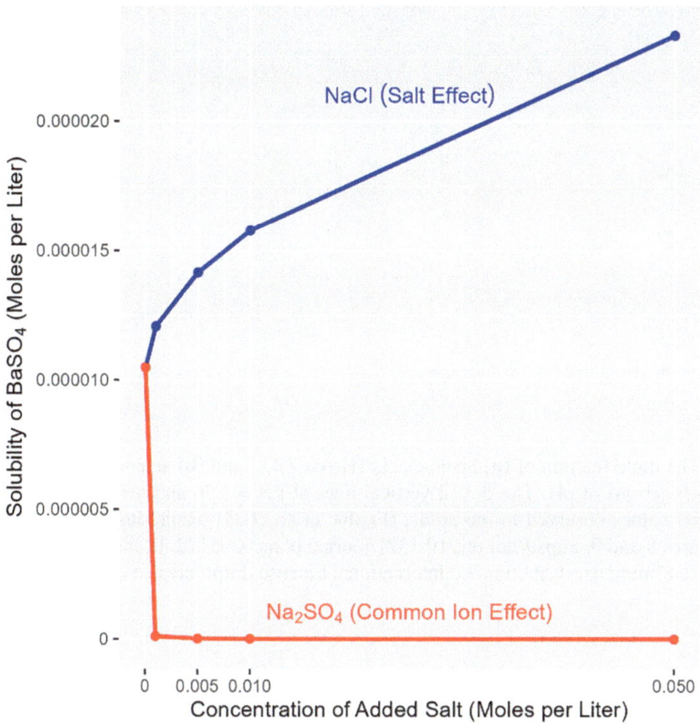

Fig. 8.2 The solubility of barium sulfate ($BaSO_4$) in moles per liter (M) versus moles per liter of added sodium chloride (NaCl; the salt effect) and moles per liter of added sodium sulfate (Na_2SO_4; the common ion effect). (Kielland 1937) (Adapted with permission from Kielland 1937. Individual activity coefficients of ions in aqueous solutions. Table II. *J Am Chem Soc* (59)9:1675–1678. https://doi.org/10.1021/ja01288a032. Copyright 1937 American Chemical Society)

In contrast, Fig. 8.3 panel b on the right suggests that from the minimum observed pH at 3.90 to the maximum observed pH at 7.96, most of the As(III) is electrically neutral. This $H_3As(III)O_3$ is not involved in the salt effect, and $H_2As(III)O_3^{-1}$ likely plays a minor role in the salt effect.

The salt effect has two processes. The first process of the salt effect is ion exchange, the displacement of adsorbed ions on solid surfaces by dissolved ions in water. For example, aqueous chloride ions ($Cl^-_{(aq)}$) from saline water exchange with adsorbed arsenate −1 ions ($H_2As(V)O_4{}^-{}_{(ex)}$) on the surfaces of solids ([Solid Surface]$^+$; Eq. 8.5). The phase designation "ex" means "exchangeable." This ion exchange releases arsenic from sediments into drinking well water (Frisbie et al. 2024; Bohn et al. 1979).

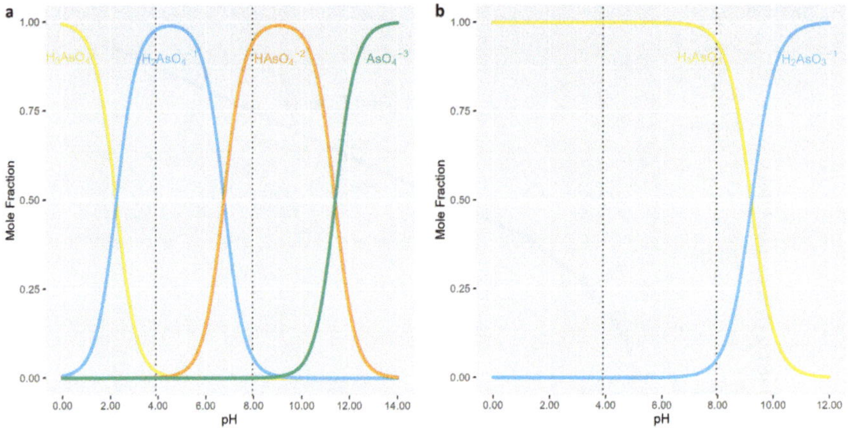

Fig. 8.3 The mole fraction of (**a**) arsenic acid ($H_3As(V)O_4$) and (**b**) arsenous acid ($H_3As(III)O_3$) species as functions of pH. The dotted vertical lines at pH = 3.90 and pH = 7.96 demarcate the range of pH values observed in our study. (Frisbie et al. 2024) (Reproduced from Frisbie et al. 2024, Figures 8 and 9. https://doi.org/10.1371/journal.pone.0295172, licensed under the terms of the Creative Commons Attribution 4.0 International License (http://creativecommons.org/licenses/by/4.0/))

$$[\text{Solid Surface}]^{+}\,H_2As(V)O^{-}_{4(ex)} + Cl^{-}_{(aq)} \rightleftarrows [\text{Solid Surface}]^{+}\,Cl^{-}_{(ex)} + H_2As(V)O^{-}_{4(aq)} \quad (8.5)$$

The second process of the salt effect is ion pairing. The uncommon ions from a dissolved salt are electrostatically attracted to and pair with the oppositely charged dissolved ions from an ionic solid. For example, in Step 1 of Eq. 8.6, the model ionic solid $CaHAs(V)O_4(s)$ is in equilibrium with its aqueous $Ca^{+2}_{(aq)}$ and $HAs(V)O_4^{-2}{}_{(aq)}$ ions. Then, in Step 2, aqueous sodium ions ($Na^{+}_{(aq)}$) from saline water pair with oppositely charged arsenate −2 ions ($HAs(V)O_4^{-2}{}_{(aq)}$) in solution to make the ion pair $NaHAs(V)O_4^{-}{}_{(aq)}$. Simultaneously, aqueous chloride ions ($Cl^{-}_{(aq)}$) from saline water pair with oppositely charged calcium ions ($Ca^{+2}_{(aq)}$) in solution to make the ion pair $CaCl^{+}_{(aq)}$. This ion pairing removes products, shifts these equilibria to the right, and releases arsenic from sediments into drinking well water (Eq. 8.6; Frisbie et al. 2024; Winstein et al. 1954).

$$\text{Step 1}: CaHAs(V)O_{4(s)} \rightleftarrows Ca^{+2}_{(aq)} + HAs(V)O^{-2}_{4(aq)}$$
$$\text{Step 2}: Ca^{+2}_{(aq)} + HAs(V)O^{-2}_{4(aq)} + Na^{+}_{(aq)} + Cl^{-}_{(aq)} \rightleftarrows CaCl^{+}_{(aq)} + NaHAs(V)O^{-}_{4(aq)} \quad (8.6)$$

4.2 *Case Study: Sea Level Rise and the Release of Arsenic into Bangladesh's Coastal Drinking Water Aquifer by the Salt Effect*

As sea levels continue to rise and floods and cyclones in Bangladesh increase in area and duration, the underlying aquifer can become more saline. The floodwater from the Indian Ocean increases the salinity of Bangladesh's coastal aquifer, which adds dissolved ions to this aquifer. The major dissolved ions in seawater are chloride ($Cl^{-}_{(aq)}$; 19.354 grams per kilogram (g/kg)), sodium ($Na^{+}_{(aq)}$; 10.77 g/kg), sulfate ($SO_4^{-2}{}_{(aq)}$; 2.712 g/kg), magnesium ($Mg^{+2}_{(aq)}$; 1.29 g/kg), calcium ($Ca^{+2}_{(aq)}$; 0.4121 g/kg), potassium ($K^{+}_{(aq)}$; 0.399 g/kg), bicarbonate ($HCO_3^{-}{}_{(aq)}$; 0.1424 g/kg), strontium ($Sr^{+2}_{(aq)}$; 0.0079 g/kg), bromide ($Br^{-}_{(aq)}$; 0.0673 g/kg), and fluoride ($F^{-}_{(aq)}$; 0.0013 g/kg). Dissolved ions can allow the salt effect to increase the release of arsenic from sediments into Bangladesh's drinking well water (Frisbie et al. 2024; Stumm and Morgan 1981).

Maps of the arsenic concentration and the specific conductance (SC) of Bangladesh's drinking well water are shown in Fig. 8.4. Specific conductance is used to estimate the salinity of water. In general, the higher the specific conductance, the higher the salinity of water. A comparison of these maps suggests a positive relationship; relatively high specific conductances are associated with relatively high concentrations of arsenic. In addition, Bangladesh's drinking well water has higher specific conductances in the south near the Indian Ocean and lower specific

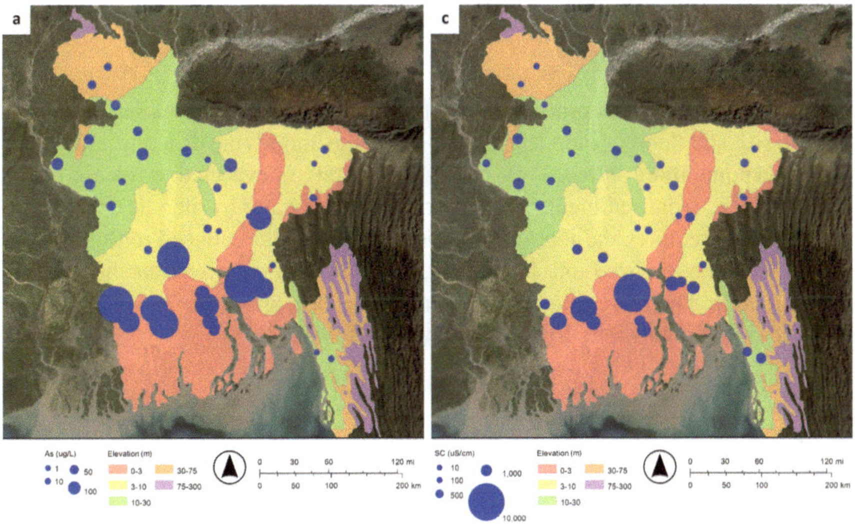

Fig. 8.4 Proportional symbol maps of (**a**) arsenic (As) in micrograms per liter (µg/L) and (**c**) specific conductance (SC) in microsiemens per centimeter (µS/cm) of Bangladesh's drinking well water. (Frisbie et al. 2024) (Reproduced from Frisbie et al. 2024, Figures 2 and 10. https://doi.org/10.1371/journal.pone.0295172, licensed under the terms of the Creative Commons Attribution 4.0 International License (http://creativecommons.org/licenses/by/4.0/))

conductances in the north near the Himalayan mountains; this observation is consistent with floods and storm surges containing seawater intruding from the south to the north (Fig. 8.4). The arsenic concentration and specific conductance of drinking well water in this study ranged from not detected to 448 μg/L and from 40 microsiemens per centimeter (μS/cm) to 5360 μS/cm, respectively. As a comparison, the conductance of seawater at the ocean's surface is about 50,000 μS/cm (Frisbie et al. 2024; APHA, AWWA, WEF 1995; Tyler et al. 2017).

The linear regression equation for arsenic concentration versus specific conductance is shown in Eq. 8.7.

$$\mathrm{As}\left(\frac{\mu g}{\mathrm{L}}\right) = 53.8 + 0.0364\left(\mathrm{SC}\left(\frac{\mu S}{\mathrm{cm}}\right)\right) \tag{8.7}$$

This linear regression gives a statistically significant positive slope (0.0364 μg cm/L μS; p-value = 0.023); this suggests that arsenic is released from sediments into Bangladesh's drinking well water by the salt effect. More specifically, this positive relationship between arsenic concentration and specific conductance suggests that any process that increases salinity, such as annual flooding, is expected to increase the release of arsenic oxyanions from sediments into Bangladesh's drinking well water by the salt effect (Frisbie et al. 2024; Castellan 1983).

4.3 Global Implications

The salt effect is a global process; it is not limited to Bangladesh. As sea levels continue to rise and flood or inundate coastal areas, the underlying drinking water aquifer can become more saline. This increase in salinity is expected to release a wide variety of ions from sediments into the drinking well water by the salt effect.

Reduction requires the ion to be in the aquifer and the reduced ion to be more soluble than the oxidized ion. In contrast, the salt effect only requires the ion to be in the aquifer. Therefore, the salt effect might release all of the 76 ions in crustal rocks from sediments into drinking well water (Table 8.1).

5 Conclusions

The chemical processes that release arsenic from Bangladesh's sediments and into its coastal drinking water wells are not limited to Bangladesh and arsenic. For example, if arsenic is in the coastal sediments in certain parts of California, these processes can release arsenic into its drinking water wells. If manganese is in the coastal sediments in certain parts of Florida, these processes can release manganese into its drinking water wells.

However, if private well owners, municipalities, and public health officials know of this problem, they can monitor the drinking well water and, if necessary, treat it to make it safe. Fortunately, if people know how to monitor coastal drinking water wells for unsafe concentrations of ions, this aspect of climate change has a relatively easy solution.

6 Discussion Questions

1. $^{137}Cs^+$ is a highly toxic isotope of ionic cesium that was released into the environment by the Chernobyl nuclear accident in 1986. Explain why the release of $^{137}Cs^+$ into drinking water aquifers by reduction is not possible. (Hint: Cs^0 is violently oxidized by water, while Cs^+ is stable in water.)
2. High fluoride ion concentrations (F^-) in drinking water can cause skeletal fluorosis, a devastating disease that significantly weakens bone tissue. A person with severe skeletal fluorosis easily suffers from broken bones.
   (a) Explain why the release of F^- into drinking water aquifers by reduction is not possible.
   (b) Explain why the release of F^- into drinking water aquifers by the salt effect is possible.
3. Arsenite, As(III), is the species of arsenic released from sediments into drinking well water by reduction. In contrast, arsenate, As(V), is the most common species of arsenic released from sediments into drinking well water by the salt effect. Explain the reasons for this difference.

Competing Interests The author has no competing interests to declare.

References

Alexander M (1977) Introduction to soil microbiology, 2nd edn. John Wiley & Sons, New York, p 143,144

American Public Health Association (APHA), American Water Works Association (AWWA), Water Environment Federation (WEF) (1995) Standard methods for the examination of water and wastewater, 19th edn. APHA, Washington, DC. parts 2000, 3000, 4000, 5000

Bacquart T, Bradshaw K, Frisbie SH, Mitchell EJ, Springston G, Defelice J, Dustin H, Sarkar B (2012) A survey of arsenic, manganese, boron, thorium, and other toxic metals in the groundwater of a West Bengal, India neighbourhood. Metallomics 7:653–650. https://doi.org/10.1039/c2mt20020a

Bhattacharya P, Jacks G, Frisbie SH, Smith E, Naidu R, Sarkar B (2002) Arsenic in the environment: a global perspective. In: Sarkar B (ed) Heavy metals in the environment. Marcel Dekker, New York, pp 147–215

Bohn HL, McNeal BL, O'Connor GA (1979) Soil chemistry. John Wiley & Sons, New York, pp 141–168. 301–314

Castellan GW (1983) Physical chemistry, 3rd edn. Addison-Wesley Publishing Company, Reading, MA, pp 366–367
Dewan AM, Nishigaki M, Komatsu M (2003) Floods in Bangladesh: a comparative hydrological investigation on two catastrophic events. J Fac Environ Sci Technol Okayama Univ 8(1):53–62. https://core.ac.uk/reader/12549250
Frisbie SH, Maynard DM, Hoque BA (1999) The nature and extent of arsenic-affected drinking water in Bangladesh. In: Sarkar B (ed) Metals and genetics. Kluwer Academic / Plenum Publishers, New York, pp 67–85
Frisbie SH, Mitchell EJ, Molla AR (2024) Sea level rise from climate change is expected to increase the release of arsenic into Bangladesh's drinking well water by reduction and by the salt effect. PLoS One 19(1):e0295172. https://doi.org/10.1371/journal.pone.0295172
Greenwood NN, Earnshaw A (1984) Chemistry of the elements. Pergamon Press, New York, pp 31–33. 1496
Kielland J (1937) Individual activity coefficients of ions in aqueous solutions. J Am Chem Soc 59(9):1675–1678. https://doi.org/10.1021/ja01288a032
Lide DR (1992) CRC handbook of chemistry and physics, 73rd edn. CRC Press, Boca Raton, FL, pp 8–17–8-27. 8–41
McMurry J, Fay RC (2004) Chemistry, 4th edn. Pearson Education, Upper Saddle River, NJ, p 721–762, 618–623
Mitchell EJ, Frisbie SH (2023) A comprehensive survey and analysis of international drinking water regulations for inorganic chemicals with comparisons to the World Health Organization's drinking-water guidelines. PLoS One 18(11):e0287937. https://doi.org/10.1371/journal.pone.0287937
Monan J (1995) Bangladesh the strength to succeed. Oxfam, Oxford, UK. p 4,5,63
Pethick J, Orford JD (2013) Rapid rise in effective sea-level in Southwest Bangladesh: its causes and contemporary rates. Glob Planet Change 111:237–245. https://doi.org/10.1016/j.gloplacha.2013.09.019
Petrucci RH (1985) General chemistry, 4th edn. Macmillan Publishing Company, New York, pp 568–570
Schwartz BS, Harris JB, Khan AI, Larocque RC, Sack DA, Malek MA, Faruque ASG, Qadri F, Calderwood SB, Luby SP, Ryan ET (2006) Diarrheal epidemics in Dhaka, Bangladesh, during three consecutive floods: 1988, 1998, and 2004. Am J Trop Med Hyg 74(6):1067–1073
Sherif MM, Singh VP (1999) Effect of climate change on sea water intrusion in coastal aquifers. Hydrol Process 13(8):1277–1287. https://doi.org/10.1002/(SICI)1099-1085(19990615)13:8<1277::AID-HYP765>3.0.CO;2-W
Spencer JN, Bodner GM, Rickard LH (2008) Chemistry structure and dynamics, 4th edn. John Wiley & Sons, Hoboken, NJ, pp 66–83. 310–317
Stumm W, Morgan JJ (1981) Aquatic chemistry. John Wiley & Sons, New York, pp 465–469. 569
Tyler RH, Boyer TP, Minami T, Zweng MM, Reagan JR (2017) Electrical conductivity of the global ocean. Earth Planets Space 69:156. https://doi.org/10.1186/s40623-017-0739-7
Winstein S, Clippinger E, Fainberg AH, Robinson GC (1954) Salt effe and ion-pairs in solvolysis. J Am Chem Soc 76(9):2597–2598. https://doi.org/10.1021/ja01638a093
World Health Organization (WHO) (2022) Guidelines for drinking-water quality, 4th edn. WHO, Geneva, Switzerland. p 125–166, 249–309, 361. https://www.who.int/publications/i/item/9789240045064. Accessed 19 Mar 2025
Zahid A, Qumrul Hassan M, Balke KD, Flegr M, Clark DW (2008) Groundwater chemistry and occurrence of arsenic in the Meghna floodplain aquifer, southeastern Bangladesh. Environ Geol 54:1247–1260. https://doi.org/10.1007/s00254-007-0907-3

Chapter 9
Evaluating the Physical, Chemical, and Bacteriological Aspects of Water Quality Related to Climate Change

Matthew J. Weinburke

1 Learning Objectives

1. Describe the physical aspects of water quality related to climate change.
2. Describe the chemical aspects of water quality related to climate change.
3. Describe the bacteriological aspects of water quality related to climate change.

2 Introduction

Although an available and adequate water supply is important, the quality of the water is just as critical. Water quality is measured by the usability of the physical, chemical, and bacteriological characteristics of the water and its source (Forari and Griggs 2017). Climate change not only affects the supply of water but also affects the quality of water. Extreme meteorological events due to climate change have impacted water quality resulting in human health and ecosystem impacts (Delpla et al. 2009). Climate change affects the physical, chemical, and bacteriological aspects of water in many ways and is significantly linked to safe and available drinking water (UN 2024; USEPA 2024a). Through shrinking glaciers, rising sea levels, floods, and droughts, climate change has a negative impact and worsens the quality of drinking water. This chapter discusses and evaluates how climate change affects water quality within the context of the physical, chemical, and bacteriological aspects of freshwater.

M. J. Weinburke (✉)
Independent Public Health Consultant, Mariposa, CA, USA

A. R. Molla et al. (eds.), *Climate Change, Drinking Water Security, and Public Health*, https://doi.org/10.1007/978-3-032-12074-8_9

2.1 Water Cycle

The normal cycle of water includes evaporation, precipitation, surface runoff, and stream flow, which is represented in Fig. 9.1. The normal water cycle is also known as the hydrologic cycle. Clouds and atmospheric vapors condense to form rain, snow, and other forms of precipitation. Some of this moisture eventually evaporates back into the atmosphere. Water comes from several sources including surface water, groundwater, frozen water, and snowpack. Surface water can come from rivers, streams and lakes while streams, creeks, lakes, and swamps absorb the moisture which can also percolate into the groundwater. Groundwater comes from springs and underground water sources such as aquifers. Frozen water and snowpack are extremely important for providing water after it melts, typically after the winter season. During heavy precipitation, flooding, and other extreme weather events, water that accumulates in streams or lakes can interact with sewage or wastewater, industrial water, or other contaminants (CDC 2024a).

The normal hydrologic cycle can include disruptions and severe weather events such as droughts and flooding. Climate change will continue to have a great effect on the hydrological cycle; this will result in the increased risks of drought, flooding, and reduced and degraded water supplies for ecosystems, wildlife, and humans (Fecht 2019; USGCRP 2024).

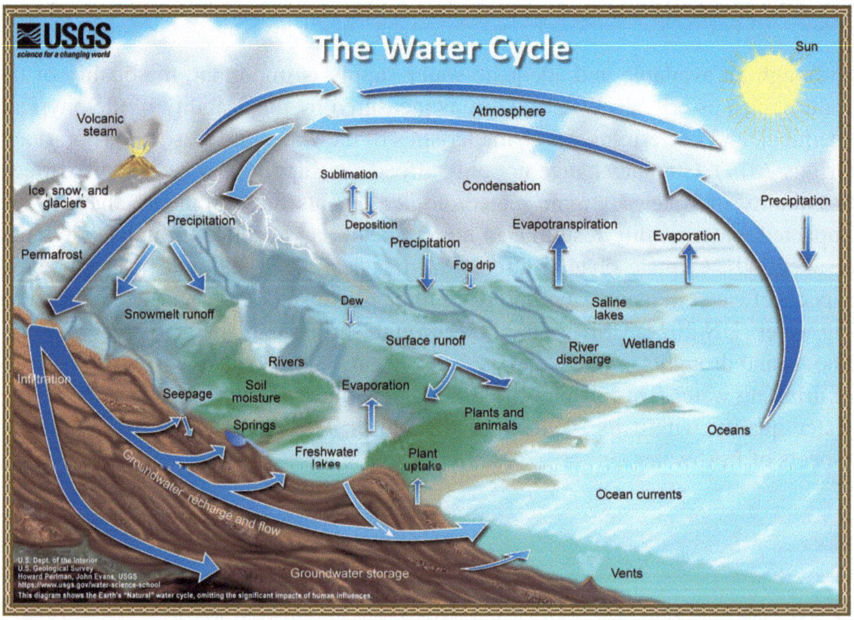

Fig. 9.1 The water or hydrologic cycle. This image provides details on the hydrologic cycle including precipitation, flow of water onto the surface and into the groundwater, and evaporation. (Reproduced from USGS 2019)

2.2 Water Quality Parameters

Around 0.01% of water on Earth is usable or considered freshwater. This is the water available for human use, and effects of climate change are resulting in a potential scarcity of water that can be exacerbated by poor water quality. The availability of safe and fresh water is important for human and other animal consumption, agriculture, industry, and recreation. Water quality is also influenced by anthropogenic (i.e., human made) activities and natural factors. These anthropogenic activities can lead to contamination of freshwater. Contaminated water can lead to a variety of problems including waterborne diseases, cancer, reproductive health issues, exposure to toxic chemicals in fish and food, a buildup of algae blooms and toxic cyanobacteria (Sharma and Bhattacharya 2017).

Water quality measures how suitable and safe water is for various uses including drinking, washing, and bathing. The three basic characteristics of water quality are physical, chemical, and bacteriological. Closely monitoring these characteristics enable the ability to evaluate, assess, and maintain water quality for human health, ecosystems, and communities. Physical aspects of water quality refer to the characteristics of water that can be observed using the senses, such as sight, smell, taste, and touch. These include attributes such as odor, color, turbidity, electrical conductivity, and the presence of solids. Chemical aspects of water quality can include pH, acidity, alkalinity, chlorine content, hardness, dissolved oxygen (DO), and biological oxygen demand. Additionally, radiological aspects are included as a subset of chemical water quality. Bacteriological aspects of water quality include bacteria, algae, protozoa, and viruses. Understanding the different aspects of water quality is important, because of the different requirements that various water interventions and applications need to be safe for human consumption (Forari and Griggs 2017).

Providing safe drinking water requires that the water is not contaminated by bacteriological, physical, or chemical contamination. This may necessitate the protection of water sources, treatment of water to remove contaminants, and the provision of water distribution systems capable of preventing contamination. Climate change effects on the water cycle's historical patterns have made it challenging to adapt and respond to drought, heavy rainfall, and other climate change issues (USGCPA 2024). Understanding these water quality aspects are very important to better managing drinking water.

2.3 Climate Change Impacts on Water Quality

As the Earth's temperature rises, surface water temperatures in lakes and oceans also rise. Climate change is expected to continue to bring an increase in floods, extreme precipitation events, droughts, higher temperatures, and sea level rises. Additionally, climate change significantly impacts the quality of drinking water by causing fluctuations in water availability due to increased droughts and extreme weather events like floods and hurricanes (UN 2024). Availability of water will not affect all countries and regions the same, but due to growing population and economic demands, and the impact of climate change, this may exacerbate the effects. Furthermore, these fluctuations in water will likely have an effect on the potential contamination of water sources with pollutants, pathogens, chemicals, and increased salinity, ultimately affecting the quality and safety of drinking water for humans throughout the world (CDC 2024b; Forari and Griggs 2017).

Climate change jeopardizes the quality and safety of our water supply. Warmer weather causes more water to evaporate, allowing the air to hold more water. Increases in air and water temperatures can also increase the risk of pathogenic waterborne diseases. This sets the stage for heavier rainfall and flooding, which decreases the quality of our water and increases health risks (APHA 2024). Heavier rainfall also can make it more likely that pathogenic organisms contaminate surface water and groundwater due to run off combined with wastewater contamination. Sea level rises can cause the intrusion of saline water into groundwater, aquifers, and surface waters. Lastly, an increase in air and water temperature can cause an increase in cyanobacterial algae blooms (EEA 2024). The following table depicts some of the climate change impacts on water quality (Table 9.1).

Table 9.1 Climate change impacts water quality. (USEPA 2024b; APHA 2024; EEA 2024)

Climate change issue	Climate change impact
Reduced water supply	Warmer temperatures lead to increased evaporation, potentially causing droughts and reduced water levels in reservoirs, impacting drinking water availability. (UN 2024)
Increased flooding	More intense storms can cause flooding, leading to sewage overflows and contamination of freshwater sources with pathogens like bacteria and parasites. (APHA 2024)
Algal blooms	Higher water temperatures can stimulate harmful algal growth, impact water quality, and make water unsafe for consumption. (USEPA 2024b)
Saltwater intrusion	Rising sea levels can push saltwater into coastal freshwater aquifers, contaminating groundwater sources. (USEPA 2024b)
Sedimentation	Increased runoff from heavy rainfall can carry sediment and pollutants into water sources, affecting water clarity and treatment processes. (USEPA 2024b)
Changes in precipitation patterns	Shifts in rainfall distribution, with longer dry periods and shorter intense rain events, can disrupt water management systems
Storm water runoff	Climate change has exacerbated the potential of the severity of storm water runoff. This, in turn, has affected the physical, chemical, and biological aspects of water quality
Shrinking glaciers and snow fields	Glaciers and snow fields are rapidly disappearing. The water from melting glaciers can contaminate freshwater in lowland areas

3 Physical, Chemical, and Bacteriological Aspects of Water Quality Related to Climate Change

3.1 Physical Aspects of Water Quality Related to Climate Change

Physical parameters of water quality are those that are determined by the senses of sight, smell, taste, and touch. These physical parameters include turbidity, color, temperature, taste, and odor.

Turbidity refers to the cloudiness of the water and measures light passing through. It is caused by suspended particles such as organic matter, clay, silt, and particulates. High turbidity makes it difficult to treat water. Pure water is typically clear with a slight blue tint, but organic matter, sediment, and vegetation in water can appear brown or green. Temperature can affect the palatability, viscosity, solubility, and odor of water. The temperature of the water also affects the ability to disinfect, chlorination, the biological oxygen demand, and the ability to remove heavy metals. Taste and odor are affected by foreign matter in water and can be indicative of the presence of some contaminants such as bacteria and harmful chemicals (Sharma and Bhattacharya 2017).

Additional physical water quality parameters include total solids, total dissolved solids (TDS), and total suspended solids (TSS). Solids represent minerals in water affecting turbidity, temperature, color, taste, odor, electrical conductivity, and DO. Electrical conductivity (EC) is another physical water quality parameter. EC measures ionic concentration of water and its ability to carry and conduct electricity. Higher conductivity means more solids in the water (Nemerow et al. 2009). All of these parameters can be negatively affected by climate change events such as flooding, drought, temperature rise, saltwater intrusion, and extreme weather events (Table 9.2).

3.2 Chemical Aspects of Water Quality Related to Climate Change

The presence of contaminants can also be measured by their chemical parameters (Sharma and Bhattacharya 2017). Chemical parameters of water quality are measures of those characteristics which reflect the environment with which water has contact. These chemical parameters can measure pH, hardness, amount of dissolved oxygen (DO), biochemical oxygen demand (BOD), calcium and magnesium, as well as organic chemicals, inorganic chemicals, and radioactive substances (Nemerow et al. 2009).

Acidity or alkalinity in water is measured by pH. Excessively high or low pH (<4 or >11) is detrimental to the quality of the water's taste and the ability to disinfect

Table 9.2 Physical drinking water attributes, contaminants, and potential health effects from climate change. (Nemerow et al. 2009; USEPA 2024d)

Contaminant	Potential health effects from climate change events
Odor	Odor should be absent or faint. Increase in odors can occur from flooding, extreme weather events, cross contamination from wastewater, decaying organic material, and sulfur or rotten egg odor from bacterial contamination
Taste	The taste of water should not be objectionable. Algae, decomposing organic matter, and dissolved gases, which can be caused by drought and increased temperatures, can cause drinking water to be objectionable
Turbidity	Turbidity can be due to suspended materials such as clay, silt, inorganic or organic materials. Turbidity is measured by nephelometric turbidity units (NTU). Drinking water regulations in the United States require that turbidity not exceed 0.5 NTU. An increase in turbidity can also be associated with pathogenic organisms such as bacteria, viruses, and protozoa. Drought, extreme weather events, and flooding can increase turbidity (Nemerow et al. 2009)
Color	Color should be less than 15 true color units. Water drained through decomposing organic material, other organics, and acids can be greater than 15 true color units. Flooding, increased weather events, and other climate change events can affect the color of the water. (Nemerow et al. 2009)
Temperature	The temperature of the water should be less than 60 °F (16 °C). Increased temperatures due to climate change can cause high water temperatures. High water temperatures accelerate the growth of nuisance organisms such as bacteria, viruses, and protozoa (Nemerow et al. 2009)

water. Additionally, a lower pH (acidic conditions) increases the solubility of heavy metals causing more risks to human health and ecosystems. Hardness refers to the mineralization of water such as dissolved materials like calcium and magnesium. Calcium and magnesium can cause mineral buildup in hot water pipes and later difficulty with dissolving and effectively utilizing soap. Calcium Carbonate ($CaCO_3$) greater than 500 mg can cause laxative properties (Sensorex 2024).

Dissolved oxygen is an indirect measure of water pollution in water ways. The lower the DO, the worse the water is for aquatic life because it increases the potential for disease-causing organisms to proliferate. DO is among the most important parameters when measuring a river's water quality. The amount of DO in the water dictates how polluted the water sample is. Low amounts of DO indicate that the water is highly polluted and that organic contaminants are consuming the DO. A low DO and a BOD means a potential increase for waterborne diseases from bacteria, algae, viruses, and protozoa.

Inorganic chemicals that can affect water quality include arsenic, asbestos, cadmium, chromium, copper, fluoride, lead, mercury, and nitrates. Several of these can naturally occur in water while all can be anthropogenically (human) caused. For example, fluoride can be both naturally occurring and/or anthropogenic. Arsenic contamination worldwide is a significant issue that causes some of the biggest health problems (WHO 2022). Mercury contamination is also an issue that can cause health issues such as neurological disorders and issues with the endocrine system (Sharma and Bhattacharya 2017). Organic chemical contamination can contain a variety of chemicals and come from a variety of human-caused sources (e.g., pesticides, industrial wastes). Contamination through organic chemical sources such as benzene, chlordane, and trihalomethanes, can cause liver and nervous system disorders and other serious health problems.

Tables 9.3 and 9.4 identify chemical contaminants, provide the sources for these inorganic and organic contaminants, and describe the potential health effects from long-term exposure in drinking water. The chart utilizes data from the US Environmental Protection Agency (EPA). Although the chart provides data on drinking water, it is generalized to freshwater due to its similarities. Furthermore, radiological contaminants in freshwater (i.e., drinking water) will be exacerbated by climate change.

3.3 Radiological Aspects of Water Quality Related to Climate Change

Although radiological aspects of water quality are a subset of chemical aspects of water quality, there is a separate section provided to discuss the radiological aspects of water quality. The presence of radionuclides in water is likely due to radioactive decay products from Uranium-238, Uranium-235, Thorium 232, and Potassium-40 (Pintilie-Nicolov et al. 2021). Radioactive substances decay to emit beta, alpha, and gamma radiation. Radiologic measures usually are alpha, beta, radium, and uranium. Radon is considered a radionuclide. Radon, which is a radioactive element that undergoes radioactive decay, is a naturally occurring radioactive gas often found in the environment, originating from the decay of uranium and radium in rocks and soil (USEPA 2012b; National Research Council 1999).

Radon concerns have focused primarily on airborne radon (radon that comes from soil and rocks surrounding a building foundation); radon in drinking water is not something that is often considered. Radon does not have an odor, is not visible, and does not have a taste. Radon in drinking water usually originates in water wells that are drilled into bedrock containing radon gas (USEPA 2012a). These wells could be private or utilized by a public water supply system.

Dissolved radon in groundwater will escape into indoor air during showering, laundering, and dishwashing. Estimates are that indoor air concentrations increase by approximately 1 picocuries per liter (pCi/L) for every 10,000 pCi/L of radon in water. For example, a water well containing 2000 pCi/L of radon would be expected to contribute 0.2 pCi/L to the indoor air radon concentration. Based on the potential for cancer, the EPA suggests that indoor air should not exceed 4 pCi/L. Radon, as a radioactive decay product in drinking water, can have health effects. This can affect the hematopoietic, gastrointestinal, reproductive, and nervous systems. According to the United States' Environmental Protection Agency (EPA) and the Surgeon General, radon is the second leading cause of lung cancer in the United States today. Climate change has the potential to increase radionuclides in water which will increase health issues and health concerns (USEPA 2012b; MDOH n.d.; PDOEP 2024).

Table 9.3 National Primary Drinking Water Regulations chart of select inorganic chemical contaminants and potential health effects from climate change. (USEPA 2024d)

Contaminant	MCLG (Maximum Contaminant Level Goal) (mg/L)	Potential health effects from long-term exposure above the MCL (Maximum Contaminant Level)	Sources of contaminant in drinking water
Arsenic	0	Skin damage or problems with circulatory system, and may have increased the risk of cancer	Erosion of natural deposits; runoff from orchards, runoff from glass and electronics production waste
Asbestos (fiber > 10 micrometers)	7 million fibers per liter (MFL)	Increased risk of developing benign intestinal polyps	Decay of asbestos or cement in water mains; erosion of natural deposits
Barium	2	Increase in blood pressure	Discharge of drilling wastes; discharge from metal refineries; erosion of natural deposits
Cadmium	0.005	Kidney damage	Corrosion of galvanized pipes; erosion of natural deposits; discharge from metal refineries; runoff from waste
Chromium (total)	0.1	Allergic dermatitis	Discharge from steel and pulp mills; erosion of natural deposits
Copper	1.3	Short term exposure: Gastrointestinal distress. Long term exposure: Liver or kidney damage	Corrosion of household plumbing systems; erosion of natural deposits
Cyanide (as free cyanide)	0.2	Nerve damage or thyroid problems	Discharge from steel/metal factories; discharge from plastic and fertilizer factories
Fluoride	4.0	Bone disease (pain and tenderness of the bones); children may get mottled teeth	Water additive which promotes strong teeth; erosion of natural deposits; discharge from fertilizer and aluminum factories
Lead	0	Infants and children: Delays in physical or mental development. Adults: Kidney problems; high blood pressure	Corrosion of household plumbing systems; erosion of natural deposits
Mercury (inorganic)	0.002	Kidney damage	Erosion of natural deposits; discharge from refineries and factories; runoff from landfills and croplands

(continued)

Table 9.3 (continued)

Contaminant	MCLG (Maximum Contaminant Level Goal) (mg/L)	Potential health effects from long-term exposure above the MCL (Maximum Contaminant Level)	Sources of contaminant in drinking water
Nitrate (measured as nitrogen)	10	Infants below the age of 6 months could become seriously ill and may die of blue-baby syndrome	Runoff from fertilizer use; leaking from septic tanks, sewage; erosion of natural deposits
Selenium	0.05	Hair or fingernail loss; numbness in fingers or toes; circulatory problems	Discharge from petroleum refineries; erosion of natural deposits; discharge from mines

Table 9.4 National Primary Drinking Water Regulations chart of select organic chemical contaminants and potential health effects from climate change (e.g., flooding, drought, temperature rise, extreme weather events). (USEPA 2024d)

Contaminant	MCLG (Maximum Contaminant Level Goal) (mg/L)	Potential health effects from long-term exposure above the MCL (Maximum Contaminant Level)	Sources of contaminant in drinking water
Benzene	0	Anemia; decrease in blood platelets; increased risk of cancer	Discharge from factories; leaching from gas storage tanks and landfills
Carbon tetrachloride	0	Liver problems; increased risk of cancer	Discharge from chemical plants and other industrial activities
Chlordane	0	Liver or nervous system problems; increased risk of cancer	Residue of banned termiticide
Chlorobenzene	0.1	Liver or kidney problems	Discharge from chemical and agricultural chemical factories
1,2-Dichloroethane	0	Increased risk of cancer	Discharge from industrial chemical factories
1,1-Dichloroethylene	0.007	Liver problems	Discharge from industrial chemical factories
Polychlorinated biphenyls (PCBs)	0	Skin changes; thymus gland problems; immune deficiencies; reproductive or nervous system difficulties; increased risk of cancer	Runoff from landfills; discharge of waste chemicals

(continued)

Table 9.4 (continued)

Contaminant	MCLG (Maximum Contaminant Level Goal) (mg/L)	Potential health effects from long-term exposure above the MCL (Maximum Contaminant Level)	Sources of contaminant in drinking water
1,1,1-Trichloroethane	0.20	Liver, nervous system, and/or circulatory problems	Discharge from metal degreasing sites and other factories
Trichloroethylene	0	Liver problems; increased risk of cancer	Discharge from metal degreasing sites and other factories
Vinyl chloride	0	Increased risk of cancer	Leaching from PVC pipes; discharge from plastic factories

Table 9.5 National primary drinking water regulations chart of select radionuclide contaminants and potential health effects from climate change (e.g., flooding, drought, temperature rise, and extreme weather events). (USEPA 2024d)

Contaminant	MCLG (Maximum Contaminant Level Goal) (mg/L)	Potential health effects from long-term exposure above the MCL (Maximum Contaminant Level)	Sources of contaminant in drinking water
Alpha particles	0	Increased risk of cancer	Erosion of natural deposits of certain minerals that are radioactive and may emit a form of radiation known as alpha radiation
Uranium	0	Increased risk of cancer, kidney toxicity	Erosion of natural deposits

Table 9.5 identifies the radiological contaminants in drinking water, provides the sources of these contaminants, and describes the potential health effects from long term exposure in drinking water. The chart utilizes data from the EPA. Although the chart provides data on drinking water, it is generalized to freshwater due to its similarities. Furthermore, radiological contaminants in freshwater (i.e., drinking water) will be exacerbated by climate change.

3.4 Bacteriological Aspects of Water Quality Related to Climate Change

Pathogen-associated water diseases have become an important public health threat, which has been exacerbated by the effects of climate change events. Bacteriological contaminants have become the greatest drinking water hazard globally (Forari and Griggs 2017). Bacteriological aspects of water quality are measurements that reflect the number of bacteria, algae, viruses, and protozoa that are present in the water and water distribution system. These microorganisms are evaluated by collecting and analyzing water samples to estimate the number of bacteria present. This process can help determine if the water is safe to consume, and if it contains bacteria and other disease-causing microorganisms.

It is critical to utilize specific microorganism indicators as measures for monitoring and determining groundwater vulnerability and surface water quality. E. coli and total coliforms are important bacterial indicators that are used to assess the quality of the drinking water. E. coli is used as a specific indicator to determine fecal contamination, whereas testing for total coliforms is less specific than testing for fecal coliforms or E. coli. Testing for total coliforms is used to identify if disease causing organisms may be present. Furthermore, testing for total coliforms and fecal coliforms can verify that the water has been adequately treated and is safe for distribution (Nemerow et al. 2009; GOC 2021).

Conducting source water assessments is extremely important for identifying enteric protozoa and enteric virus contamination and minimizing waterborne illnesses. Common enteric protozoa include *Giardia* and *Cryptosporidium,* and common enteric viruses include noroviruses, rotaviruses, enteroviruses, and hepatitis A virus. Both enteric protozoa and enteric viruses can cause gastroenteritis symptoms including vomiting, abdominal cramping, and diarrhea. These water assessments include testing to identify total and fecal coliform contamination in all water sources (e.g., watershed, aquifers, rivers, lakes). Identifying and evaluating pathways (both confirmed and potential) by which the pathogens can make their way into the source water, assessing runoff from precipitation events, as well as identifying and addressing factors such as rainfall and other periods of elevated precipitation that may contribute to increased high concentrations of microbiological pathogens, are important components of source water assessments. Lastly, source water monitoring for enteric protozoa, enteric viruses, and other microbiological pathogens can help prevent contamination and outbreaks while ensuring the health and safety of drinking and recreational waters (Nemerow et al. 2009; GOC 2021).

Monitoring enteric pathogens is essential to prevent contamination of drinking water and to protect the health and safety of the public. Source water contamination by enteric pathogens (protozoa, viruses, and/or bacteria) is a serious public health concern. The monitoring of E. coli, Giardia, Cryptosporidium, enteric viruses, and other protozoa should reflect both normal pathogen loads in the water source and peak events. Historical data on precipitation, river and stream flow, snowmelt, and the turbidity of the water source can show trends and patterns that help to predict when peak pathogen loads are likely to occur. Analyzing these factors, along with a variety of other data and assessments, will help to ensure that an effective mitigation plan is implemented (GOC 2021).

Climate change through flooding, drought, temperature rise, saltwater intrusions, extreme weather events, and other climate triggered events can increase and exacerbate bacteriological organisms increasing the potential for health risks. The following table identifies the bacteriological contaminants in drinking water, provides the sources for these contaminants, and describes the potential health effects from long term exposure in drinking water. The chart utilizes data from the U.S. Environmental Protection Agency. Although the chart provides data on drinking water, it is generalized to freshwater due to its similarities. Furthermore, bacteriological contaminants in freshwater (i.e., drinking water) will be exacerbated by climate change (Table 9.6).

Table 9.6 Select National Primary Drinking Water bacteriological contaminants and potential health effects from climate change. (USEPA 2024d)

Contaminant	MCLG (Maximum Contaminant Level Goal)[1]	Potential health effects from long-term exposure above the MCL (Maximum Contaminant Level)	Sources of contaminant in drinking water
Cryptosporidium	0	Gastrointestinal illness (such as diarrhea, vomiting, and cramps)	Human and animal fecal waste
Giardia lamblia	0	Gastrointestinal illness (such as diarrhea, vomiting, and cramps)	Human and animal fecal waste
Legionella	0	Legionnaire's disease, a type of pneumonia	Found naturally in water; multiplies in heating systems
Total coliforms (including fecal coliform and *E. Coli*)	0	Not a health threat in itself; it is used to indicate whether other potentially harmful bacteria may be present[2]	Coliforms are naturally present in the environment; as well as feces; fecal coliforms and *E. Coli* only come from human and animal fecal waste[2]
Turbidity	N/a	Higher turbidity levels are often associated with higher levels of disease causing microorganisms such as viruses, parasites, and some bacteria. These organisms can cause symptoms such as nausea, cramps, diarrhea, and associated headaches	Soil runoff
Viruses (enteric)	0	Gastrointestinal illness (such as diarrhea, vomiting, and cramps)	Human and animal fecal waste

4 Case Studies

4.1 *Case Study: Effects of Drought on Drinking Water Contaminants in California (Sum 2024)*

In the last 20 years, droughts in California related to climate change have increased the severity of drinking water disparities among underserved and marginalized communities (Sum 2024). Droughts have been found to cause critical threats to drinking water systems, and also threats to vulnerable communities, especially those in resource poor Community Water Systems (CWS). CWSs are small water systems which provide over 90% of drinking water to residents in the United States. Not only have there been disparities among vulnerable marginalized populations, but drought conditions have also created an increase in issues of drinking water contaminants exacerbating the existing disparities. Droughts have created an unequal exposure to drinking water contaminants among marginalized populations and specifically among Latino/Latina communities in California (Sum 2024).

Impending droughts, fueled by climate change, may exacerbate chemical and bacteriological contaminants in drinking water especially in groundwater sources (Sum 2024). The smaller and more resource poor CWSs serve a significant amount of Latino/Latina residents who utilize drinking water from groundwater sources. During droughts in California, there are large scale changes in the sources of drinking water and that increases the potential for contaminants, especially arsenic and nitrates. These drought conditions are particularly apparent in the Central Valley area of California. There has been excessive and intensified groundwater pumping and an acceleration in the withdrawal of the more shallow and contaminated water leading to an increase in nitrates in the drinking water. Arsenic, however, has been found to increase in surface water in California during drought conditions (Sum 2024).

Arsenic and nitrate are examples of contaminants that have been found exceeding the USEPA limits greater among the California Latino/Latina communities than other communities (USEPA 2024c, USEPA 2024d). Exposure to arsenic and nitrates can cause a variety of health problems including cancers and cardiovascular diseases. When nitrates are consumed, they are converted into nitrites in the blood causing an oxygen deprivation condition called "Blue Baby Syndrome" or methemoglobinemia (CDC 2024b). Small CWSs, rather than large CWSs, showed a more significant arsenic and nitrate contamination issue among Latino/Latina communities in the Central Valley of California.

Droughts fueled by climate change have exacerbated chemical and bacteriological contaminants in the drinking water, especially in groundwater nitrates and surface water arsenic (Sum 2024). These smaller, predominantly minority and lower socioeconomic communities, have less resources to mitigate these changes and are more vulnerable and at risk to nitrate and arsenic contamination (Sum 2024).

This case study shows that climate change not only affects the overall quality of drinking water in the general population, but disproportionately affects vulnerable populations sensitive to the impacts of poor water quality and reduced water capacity and quantity. These communities are more affected than the general population due to reduced resources and funding. Additionally, more issues may surface that are not known due to the increasing effects of climate change. These issues will affect vulnerable and marginalized communities at an increasing rate due to the worsening of climate change concerns.

4.2 Case Study: Depletion of Groundwater and Declining Surface Water Quality (Lall et al. 2018)

Water systems throughout the world face an increasing risk of depletion and contamination due to the effects of climate change. Long term drought will affect the quantity and quality of surface water and groundwater sources. Since groundwater can also be composed of more dissolved solids or minerals than surface water, it is also important for the nourishment of vegetation. The strata of groundwater may be unconsolidated such as sand, clay, and gravel or consolidated such as granite and limestone. Over 40% of water used for agriculture and drinking water in the world comes from groundwater (Nature Conservancy 2024). Persistent dry periods have affected North America in the last 20 years (IPCC 2008). Water quantity issues have affected every area of the United States. Climate change issues are expected to persist with variable precipitation, rising temperatures, extreme weather events, drought, and reduced snowpack. Climate change has affected both the quality and quantity of surface water and groundwater. This has resulted in negative consequences to the physical, bacteriological, and chemical aspects of water quality.

Throughout the world, climate change has had negative consequences to the quality and quantity of water. In the United States, climate change has affected water quality and quantity in a variety of regions. Northern and Southern plants have experienced a large amount of flooding causing a variety of issues including agricultural damage. In the Northeast, rising sea levels and heavy precipitation have affected and exacerbated the infrastructure problems. The Midwest has experienced numerous issues due to increasing flood risks and precipitation. Warmer temperatures have created other water quality issues including harmful algae blooms introducing toxic cyanobacteria into drinking and recreational waters. In the Northwest and Alaska, habitat loss and damage to fishing and fisheries have occurred through reduced snowpack, increasing flooding, and increasing extreme weather events. In the Southwest, droughts have affected water supply and water quality issues. Hawaii, United States, Pacific Islands, and Caribbean islands have experienced rising temperatures, rising sea levels, and saltwater intrusion. All these regions have incurred issues on their quality and quantity of drinking water and water available for agriculture, recreation, wildlife, and fishing (Lall et al. 2018; NOAA NCEI 2018).

Groundwater has been greatly affected by climate change. Groundwater provides over 38% of drinking water in the United States, and over 40% of water used for agriculture and drinking water throughout the world (Nature Conservancy 2024). Groundwater use has increased significantly throughout the world and the United States, resulting in a groundwater withdrawal rate larger than is being recharged. The increasing use of groundwater and the variability of droughts has created enormous challenges with groundwater. Groundwater supplies have been decreasing throughout the United States since 1900–2000 and has accelerated more since 2001 due to persistent droughts (Lall et al. 2018). Depletion of aquifers is a major concern for groundwater sources in the United States (Lall et al. 2018).

Surface water sources have also been greatly affected by climate change and climate change events (Bates et al. 2008). Surface water sources are dependent on seasonality, variability of streamflow, and a dependable water flow (e.g., snowpack, annual precipitation). Increases in precipitation events such as extreme weather events can increase sediment, microbiological pathogens, nutrients, and other issues to streams, lakes, rivers, estuaries, and other surface water sources (Lall et al. 2018). During periods of drought and other low flow water events, there have been challenges with aquatic life, the increase in the potential of algal blooms, reduce DO in surface water, and an increase in the biological oxygen demand of the surface waters (Lall et al. 2018). Algal blooms are expected to increase due to higher temperatures, potentially increasing the negative effects on human health and the ecosystem. Impacts of global warming on lakes and other surface waters include an extended growing period of algal blooms at high latitudes, an increase in nutrient loss, and an increase in aquatic organic material resulting in more negative effects on fish and other aquatic life forms. Consequently, the effects of climate change may result in less reliable sources of drinking water for humans and the potential of contaminated water for all wildlife and the ecosystem (Bates et al. 2008).

Heavy and variable precipitation has increased in intensity. Higher population and higher temperatures have increased water needs and uses. The continued warming will not only affect the quantity of water supplies but the physical, chemical, and bacteriological quality of water. Extreme precipitation events and long-lasting droughts are expected to cause problems for surface water quantity and quality. The Colorado River Basin faces many challenges due to persistent drought affecting this area (Lall et al. 2018).

This case study shows the problems that climate change causes on regions throughout the United States and its territories. Climate change has affected both the quality and quantity of surface water and groundwater. This has resulted in negative consequences to the physical, bacteriological, and chemical aspects of water quality.

4.3 Case Study Algae in Water (USEPA 2024e)

Algae are important to freshwater ecosystems, but under the right conditions some types of algae can be harmful to animals, humans, and their ecosystems. Algae help to make oxygen for the survival of plants, animals, and humans. However, some types of algae can be harmful if they grow too quickly or if they make toxins. Phytoplankton are a type of algae that are small single celled organisms (e.g., plants, bacteria). Cyanobacteria or blue green algae is a type of phytoplankton that can cause harmful algal blooms (HABs) in fresh water. Diatoms, dinoflagellates, or red tides are another type of phytoplankton that can cause HABs in salt water (CDC 2024c). Climate change is significantly increasing the risk of HABs in all water sources including drinking water (CDC 2024c; UN 2024; APHA 2024).

Too many nutrients, such as nitrogen and phosphorus, help HABs to grow. These can come from fertilizers from farms and lawns, sewage or wastewater contamination, and runoff from storms. Rising sea water or upwelling can increase nutrient levels. HABs tend to grow more in warm water resulting in algal blooms to occur more in the summer or fall time periods. HABs can also grow when the water is slowly moving or there is less air circulating through the water. Changes in pH or turbidity can increase nutrients which result in an increase of algal blooms. Changes in the ecosystems can also impact the growth of algae and HABs; which can be exacerbated by droughts, heavy rainfall, flooding and other climate change related events (CDC 2024c).

Climate change can increase the growth of HABs in freshwater, seawater, and brackish waters. Increases in water temperatures as a result of climate change are expected to increase the duration and severity of cyanobacteria HABs (CyanoHABs). These can grow faster than other algae at warmer temperatures. CyanoHABs can absorb sunlight and release heat making temperature in surface waters hotter. Increased carbon dioxide in the air and water can also lead to a rapid growth of algae, especially CyanoHABs. Increased rainfall from storms and extreme rainfall events can increase nutrients the runoff from land into waterbodies that can flush large amounts of nutrients in the surface water which can increase CyanoHABs. Furthermore, coastal upwelling can be increased by climate change which can increase nutrients which lead to more algal blooms and CyanoHABs (USEPA 2024e).

Coastal, oceanic, and Great Lakes' waters are increasingly impacted by HABs, however, HABs are affecting all of the United States. Warming temperatures have contributed to large CyanoHABs in many regions through the world and the United States. Lake Erie is an example of a location that has seen a significant rise in HABs. The National Oceanographic Atmospheric Association forecasts seasonal blooms of CyanoHABs, typically from July to October, when warmer water conditions favor the growth of the algae. These can last even into the early winter months. Specifically, the western portion of Lake Eric has seen an extreme increase in HABs since the early 2000s. They primarily consist of CyanoHABs posing a risk to humans, animals, ecosystems, and affecting recreational water use and businesses (NOAA NCCOS 2024; CDC 2024c).

Mitigating harmful algal blooms and CyanoHABs is going to be a challenge, especially because the effects of climate change are exacerbating the growth of the algal blooms. Reducing nutrient runoff is important. This can be done through improved agricultural practices, education of the public on the use of fertilizers, better stormwater runoff management, and the implementation of stormwater plans. Early detection of algal blooms is important especially when drought conditions or extreme weather events are affecting a community or region. Preventive and advanced water treatment is also important to remove HABs. Lastly, addressing climate change as a whole through reduction of greenhouse gas emissions is going to be extremely important to reduce the growth of algae and HABs (USEPA 2024e; NOAA NCCOS 2024; CDC 2024c; CA.GOV 2024).

5 Conclusion

This chapter discusses and evaluates how climate change affects water quality within the context of the physical, chemical, and bacteriological aspects of freshwater. Although an available and adequate water supply is important, the quality of the water is just as critical. Extreme meteorological events due to climate change have impacted water quality resulting in human health and ecosystem impacts (Delpla et al. 2009). Due to climate change, global disparities in access to safe and available clean and fresh water have become apparent. Adverse effects of water quality due to climate change are increasingly a significant issue affecting communities locally, regionally, and nationally. Safe drinking water, from all water sources in the context of climate change has an increasingly significant human health and ecosystem impact (Delpla et al. 2009).

Consequently, access to clean drinking water is crucial for public health, and its problematic disruption due to climate change can lead to increased healthcare burdens. Water quality and the adverse health and ecological effects caused by contamination, pollution, and climate change can be effectively managed. This includes effectively treating water, addressing climate change, preventing the contamination of water sources, identifying alternative water sources, and implementing best public health practices.

6 Discussion Questions

1. How do physical, chemical, and bacteriological aspects of water quality differ? How will climate change affect these aspects of water quality? Provide an example of each.
2. Utilizing examples presented in this chapter or from other cases, what is a real-world example of climate change impacting human health through the physical, bacteriological, and chemical aspects of water quality?
3. Did you observe any patterns or trends between the source contaminants of drinking water? Utilizing the case study on Latino/Latina communities in the Central Valley of California, what other populations do you think would be at risk of the effects of climate change on drinking water quality? Explain your answer?

References

American Public Health Association (APHA) (2024) Climate changes health: water quality and accessibility. https://www.apha.org/topics-and-issues/climate-health-and-equity/water-quality#:~:text=Unfortunately%2C%20climate%20change%20jeopardizes%20the,water%20and%20increases%20health%20risks. Accessed 2 Oct 2024

Bates BC, Kundzewicz ZW, Wu S (2008) In: Palutikof JP (ed) Climate change and water. Technical paper of the intergovernmental panel on climate change. IPCC Secretariat, Geneva, p 210. https://archive.ipcc.ch/pdf/technical-papers/climate-change-water-en.pdf. Accessed 7 Nov 2024

California.Gov (CA.GOV) (2024) Are harmful algal blooms affecting our waters? https://mywaterquality.ca.gov/habs/what/index.html. Accessed 10 Nov 2024

Centers for Disease Control and Prevention (CDC) (2024a) Drinking water sources: an overview. https://www.cdc.gov/drinking-water/about/drinking-water-sources-an-overview.html#:~:text=Source%20water%20refers%20to%20bodies,and%20water%20under%20the%20ground. Accessed 12 Nov 2024

Centers for Disease Control and Prevention (CDC) (2024b) Drinking water: what causes tap water concentration. https://www.cdc.gov/drinking-water/causes/index.html. Accessed 7 Nov 2024

Centers for Disease Control and Prevention (CDC) (2024c) Harmful algal blooms: contribution factors and impacts. https://www.cdc.gov/harmful-algal-blooms/about/harmful-algal-blooms-contributing-factors-and-impacts.html. Accessed 7 Nov 2024

Delpla I, Jung AV, Baures E, Clement M, Thomas O (2009) Impacts of climate change on surface water quality in relation to drinking water production. Environ Int 35(8):1225–1233. https://doi.org/10.1016/j.envint.2009.07.001

European Environment Agency (EEA) (2024) European climate risk assessment, EEA report no 1/2024. https://www.eea.europa.eu/publications/european-climate-risk-assessment. Accessed 7 Nov 2024

Fecht, S (2019) Columbia climate school. State of the planet. How climate change impacts our water. https://news.climate.columbia.edu/2019/09/23/climate-change-impacts-water/. Accessed 7 Oct 2024

Forari JA, Griggs A (2017) Water contamination. In: Levy BS, Wegman DH, Baron SL, Sokas RK (eds) Occupational and environmental health. Oxford University Press, Oxford, UK, pp 365–380

Government of Canada (GOC) (2021) Overview of the microbiological aspects of drinking water quality. https://www.canada.ca/content/dam/hc-sc/documents/services/environmental-workplace-health/reports-publications/water-quality/micro-overview-eng.pdf. Accessed 2 Oct 2024

Intergovernmental Panel on Climate Change (IPCC) (2008) IPCC sixth assessment report. https://www.ipcc.ch/report/ar6/wg2/chapter/chapter-14/. Accessed 7 Nov 2024

Lall U, Johnson T, Colohan P, Aghakouchak A, Brown C, McCabe G, Pulwarty R, Sankarasubramanian A (2018) Water. In: Reidmiller DR, Avery CW, Easterling DR, Kunkel KE, Lewis KLM, Maycock TK, Stewart BC (eds) Impacts, risks, and adaptation in the United States: fourth national climate assessment, volume II. United States Global Change Research Program, Washington, DC, USA, pp 145–173. https://doi.org/10.7930/NCA4.2018.CH3. https://repository.library.noaa.gov/view/noaa/19487

Minnesota Department of Health (MDOH) (n.d.) Naturally occurring radionuclides in drinking water are an underestimated public health risk. Naturally Occurring Radionuclides in Drinking Water…poster. Accessed 2 Oct 2024

National Research Council (1999) Risk assessment of radon in drinking water. The National Academies Press, Washington, DC, p 296. http://books.nap.edu/catalog.php?record_id=6287. Accessed 2 October 2024

National Oceanic and Atmospheric Association (NOAA) National Centers for Environmental Information (NCEI) (2018) Billion-dollar weather and climate disasters. https://www.ncei.noaa.gov/access/billions/events/US/1980-2017. Accessed 9 Jan 2025

National Oceanic and Atmospheric Association (NOAA) National Centers for Coastal Ocean Science (NCCOS) (2024) Lake Erie harmful algal bloom forecast. https://coastalscience.noaa.gov/science-areas/habs/hab-forecasts/lake-erie/. Accessed 7 Nov 2024

Nature Conservancy (2024) Groundwater: our most valuable hidden resource https://www.nature.org/en-us/what-we-do/our-insights/perspectives/groundwater-most-valuable-resource/#:~:text=38%25,of%20all%20drinking%20water%20worldwide. Accessed 30 Nov 2024

Nemerow NL, Agardy FJ, Sullivan PJ, Salvato JA (2009) Environmental engineering: water, wastewater, soil and groundwater treatment and remediation. John Wiley & Sons, New York

Pennsylvania Department of Environmental Protection (PDOEP), Bureau of Radiation Protection, Pennsylvania Radon Information Center (2024) Pennsylvania Consumer's Guide to Radon Reduction. .Accessed 20 Jan 2025

Pintilie-Nicolov V, Georgescu PL, Iticescu C, Moraru DI, Pintilie AG (2021) The assessment of the annual effective dose due to ingestion of radionuclides from drinking water consumption: calculation methods. J Radioanal Nucl Chem 327:49–58. https://doi.org/10.1007/s10967-020-07438-5

Sensorex (2024) Three main types of water quality parameters explained. https://sensorex.com/three-main-types-of-water-quality-parameters-explained/#Physical_Water_Quality_Parameters. Accessed 7 Oct 2024

Sharma S, Bhattacharya A (2017) Drinking water contamination and treatment techniques. Appl Water Sci 7:1043–1067. https://doi.org/10.1007/s13201-016-0455-7

Sum S (2024) Impact of droughts on served drinking water disparities in California, 2007-2020. Am J Public Health 114(9):935–945. https://doi.org/10.2105/ajph.2024.307758

United Nations (UN) (2024) Climate action water at the center of the climate crisis. https://www.un.org/en/climatechange/science/climate-issues/water#:~:text=Sea%2Dlevel%20rise%20is%20projected%20to%20extend%20salinization,humans%20and%20ecosystems%20in%20coastal%20areas%20(IPCC).&text=Water%20quality%20is%20also%20affected%20by%20climate,from%20sediments%20to%20pathogens%20and%20pesticides%20(IPCC). Accessed 2 Oct 2024

United States Environmental Protection Agency (USEPA) (2012a) A citizen's guide to radon: the guide to protecting yourself and your family from radon. https://www.epa.gov/sites/default/files/2016-12/documents/2016_a_citizens_guide_to_radon.pdf. Accessed 2 Oct 2024

United States Environmental Protection Agency (USEPA) (2012b) Proposed radon in drinking water regulation. http://water.epa.gov/lawsregs/rulesregs/sdwa/radon/regulations.cfm. Accessed 2 Oct 2024

United States Environmental Protection Agency (USEPA) (2024a) Climate adaptation and saltwater intrusion. https://www.epa.gov/arc-x/climate-adaptation-and-saltwater-intrusion. Accessed 2 Oct 2024

United States Environmental Protection Agency (USEPA) (2024b) Climate impacts on water quality. https://www.epa.gov/arc-x/climate-impacts-water-quality. Accessed 2 Oct 2024

United States Environmental Protection Agency (USEPA) (2024c) Types of drinking water contaminants. https://www.epa.gov/ccl/types-drinking-water-contaminants. Accessed 12 Nov 2024

United States Environmental Protection Agency (USEPA) (2024d) National primary drinking water regulations. https://www.epa.gov/ground-water-and-drinking-water/national-primary-drinking-water-regulations. Accessed 12 Nov 2024

United States Environmental Protection Agency (USEPA) (2024e) Climate change and freshwater harmful algal blooms. https://www.epa.gov/habs/climate-change-and-freshwater-harmful-algal-blooms. Accessed 7 Nov 2024

United States Geological Survey (USGS) (2019) The natural water cycle. https://www.usgs.gov/media/images/natural-water-cycle-jpg. Accessed 8 Jan 2025

United States Global Change Research Program (USGCRP) (2024) National climate assessment: fifth national climate assessment: chapter 4: water. https://nca2023.globalchange.gov/chapter/4/. Accessed 28 Nov 2024

World Health Organization (WHO) (2022) Guidelines for drinking-water quality, 4th edn. WHO, Geneva, Switzerland, pp 341–343. https://apps.who.int/iris/rest/bitstreams/1414381/retrieve. Accessed 30 Oct 2024

Part III
Cost of Climate Change and Drinking Water

Chapter 10
Climate Change Vulnerability Assessment, Risk Management, and Risk Assessment

Lakshmi Raghu Nagendra Prasad Rentachintala

1 Learning Objectives

1. Review available literature on climate change vulnerability, risk assessment, and risk management and associated methodologies.
2. Compare and evaluate existing methodologies on vulnerability, risk assessment, and risk management of climatic variation.
3. Provide critical remarks and future research needs for vulnerability, risk assessment, and risk management of climatic variation.

2 Introduction

Climate change would more exacerbate various hazards in numerous ways like happening of new risks, escalating the prevailing unsafe occasions (IPCC 2014) and has more raised probability of peak climate incidents (Eckstein et al. 2018) with the potential for activating hazards (Brooks 2003). Multiple-hazard with climatic variation will mark the value, occurrence, circulation, variety and severity of impacts and the critical possibility of peak occasions, though negotiating and decreasing the resilience of community and atmosphere. Thus, climatic change serves as an aggravating representative for deteriorating the impact of the hazards (IPCC 2014).

Collective assessment of risk for multiple-peril postures a diversity of tests, such as uniting impacts of every discrete threat for combined impact; considering the severities of various hazards, insertion of occurrences of those threats and dealing

L. R. N. P. Rentachintala (✉)
Department of Civil Engineering, Bapatla Engineering College,
Bapatla, Andhra Pradesh, India
e-mail: rlr.nagendraprasad@becbapatla.ac.in

A. R. Molla et al. (eds.), *Climate Change, Drinking Water Security, and Public Health*, https://doi.org/10.1007/978-3-032-12074-8_10

with the connections between diverse perils (Carpignano et al. 2009). Also, the recognition of proper suitable pointers for collective assessment of risk of multiple-threat is questionable, intricate and perilous (Delmonaco et al. 2006).

The present study aims at a review of available literature on climatic change impacts, vulnerability, assessment, and management of associated risks especially in the context of drinking water. Also, comparison and evaluation of existing methodologies, critical remarks, and future research needs are identified and presented.

3 Literature Review

Various attributes of climate change are temperature governing variations, melting of frost, and the high level of specialty of both normal and anthropological systems, such as mountain species, density of valley residents, and economy on tourism basis (Terzi et al. 2019). These climate change characteristics need the utilisation of risk evaluation methodologies capable of describing the intricate collaborations among multiple perils, biophysical and socio-monetary systems for adaptation of climatic variation (Terzi et al. 2019).

Various scenarios are developed to interpret plausible impacts of climatic change by Intergovernmental Panel on Climate Change (IPCC) to denote worldwide growth pathways and resultant emissions to cover a vital portion of uncertainties in emissions (Calow et al. 2011).

Societal and environmental stresses, including climate change on water, are different across the regions and sectors (Iglesias et al. 2007). A framework to manage water stress risk is for preparing a method instead of a crisis technique. Significant local level managing is key with multiple institutions and multiple stakeholder association (Iglesias et al. 2007).

3.1 Challenges in Climatic Change Assessments

Still, there remain knowledge gaps on analysis of climate change associated risks to apply the framework of chain of impacts, stakeholder contribution, socio-economic approaches and community exposure and transboundary impacts (Aall 2019).

There remains negligence to extend capability and resilience which is a condition to manage climate change risks and it can mitigate meagre vulnerability (O'Brien et al. 2006). Mitigation of vulnerability is a vital feature in the reduction of climate change risk. Also, a growth which extends authority and resilience as a requirement to manage risks related to climatic change is needed (O'Brien et al. 2006).

About 40–50% cities would be subjected to the tests to improved handling with flooding or/and famine threats attributed to susceptible infrastructure (Dong et al. 2020). With no spatial distribution outline, the vulnerability approach at the city scale is multifactor driven and diverse, while dissimilar features of sensible and adaptive capability may govern distinctly or unnaturally (Dong et al. 2020). These outcomes specify the need of more city scale evaluations and city-specific replies of urban infrastructure. The evaluation method could act as a hands-on tool to enable more interpretation of climate vulnerabilities and advance of efficient policies for adaptation (Dong et al. 2020).

3.2 Impacts of Climate Change on Water Quality, Health, and Environment

Climatic variation can impact the quality of water resources, for instance, lakes and rivers, modifying the values of physical-chemical and biological parameters, micropollutants (Delpla et al. 2009). Climate change led to more heating and vastly varying hydrologic cycle. Climate change impacts communities as well at a vital scale and drive to drought. Drought can cause vulnerability on water quality as well (Kim et al. 2019).

Climatic change is predicted to enhance the exposure level of various environment related contaminants as straight and ancillary impacts on the utilisation and conveyance and fate of chemicals (Balbus et al. 2013). Variations in human attitude will also impact the way of contact with contaminated air, water, and food (Balbus et al. 2013). Dietary variations, psycho-social stress, and co-exposure to stressors like more heating are probably to rise the vulnerability of humans to chemicals (Balbus et al. 2013). These variations are probable to have vital implications for present exercises for chemical assessment. Prevailing monitoring approaches need to be robust enough to perceive adverse episodic alterations in acquaintance (Balbus et al. 2013). Institutions accountable to assess and manage health risks of chemicals, thus essential to be further initiative and account for the inferences of climate change for the techniques and methods (Balbus et al. 2013).

Rural drinking water sustainability under climatic variation can be viewed on the wide examination of rural water protection (Huang et al. 2020). Also, the focus is on subjects of water quality evaluations, quality and quantity of drinking water enhancement, system conservation and community managing, and making of decision in rural regions across the world (Huang et al. 2020).

Most of human health risk assessments (HHRAs) (86%) used deterministic techniques with 14% applied probabilistic and stochastic approaches (Ford et al. 2017). 47% of studies recognised no less than one literature gap in the domains of investigation, risk management, and civic acquaintance (Ford et al. 2017).

World Health Organization (WHO) prepared a Water Safety Plan (WSP), a Risk Assessment and a Risk Management (RA/RM) framework (van den Berg et al. 2019). WSP necessitates a general risk assessment to identify all perils and perilous occurrences from source to tap, while the six legally needed RA/RM respectively focus on definite risks at the next level. Each risk assessment offers information on exact perils and perilous occurrences that cover a portion of the water supply chain (van den Berg et al. 2019). Thus, legal prerequisites are accompanied with extra RA/RM events at sector and water company level like codes of practices and standard operating methods. The results of all RA/RM methods united offer data from source to tap (van den Berg et al. 2019). With the use of multiple RA/RM methods, it is critical to segment and unite evidence obtained from the various events (van den Berg et al. 2019).

3.3 Various Climatic Change Risk Assessment Studies

The multiple-hazard risk index was recommended as a pointer from various regions of Indian Himalayan Region (IHR) (Sekhri et al. 2020). Exposure, capacity of adaptation, sensitivity, and managing capacity were well-defined with inclusive and sub-local pointers recognised over additive and subtractive methods (Sekhri et al. 2020). The outcomes specified that the varying risks amongst the regions of IHR were driven by the diversity of factors like demography, facilities, usual capital, association, skill, and spatial specifics of the regions.

A framework to evaluate vulnerability of climatic change is vital with a holistic perception. It will unite drivers which can impact the geography considering socio-cultural aspects and timelines post disasters (Villegas-González et al. 2017). This framework will have the capacity to evaluate geographical attributes and develops local acquaintance to prevent, mitigate, and reply to climatic change impacts (Villegas-González et al. 2017).

Potable water systems comprise risks and risks need to be managed. Risk management can be carried with a framework methodology and well-informed decision making (Lindhe 2008). The framework needs to consist of integrated manner with best use of resources for risk mitigation (Lindhe 2008). Also, a probabilistic approach is to be considered to deal with uncertainties (Lindhe 2008). The suggested integrated management of risk in the setting of drinking water system is to be efficient to model the whole system from source to tap (Lindhe 2008). The system then offers information on levels of risks and dynamic response of the system in regard to the rate of failure and their periods (Lindhe 2008). It allows results verification with goals of performance and level of acceptance of risks. Thus, the integrated risk management approach needs to consist of a set of tools and sub-optimisation of risk mitigation choices (Lindhe 2008).

Within the European project Life+ TRUST (Tool for Regional-scale assessment of groundwater Storage improvement in adapting to climaTe variation), a Regional Risk Assessment (RRA) method was established to recognise the impact under climatic change on groundwater and related ecology, such as surface waters, agrarian areas, environment and to rank areas and receptors at risk (Pasini et al. 2012). This RRA framework was considered to comply with the Sources–Pathway–Receptor–Consequence (SPRC) approach. Related impacts on groundwater and surface waters were chosen and analysed with hazard state, revelation, susceptibility, and assessment of risk (Pasini et al. 2012).

There is a novel approach called Regional Risk Assessment (RRA)-based approach which will not just be based on the notion of risk as a blend of exposure, hazard, and vulnerability, but explains thorough (technical and concept based) forming and computation describing the features in terms of the anticipated equations and classification methods, to develop maps of risk spatial form and associated statistics differentiating the most crucial areas (hot spots of risk) (Ronco et al. 2017).

An efficient operational tool was developed which could aid decision makers to raise risk consciousness and readiness and, perhaps, to ease management of emergency. The anticipated tool is on Bayesian Belief Networks (BBN), a probabilistic method which has verified more capacity to integrate various knowledge sources (Pagano et al. 2014).

Community risk assessments (CRAs) refer to participating approaches to evaluate perils, vulnerabilities, and capabilities to assist community-based disaster risk mitigation, utilised by many NGOs, community-based institutions, and the Red Cross/Red Crescent (Van Aalst et al. 2008). CRAs can aid to address the climate change associated contests by nurturing public involvement in climatic risk mitigation (Van Aalst et al. 2008). Yet, a vital test is to retain CRAs modest for widespread use (Van Aalst et al. 2008). This needs special care to amend CRA tools in the contextual materials and trainings for CRA organisers, and in the direction to interpret CRA results (Van Aalst et al. 2008). Another test is the use of a restricted set of CRA outcomes in guiding risk mitigation in different societies and to apprise national and international adapting policies (Van Aalst et al. 2008). This needs exact consideration to sample and upkeep to scale up quality findings. Thus, robust relations are essential between institutions enabling CRAs and suppliers of climatic data to address in specific terms for the translation of climate information to the public (Van Aalst et al. 2008).

3.4 Various Climate Change Vulnerability Assessment Studies

Climatic change impacts need to be assessed in terms of assessments of vulnerability to identify, prioritise, and acclimate to risk management (Howe et al. 2013). A methodology of risk ranking leads to a participating vulnerability scoping diagram, which helps quick evaluation and amalgamation of various participant perceptions for vulnerability. This methodology is on qualitative contextual ranking of risk as a first step in assessing vulnerability (Howe et al. 2013).

Another novel tool was developed that can transform theoretical perception (vulnerability of water resources (VWR)) to a framework of operative under varying heating and precipitation, and population change (smaller level) (Anandhi and Kannan 2018). This device had three phases and used a novel system thoughtful method. Stage-1: Transforming theoretic notion to features recognised; Stage-2: Operationalising features to methodology in VWR; Stage-3: Utilisation of the procedure to develop a concept-based modelling instrument for VWR: WR-VISTA (Water Resource Vulnerability assessment conceptual model using Indicators chosen by System's Thinking Approach) (Anandhi and Kannan 2018). The specific innovations were: (1) The key attributes in VWR were recognised in Stage-1 (aiming system, scheme elements, scale, level of details, data sources, framework, and indicators); (2) WR-VISTA united two vulnerability assessment contexts: the European's Driver–Pressure–State–Impact–Response framework (DPSIR) and the Intergovernmental Panel on Climate Change's framework (IPCC's); and (3) utilised system thoughtful methods in VWR for pointer choice (Anandhi and Kannan 2018).

The Vulnerability Scoping Diagram presents to provide the relation of assessments with diverse methods. The diagram is depicted with latest study on famine vulnerabilities, highlighting that mutual perceptions into vulnerability may arise if distinct study methods use a similar structure to organise data about exposure, sensibility, and adaptation capacity with difference between assessments. This approach comprises "Eight Steps" methodical protocol, will permit a vulnerability meta-analysis, the messages with which may allow places to recognise useful mitigation or adaptation or choices with no mandate to conduct at the start their particular vulnerability assessments (Polsky et al. 2007).

Water vulnerability can be assessed using parameters, such as exposure, sensibility, and adaptive capability (Chauhan et al. 2021). Edaphic (aspect, elevation) and climatic (land surface heating) with few water infrastructure (distance to water delivery) and people can be considered to develop vulnerability index with Analytical Hierarchy Process to assign weights (Chauhan et al. 2021). These pointers are modest to derive with secondary causes and can consider the variability at micro-level. Moreover, the present adapting approaches for water safety can also be obtained by organising surveys in a random manner (Chauhan et al. 2021). The survey outcomes that the adaptation approach must be manageable at specific and institution level (Chauhan et al. 2021). Policy techniques like optimal usage of water, grey water recycle, spring rejuvenating, stormwater harvesting, and leak proof set-up with involvement of novel methods may be adapted to mitigate the water vulnerability in urban areas with community involvement (Chauhan et al. 2021).

A model is developed which works as a decision support system (DSS) to direct users with a participant-based evaluation of vulnerability of water systems for the community (Coletti et al. 2016). The DSS unified data sources as an online atmosphere which enables professed risks that are well-defined and ordered in a qualitative way by users which can be related against the impacted the former incidents experienced by the community (Coletti et al. 2016). A model with a prototype design apt for semantic web claims in which the numerous elements are linked to make DSS more efficient (Coletti et al. 2016).

The Vulnerability and Risk Assessment (VRA) aims to activate a sagacity of authorisation and alliance among stakeholders (Morchain and Kelsey 2016). This is an intricate procedure in which chaos can ascend. The VRA method invites this and responds to it with candidness and investigation, while endorsing contribution from root level (Morchain and Kelsey 2016).

An emphasis on water security is a prior adaptation strategy to deliver instant profits to vulnerable and unexpected population (Sadoff and Muller 2009). It advances the Millennium Development goals, though solidification methods and capacity for long term climate risk management (Sadoff and Muller 2009).

A framework for climate change risk assessment as GIS-based, modelling of geographical variance in both climate and asset susceptibility in a country is described (Hawchar et al. 2020). This allows for the recognition of plausible climatic change risk hotspots crossways a variety of critical infrastructure segments (Hawchar et al. 2020). The approach permits to develop multi-sector semi-quantitative maps of risk ranking that count the geographical vicinities of many assets from diverse critical infrastructure domains which are susceptible to a definite climatic peril (Hawchar et al. 2020).

The Freshwater Vulnerability Index (FVI) raised to a range of 0.38–0.44 signifying a move from modest to high vulnerability area and the results also designate that the country is probable to arrive at a water scarceness zone (water availability of <1000 m^3 /capita) by 2030 and experience overutilisation of resources of water (water utilisation rate > 100%) by 2040 (Boojhawon and Surroop 2021). The resource stress group exhibited the maximum influence on the computed FVI with a portion of share of 33–38% to the present FVI, 34–38% to the projected FVI for a as normal scenario and 40–43% to the projected FVI under the impacts of climate change (Boojhawon and Surroop 2021).

A holistic interpretation of vulnerability with a united human-environment perspective is needed under the impacts of climatic change (Karmaoui et al. 2014). Eight various indicators are recognised, grouped into four types, about land features, service inadequacies for users attributed to infrastructure letdown, conduit path features, and physical features of aqueduct pipes (Borzì 2023). Also, a graphical depiction is projected with the Kiviat chart, developing a vulnerability chart which presents an effective tool to find the key vulnerability aspects in prevailing water supply set-up, to manage interventions, to plan and design procedures of new infrastructure, and to compare diverse design results (Borzì 2023).

A robust framework is needed for the evaluation of risk and enhancing the resilience of critical infrastructure (C.I) on the principles of elasticity, variety, and work ecosystem, with short-range and enduring impacts of climatic risk (Kumar et al. 2021).

Climatic change enhances the seasonal risks of water supply uncertainty, thus increases socioeconomic and public health risks. With deep uncertainty, a pattern variation to assess climate impacts on water supply is required (Leveque et al. 2021). When concerned authorities resolve the policies to secure and enhance water supply under climate change, the existing and probable conflicts, the local water requirement, the influence to the city's greenhouse gas releases and forthcoming variations in water obtainability are aspects to consider (Martinez et al. 2015).

Stressors under climate change imply a stern menace to the societies and ecology in the basin and have intensified their prevailing vulnerabilities (Divya Mohan and Shirish Sinha 2015). There is a necessity to design and execute proper adaptation approaches to aim at the maximum vulnerable societies and ecosystem (Divya Mohan and Shirish Sinha 2015).

An integrated assessment method to assess impacts of climate change by applying chosen assessment devices and methods, comprising Bayesian Networks, Multiple Criteria Analysis, Impact Matrix, Problem Tree Analysis and Participatory Rural Appraisal (PRA) is useful to perform Climate Change Vulnerability Assessment (CCVA) at the local level (Noi and Nitivattananon 2015).

Integrated analytical framework assists and analyses vulnerability to climatic variation, while recognising and comparing adaptation choices that could lessen vulnerability, with four wide steps: (i) find probable level of exposure to climatic change, and how climate change might interact along with prevailing stresses and other forthcoming drivers of variation; (ii) assess the sensitivity of stocks of capital assets and flows of ecological amenities to climatic variation; (iii) recognise factors influencing verdicts to build and/or adopt various adapting policies on innovation or the use/substitution of prevailing assets; and (iv) recognise and determine plausible trade-offs among adapting choices (Reed et al. 2013).

A surrogate model on a machine learning method and works on Gaussian process regression was developed with climate and water supply data from various water supply companies (Starcevic 2023). Also, forthcoming water demand was assessed with climate projected data from various scenarios. An early warning system was prepared to find upcoming water balance and the countermeasures determination and depiction (Starcevic 2023).

Coastal City Flood Vulnerability Index, CCFVI provides a wide synopsis of flood vulnerability and the effect of plausible adapting choices (Balica et al. 2012). Information and familiarity of vulnerability is a valuable primary stage in the lengthy procedure of forming climatic adapting and resilient paths in the framework of sustainable urban growth (Sanchez Rodriguez and Morales Santos 2018).

4 Comparison and Evaluation of Various Methodologies

The lack of drinking water is the prime concern which has become sterner due to latest drought attributed to climatic change and poor water management (Cho 2020). Further, the eruption of diseases due to more heating is recognised as a bothering concern (Cho 2020). Risk assessment needs to be integral for action and planning to minimise the vulnerability from climate change (Cho 2020). It is essential to aid education and cognisance in developing countries to accomplish assessment of risk well and advance both problem-solving and policy-creation capabilities to adapt to a change of climate (Cho 2020).

The anticipated impacts of climatic variation on potable water quantity and quality need to be evaluated (Delpla et al. 2009). Degradation trends of drinking water quality relating to climatic change drives to enhanced risk conditions in association with probable health impacts (Delpla et al. 2009).

Risk associated with water quality attributed to climate change can translate from watershed characteristics. An approach is proposed to assess drought on risk of water quality and to observe drought with the occurrence of exceeding the water quality target with the intent of a resilient response of upbeat kind (Kim et al. 2019).

A catchment risk management programme(s) established were evaluated in terms of key elements, data requirements, and review of prevailing methods. This risk management, also, was applied to ecology and human health risk assessment with Hazard Analysis and Critical Control Points (HACCP) (Miller et al. 2009). It was found that contaminant budgeting, cause/effect relationships and size of catchment, landscape and coverage of information played a vital role in water quality risk management (Miller et al. 2009).

The proposed framework details the mitigation of risk and handling strategies (on adjusting internal and external strengths) to address risks (Sekhri et al. 2020). Risk mitigation was suggested to achieve with the resilience of habitation, natural capital augmentation, external associations, adaptation to climate change, and technologic intervention (Sekhri et al. 2020).

Climatic variation is anticipated to impact adversely on water resources accessibility, water need is anticipated to upsurge. Problems of water quality will be exaggerated, specifically attributed to land use variations and salt water intrusion (Kanakoudis et al. 2017).

The worldwide vulnerability of the system enumerates hydrologic, environmental, monetary, and social susceptibilities. The outcomes attained display that the anticipated procedure may offer a novel method to analyse susceptibility in semi-arid areas. Further, it made it conceivable to analyse and launch that the utmost present and forthcoming vulnerabilities of the scheme are the consequence of events in agrarian extents and urban areas (Orozco et al. 2020).

Climate change-induced vulnerability assessment need to frame all-inclusive adapting strategies at the policy- and decision-developing levels to aid the societies that deal with the adversative impacts of climate variation (Bera et al. 2022).

5 Critical Remarks

Change of climate, present demographic trends, monetary growth and associated land use variations have straight impacts on the growing need for freshwater resources (Gain et al. 2012). Lack of institutional harmonisation and prospects such as adaptation are the key constraints for vulnerability assessment linked to water resources (Gain et al. 2012).

A schematic framework is useful to characterise risk attributes related to climatic change especially on water resources at small to catchment scales (Sharma and Bharat 2009). This type of framework can generate concepts and leaves an open end to carry aspects to next research and application (Sharma and Bharat 2009).

Vulnerability assessment varies and depends on population, land use, and community sensitivity drivers for exposure, sensitivity, and adaptive capability system (Howe et al. 2013).

Existing methods useful to evaluate climatic change risks frequently deal specific risks distinctly and will not achieve an all-inclusive depiction of cumulative outcomes allied to various hazards (Terzi et al. 2019). Also, revolutionary multiple-layer single risk assessment (dealing various threats) is now in wide use, leading disingenuous evaluation of multiple risk process (Terzi et al. 2019). A review of five cutting-edge modelling techniques (agent-based models, Bayesian networks, system dynamic models, event and fault trees, and hybrid models), is desirable to appraise their plausible applications for multiple risk assessments and climate change adaptations in hilly regions (Terzi et al. 2019).

Also, when risk assessment is to be carried, it is essential to study natural hazards on a territorial basis (Villegas-González et al. 2017). Stakeholder involvement assists in the design of risk management with due consideration of territorial characteristics to enhance local knowledge to appraise vulnerability which can augment support mechanisms to enhance local resilience (Villegas-González et al. 2017).

A huge-scale, quick risk screening of raw water quality on catchment response to pressures as precondition to an added planned insertion of incipient risks in water resource and ecological management is needed (Vorstius et al. 2019). Forms and spatial dependences were examined with principal component analysis, idleness and cluster analyses, and regression trees (Vorstius et al. 2019). These statistical methods specify the collaboration between intrinsic catchment biophysical features, land use and climate in featuring water quality risks and recognise the emphasis to prioritise catchment interferences and mitigation of risk in the upcoming (Vorstius et al. 2019).

Risk assessment and risk management (RARM) of drinking water management including small drinking water supplies need to set aim to (i) recognise permitting environments and methods which aid execution and approval of RARM, (ii) suggest techniques to control data for better usage in RARM, (iii) develop a field- and desk-based tool to assist RARM, and (iv) distribute suggestions via journal publications (Herschan 2023).

The two tools built (a sanitary inspection form for household water treatment and an interaction matrix model) are the first of their type for this claim and are key, hands-on outputs of the activity (Herschan 2023). RARM in general organise constraints of data disintegration, methodological and economic constraints.

Climate change adaptation uncertainty can be characterised on level, source, and nature. Most of the uncertainties on impacts of climate change adaptation are epistemic (decreasable) by nature but not certainties on adapting methods are intricate, with indistinctness frequently being additional to impact not certainties (Refsgaard et al. 2013). Strategies to handle with no certainty in climatic change adapting should mimic the kind of the uncertainty basis and how they relate with risk level and decision forming: (i) epistemic noncertainties can be condensed by acquisition of more acquaintance; (ii) noncertainties associated to vagueness can be lessened by discussion and information division among the various participants; and (iii) aleatory noncertainty is, by its kind, non-decreasable (Refsgaard et al. 2013). The noncertainty cascade comprises various sources and their dissemination with methodical and socio-monetary models may augment significantly to forecast noncertainties, though they may also withdraw each other (Refsgaard et al. 2013). Thus, huge noncertainties can have slight outcomes to make decision even, because multi sources of information afford adequate information to validate deed in climate change adaptation (Refsgaard et al. 2013).

6 Future Research Needs

Multi-risk assessment is to be assessed on discrete criteria, such as dynamics of spatio-temporal nature, management of uncertainty, assessment of cross-sector, integration of adaptation amounts, data and degree of complexity (Terzi et al. 2019). Inadequate application of the chosen methods needs to address the climate and risks contest in hilly regions (Terzi et al. 2019). In specific, system dynamic and hybrid models should establish more capacity to characterise climate variation impacts on multiple-risk procedures for an efficient execution of climatic variation strategies for adaptation (Terzi et al. 2019).

Vital queries on the distinct aspects of multiple risk assessments and the existing tools and methodologies need to be dealt with in future (Terzi et al. 2019). Review and analysis outcome of various literature on risk analysis related to treatment of water systems and their risk assessment methods are as mentioned below and need further research in future (1) The methodical basis of assessment and management of risk is yet an ongoing subject; (2) principles, theories, and base methods related to systems of water supply are in unremitting development, prevailing devices are apt, and an increasing number of uses are obtainable and of prodigious attention; and (3) analysis of risk methods are in the voyage to advance the essential wide technical, society, and political receiving in the sector of water treatment (Kombo Mpindou et al. 2022).

7 Conclusions

Climate change induced vulnerability, impacts, risk evaluation and risk management methodologies are reviewed to the extent available literature. This review has performed especially considering drinking water, water resources and water quality aspects. There are various methods, techniques and approaches proposed and being implemented in various parts of the world. However, one common aspect that can be perceived is an integrated frame work that needs to be implemented and fulfilled to cater to all the requirements to deal with various vulnerabilities, associated risk assessment and management for efficient climate change adaptation strategies. However, with more region specific, accurate data, more impacting problems and key constraints that are governing capacity of any existing and future projected systems to cope with climate change need to be developed in future.

8 Discussion Questions

1. Are vulnerability, risk assessment, and risk management to be dealt with separately or combinedly in connection with climate change?
2. Whether adaptation or mitigation is the strategy to carry to deal with climatic variation or not?
3. What are the key drivers for climate change mitigation strategies?

Competing Interests This study was funded by [X] [grant number X].

References

Aall C (2019) Methods for climate change risk assessments: an. Nat Hazards Earth Syst Sci 19(10):2117–2139. https://www.vestforsk.no/sites/default/files/2021-01/D1.1-knowledge%20review%20final2%20%281%29.pdf

Anandhi A, Kannan N (2018) Vulnerability assessment of water resources–translating a theoretical concept to an operational framework using systems thinking approach in a changing climate: case study in Ogallala aquifer. J Hydrol 557:460–474. https://doi.org/10.1016/j.jhydrol.2017.11.032

Balbus JM, Boxall AB, Fenske RA, McKone TE, Zeise L (2013) Implications of global climate change for the assessment and management of human health risks of chemicals in the natural environment. Environ Toxicol Chem 32(1):62–78. https://doi.org/10.1002/etc.2046

Balica SF, Wright NG, Van der Meulen F (2012) A flood vulnerability index for coastal cities and its use in assessing climate change impacts. Nat Hazards 64:73–105. https://doi.org/10.1007/s11069-012-0234-1

Bera A, Meraj G, Kanga S, Farooq M, Singh SK, Sahu N, Kumar P (2022) Vulnerability and risk assessment to climate change in Sagar Island, India. Water 14(5):823. https://doi.org/10.3390/w14050823

Boojhawon A, Surroop D (2021) Impact of climate change on vulnerability of freshwater resources: a case study of Mauritius. Environ Dev Sustain 23:195–223. https://doi.org/10.1007/s10668-019-00574-3

Borzì I (2023) Vulnerability assessment of water supply infrastructures through multiple indicator methodology. J Water Clim Change 14(11):3967–3984. https://doi.org/10.2166/wcc.2023.148

Brooks N (2003) Vulnerability, risk and adaptation: a conceptual framework. Tyndall Centre for climate change research. Working paper 38(38):1–16. https://nickbrooksconsultancy.wordpress.com/wp-content/uploads/2012/01/tynwp38.pdf

Calow R, Bonsor H, Jones L, O'Meally S, MacDonald A, Kaur N (2011) Climate change, water resources and WASH: a scoping study. Overseas Development Institute. https://nora.nerc.ac.uk/id/eprint/15236

Carpignano A, Golia E, Di Mauro C, Bouchon S, Nordvik JP (2009) A methodological approach for the definition of multi-risk maps at regional level: first application. J Risk Res 12:513–534. https://doi.org/10.1080/13669870903050269

Chauhan D, Thiyaharajan M, Pandey A, Singh N, Singh V, Sen S, Pandey R (2021) Climate change water vulnerability and adaptation mechanism in a Himalayan City, Nainital, India. Environ Sci Pollut Res Int:1–18. https://doi.org/10.1007/s11356-021-15713-5

Cho H (2020) Climate change risk assessment for Kurunegala, Sri Lanka: water and heat waves. Climate 8(12):140. https://doi.org/10.3390/cli8120140

Coletti A, De Nicola A, Villani ML (2016) Building climate change into risk assessments. Nat Hazards 84(2):1307–1325. https://doi.org/10.1007/s11069-016-2487-6

Delmonaco G, Margottini C, Spizzichino D (2006) New methodology for multi-risk assessment and the harmonization of different natural risk maps. Deliverable 3(1), Armonia project: Rome, Italy. https://www.researchgate.net/publication/317957266_New_methodology_for_multi-risk_assessment_and_the_harmonisation_of_different_natural_risk_maps

Divya Mohan, Shirish Sinha (2015) Assessing vulnerability to climate change in the Ganges Basin using a combined macro- and micro-level approach. Clim Dev. https://doi.org/10.1080/17565529.2015.1064810

Dong X, Jiang L, Zeng S, Guo R, Zeng Y (2020) Vulnerability of urban water infrastructures to climate change at city level. Resour Conserv Recycl 161:104918. https://doi.org/10.1016/j.resconrec.2020.104918

Eckstein D, Künzel V, Schäfer L (2018) Who suffers Most from extreme weather events? Weather-related loss events in 2016 and 1997 to 2016. Global climate risk index, Germanwatch, Germany. https://www.germanwatch.org/sites/default/files/publication/20432.pdf

Ford L, Bharadwaj L, McLeod L, Waldner C (2017) Human health risk assessment applied to rural populations dependent on unregulated drinking water sources: a scoping review. Int J Environ Res Public Health 14(8):846. https://doi.org/10.3390/ijerph14080846

Gain AK, Giupponi C, Renaud FG (2012) Climate change adaptation and vulnerability assessment of water resources systems in developing countries: a generalized framework and a feasibility study in Bangladesh. Water 4(2):345–366. https://doi.org/10.3390/w4020345

Hawchar L, Naughton O, Nolan P, Stewart MG, Ryan PC (2020) A GIS-based framework for high-level climate change risk assessment of critical infrastructure. Clim Risk Manag 29:100235. https://doi.org/10.1016/j.crm.2020.100235

Herschan JB (2023) Techniques and tools to support the practical implementation of risk assessment and risk management in small drinking-water supplies (Doctoral dissertation, University of Surrey). https://doi.org/10.15126/thesis.900808

Howe PD, Yarnal B, Coletti A, Wood NJ (2013) The participatory vulnerability scoping diagram: deliberative risk ranking for community water systems. Ann Assoc Am Geogr 103(2):343–352. https://doi.org/10.1080/00045608.2013.754673

Huang WW, Chen XJ, Fan YR, Li YP (2020) Management of drinking water source in rural communities under climate change. J Environ Inf 39(2):136–151. https://doi.org/10.3808/jei.202000431

Delpla I, Jung AV, Baures E, Clement M, Thomas O (2009) Impacts of climate change on surface water quality in relation to drinking water production. *Environ Int 35*(8):1225–1233. https://doi.org/10.1016/j.envint.2009.07.001

Iglesias A, Garrote L, Flores F, Moneo M (2007) Challenges to manage the risk of water scarcity and climate change in the Mediterranean. Water Resour Manag 21:775–788. https://doi.org/10.1007/s11269-006-9111-6

IPCC (2014) Intergovernmental panel on climate change–climate change 2014, impacts, adaptations, and vulnerability. Summary for policy makers. Working group II contribution of the fifth assessment report to the Intergovernmental Panel on Climate Change. https://www.ipcc.ch/report/ar5/wg2/

Kanakoudis V, Tsitsifli S, Papadopoulou A, Cencur Curk B, Karleusa B (2017) Water resources vulnerability assessment in the Adriatic Sea region: the case of Corfu Island. Environ Sci Pollut Res 24:20173–20186. https://doi.org/10.1007/s11356-017-9732-8

Karmaoui A, Messouli M, Ifaadassan I, Khebiza MY (2014) A multidisciplinary approach to assess the environmental vulnerability at local scale in context of climate change (pilot study in upper Draa Valley, South Morocco). Glob J Technol Optim 6(1). https://doi.org/10.4172/2229-8711.1000167

Kim JS, Jain S, Lee JH, Chen H, Park SY (2019) Quantitative vulnerability assessment of water quality to extreme drought in a changing climate. Ecol Indic 103:688–697. https://doi.org/10.1016/j.ecolind.2019.04.052

Kombo Mpindou GOM, Escuder Bueno I, Chordà Ramón E (2022) Risk analysis methods of water supply systems: comprehensive review from source to tap. Appl Water Sci 12(4):56. https://doi.org/10.1007/s13201-022-01586-7

Kumar N, Poonia V, Gupta BB, Goyal MK (2021) A novel framework for risk assessment and resilience of critical infrastructure towards climate change. Technol Forecast Soc Change 165:120532. https://doi.org/10.1016/j.techfore.2020.120532

Leveque B, Burnet JB, Dorner S, Bichai F (2021) Impact of climate change on the vulnerability of drinking water intakes in a northern region. Sustain Cities Soc 66:102656. https://doi.org/10.1016/j.scs.2020.102656

Lindhe A (2008) Integrated and probabilistic risk analysis of drinking water systems. https://publications.lib.chalmers.se/records/fulltext/74243.pdf

Martinez S, Kralisch S, Escolero O, Perevochtchikova M (2015) Vulnerability of Mexico City's water supply sources in the context of climate change. J Water Clim Change 6(3):518–533. https://doi.org/10.2166/wcc.2015.083

Miller R, Guice J, Deere D (2009) Risk assessment for drinking water sources. Research Rep 78. https://kh.aquaenergyexpo.com/wp-content/uploads/2023/12/Risk-Assessment-for-Drinking-Water-Sources.pdf

Morchain D, Kelsey F (2016) Finding ways together to build resilience: The vulnerability and risk assessment methodology. https://oxfamilibrary.openrepository.com/bitstream/handle/10546/593491/www.oxfam.de?sequence=1

Noi LVT, Nitivattananon V (2015) Assessment of vulnerabilities to climate change for urban water and wastewater infrastructure management: case study in Dong Nai river basin, Vietnam. *Environ Dev* 16:119–137. https://doi.org/10.1016/j.envdev.2015.06.014

O'Brien G, O'keefe P, Rose J, Wisner B (2006) Climate change and disaster management. Disasters 30(1):64–80. https://doi.org/10.1111/j.1467-9523.2006.00307.x

Orozco I, Martínez A, Ortega V (2020) Assessment of the water, environmental, economic and social vulnerability of a watershed to the potential effects of climate change and land use change. Water 12(6):1682. https://doi.org/10.3390/w12061682

Pagano A, Giordano R, Portoghese I, Fratino U, Vurro M (2014) A Bayesian vulnerability assessment tool for drinking water mains under extreme events. Nat Hazards 74:2193–2227. https://doi.org/10.1007/s11069-014-1302-5

Pasini S, Torresan S, Rizzi J, Zabeo A, Critto A, Marcomini A (2012) Climate change impact assessment in Veneto and Friuli plain groundwater. Part II: a spatially resolved regional risk assessment. Sci Total Environ 440:219–235. https://doi.org/10.1016/j.scitotenv.2012.06.096

Polsky C, Neff R, Yarnal B (2007) Building comparable global change vulnerability assessments: the vulnerability scoping diagram. Glob Environ Chang 17(3–4):472–485. https://doi.org/10.1016/j.gloenvcha.2007.01.005

Reed MS, Podesta G, Fazey I, Geeson N, Hessel R, Hubacek K, Letson D, Nainggolan D, Prell C, Rickenbach MG, Ritsema C (2013) Combining analytical frameworks to assess livelihood vulnerability to climate change and analyse adaptation options. Ecol Econ 94:66–77. https://doi.org/10.1016/j.ecolecon.2013.07.007

Refsgaard JC, Arnbjerg-Nielsen K, Drews M, Halsnæs K, Jeppesen E, Madsen H, Markandya A, Olesen JE, Porter JR, Christensen JH (2013) The role of uncertainty in climate change adaptation strategies—a Danish water management example. Mitig Adapt Strateg Glob Chang 18(2013):337–359. https://doi.org/10.1007/s11027-012-9366-6

Ronco P, Zennaro F, Torresan SILVIA, Critto A, Santini M, Trabucco A, Zollo AL, Galluccio G, Marcomini A (2017) A risk assessment framework for irrigated agriculture under climate change. Adv Water Resour 110:562–578. https://doi.org/10.1016/j.advwatres.2017.08.003

Sadoff C, Muller M (2009) Water management, water security and climate change adaptation: early impacts and essential responses. Global Water Partnership, Stockholm. https://hdl.handle.net/10535/4999
Sanchez Rodriguez RA, Morales Santos AE (2018) Vulnerability assessment to climate variability and climate change in Tijuana, Mexico. *Sustainability* 10(7):2352. https://doi.org/10.3390/su10072352
Sekhri S, Kumar P, Fürst C, Pandey R (2020) Mountain specific multi-hazard risk management framework (MSMRMF): assessment and mitigation of multi-hazard and climate change risk in the Indian Himalayan region. Ecol Indic 118:106700. https://doi.org/10.1016/j.ecolind.2020.106700
Sharma D, Bharat A (2009) Conceptualizing risk assessment framework for impacts of climate change on water resources. Curr Sci:1044–1052. http://www.jstor.org/stable/24104596
Starcevic I (2023) Risk analysis of climate change impacts on the quantitative drinking water supply. https://odr.chalmers.se/handle/20.500.12380/305948
Terzi S, Torresan S, Schneiderbauer S, Critto A, Zebisch M, Marcomini A (2019) Multi-risk assessment in mountain regions: a review of modelling approaches for climate change adaptation. J Environ Manag 232:759–771. https://doi.org/10.1016/j.jenvman.2018.11.100
Van Aalst MK, Cannon T, Burton I (2008) Community level adaptation to climate change: the potential role of participatory community risk assessment. Glob Environ Chang 18(1):165–179. https://doi.org/10.1016/j.gloenvcha.2007.06.002
van den Berg HH, Friederichs L, Versteegh JF, Smeets PW, de Roda Husman AM (2019) How current risk assessment and risk management methods for drinking water in The Netherlands cover the WHO water safety plan approach. Int J Hyg Environ Health 222(7):1030–1037. https://doi.org/10.1016/j.ijheh.2019.07.003
Villegas-González PA, Ramos-Cañón AM, González-Méndez M, González-Salazar RE, De Plaza-Solórzano JS (2017) Territorial vulnerability assessment frame in Colombia: disaster risk management. Int J Disaster Risk Reduct 21:384–395. https://doi.org/10.1016/j.ijdrr.2017.01.003
Vorstius C, Rowan JS, Brown I, Frogbrook Z, Palarea-Albaladejo J (2019) Large-scale risk screening of raw water quality in the context of drinking water catchments and integrated response strategies. Environ Sci Pol 100:84–93. https://doi.org/10.1016/j.envsci.2019.05.012

Chapter 11
On the Application of the Statistical and Fuzzy Systems Regression and Clustering to Analyze the Multivariate Chemical Composition of Groundwater: Case Study of Arsenic-Contaminated Groundwater in Bangladesh

Boris Faybishenko and Seth H. Frisbie

1 Learning Objectives

1. Identify the goal(s) of the application of traditional statistical analysis and fuzzy systems modeling methods in analyzing the chemical composition of water samples;
2. Demonstrate the application of the statistical hierarchical analysis, fuzzy regression, and fuzzy clustering to the arsenic concentration data collected by Frisbie et al. in 1997 during a national-scale survey of arsenic in Bangladesh (Frisbie et al. 2024); and
3. Demonstrate that fuzzy systems analysis can be considered as a complementary approach to statistical analysis when dealing with uncertain or qualitative information.

This chapter uses data provided by the co-author, Seth H. Frisbie, and published in PLoS ONE under the terms of the Creative Commons Attribution 4.0 International License (http://creativecommons.org/licenses/by/4.0/). The analysis of this dataset using fuzzy regression and clustering is unique and has not been published elsewhere.

B. Faybishenko (✉)
Lawrence Berkeley National Laboratory, Berkeley, CA, USA
e-mail: bafaybishenko@lbl.gov

S. H. Frisbie
Department of Chemistry and Biochemistry, College of Arts and Sciences, Norwich University, Northfield, VT, USA
e-mail: sfrisbie@norwich.edu

A. R. Molla et al. (eds.), *Climate Change, Drinking Water Security, and Public Health*, https://doi.org/10.1007/978-3-032-12074-8_11

2 Introduction

Groundwater is becoming increasingly scarce in Bangladesh due to climatic, hydrological, and socio-economic changes (Fahim et al. 2023). According to the new version of the Koppen-Geiger climate classification (Beck et al. 2018), the north-western part of Bangladesh is located within the temperate, dry winter, monsoon influenced hot summer zone (Cwa), the central part is within the tropical savannah zone (Aw), and the north-eastern and south-eastern parts are in the tropical monsoon zone (Am) (Fahim et al. 2023).

Bangladesh's groundwater contains significant concentrations of arsenic (As), which poses a significant health risk for over half the population (Khalequzzaman et al. 2005). Approximately 97% of Bangladeshis drink well water, despite the concentration of 49% of water exceeding the 10 micrograms per liter (μg/L) World Health Organization (WHO) drinking water guideline for arsenic.

Researchers have explored various approaches to address this issue. In 1997, Frisbie et al. collected drinking water well samples during a national-scale survey of arsenic contamination in Bangladesh (Frisbie et al. 1999); these results were published in 2024 (Frisbie et al. 2024). The dissolved oxygen concentration, oxidation-reduction potential (ORP), specific conductance, pH, and temperature were measured at sampling with calibrated portable electronic sensors. The arsenic concentration was measured in a laboratory using the silver diethyldithiocarbamate method. Based on these results, Frisbie et al. (2024) hypothesized that the consequences of climate change could increase the release of arsenic from sediments into Bangladesh's drinking well water. The dissolved oxygen concentration and oxidation-reduction potential results suggested that arsenic is released from sediments into Bangladesh's drinking well water by reduction. In addition, the specific conductance and pH results also suggested that arsenic is released from sediments into Bangladesh's drinking well water by the salt effect, an increase in the solubility of an ionic solid by dissolved ions that are different from those making up the solid. Rising sea levels, more extensive flooding, and more extreme cyclones in Bangladesh can cause the underlying aquifer to become more reduced and saline. These changes may increase the release of arsenic from sediments and into Bangladesh's drinking well water. In their study, Frisbie et al. (2024) used linear regression analysis to examine how the independent variables, dissolved oxygen (measured in milligrams per liter, mg/L), oxidation-reduction potential (measured in millivolts, mV), specific conductance (measured in microsiemens per centimeter, μS/cm), pH, and temperature (measured in Celsius, °C), affect the dependent variable, arsenic concentration (measured in μg/L). Further research is essential to fully understand the complex relationships between chemical changes in coastal aquifers caused by rising sea levels and increased arsenic release from sediments into Bangladesh's drinking water wells. Understanding this concept is essential for creating effective strategies to mitigate the issue.

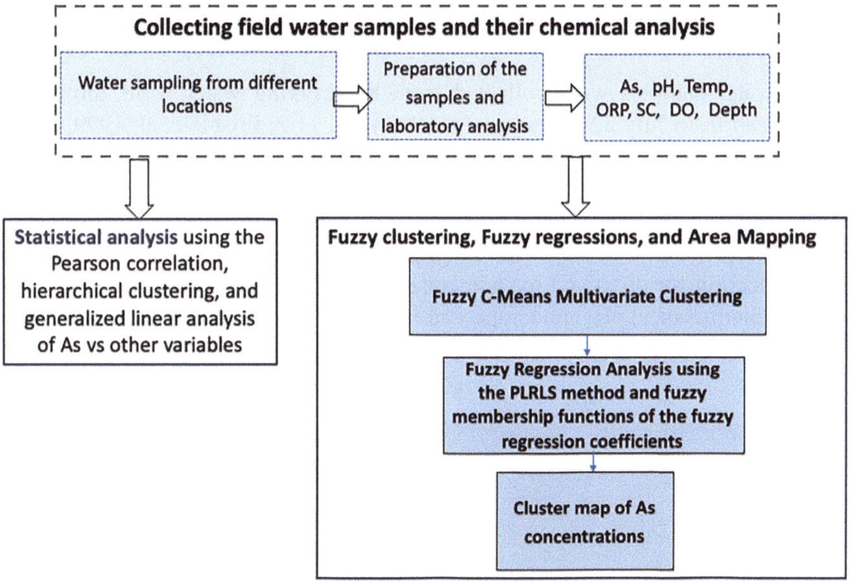

Fig. 11.1 A flowchart illustrating the methodology of combining traditional statistical analysis with fuzzy systems analysis

This chapter aims to demonstrate how combining traditional statistical and fuzzy systems analyses can describe the relationships between arsenic concentration and other variables derived from the chemical analysis of water samples collected by Frisbie et al. (2024). Additionally, it will explain how to perform fuzzy clustering zonation of the studied area in Bangladesh. The general approach to the methodology is illustrated in the flowchart in Fig. 11.1.

The structure of this chapter is organized as follows:

Section 3 describes the groundwater sampling campaign in Bangladesh and summarizes the statistical analysis presented in the paper by Frisbie et al. (2024).

Section 4 presents the results of the statistical analysis conducted in this chapter. It encompasses the findings of the data's Quality Assurance and Quality Control (QA/QC) and the application of traditional statistical correlation, Principal Component Analysis (PCA), and hierarchical clustering.

Section 5 presents the outcomes of the fuzzy clustering analysis and fuzzy regression derived from the fuzzy systems analysis.

Section 6 offers the conclusions drawn from the study.

The original and QA/QC-ed dataset files and various files demonstrating the calculation results are attached. Five Excel CSV files with input data used for calculations and resulting output data are stored at Google Drive at https://drive.google.com/drive/folders/1NsP-imLoGcz3plZecMJWSPBIOFPtg_HZ

3 Groundwater Sampling

Groundwater samples were collected from 83 drinking water wells throughout Bangladesh from July 22 to August 14, 1997 (Fig. 11.2; Frisbie et al. 1999, 2024). The wells were screened at reported depths of 4.6–370 m (15–1200 ft) below ground surface (bgs). The collection, preservation, and storage techniques ensured that each sample represented groundwater quality (APHA 1995; Stumm and Morgan 1981). Samples were analyzed with calibrated portable electronic sensors immediately after collection for dissolved oxygen concentration, oxidation-reduction potential, specific conductance, pH, and temperature. Dissolved oxygen was measured with a Hanna HI 9143 m. Oxidation-reduction potential, pH, and temperature were

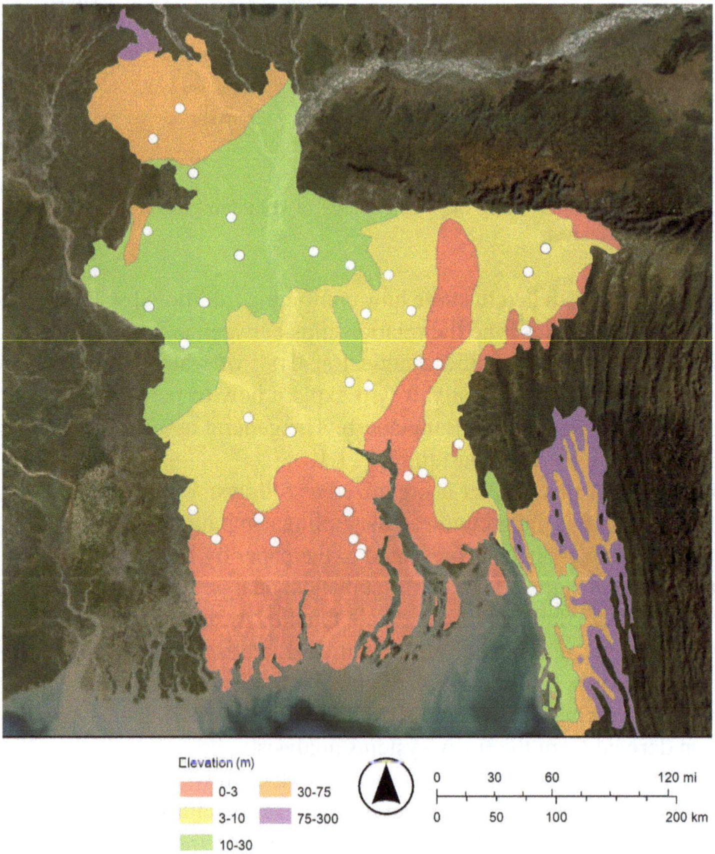

Fig. 11.2 Map of Bangladesh showing the locations of collecting groundwater samples from drinking water wells. The color legend illustrates the ground surface elevation (Frisbie et al. 2024) [Reproduced from Frisbie et al. 2024, Figure 1. https://doi.org/10.1371/journal.pone.0295172, licensed under the terms of the Creative Commons Attribution 4.0 International License (http://creativecommons.org/licenses/by/4.0/)]

measured with Cole-Parmer Digi-Sense meters. Specific conductance was measured with a Cole-Parmer Model 19,815 m. Figure 11.2 displays a map of Bangladesh that highlights the locations where groundwater samples were collected from drinking water wells. Areas with lower ground surface elevations are more susceptible to annual flooding from rivers and saltwater intrusion from the Bay of Bengal, affecting over 21% of Bangladesh's land during a typical monsoon season.

Subsamples collected for the laboratory analysis were preserved by acidification to pH < 2 with 18.6% (weight/weight) nitric acid (HNO_3) and stored in ice-packed coolers for transportation to the laboratory. The temperature of all laboratory subsamples was maintained at 0–4 °C until immediately before analysis. The concentration of arsenic was measured at the laboratory of the International Centre for Diarrhoeal Disease Research, Bangladesh, using the silver diethyldithiocarbamate method (APHA 1995). In their paper, Frisbie et al. (2024) applied a linear regression to determine the effect of the dependent variable arsenic (As) concentration in micrograms per liter (μg/L) on the independent variables: dissolved oxygen (DO) in milligrams per liter (mg/L), oxidation-reduction potential (ORP) in millivolts (mV), specific conductance (SC) in microsiemens per centimeter (μS/cm), pH, and temperature (°C). The first four regressions (As on DO, As on ORP, As on SC, and As on pH) each had a statistically significant 95% confidence level. To ensure the integrity of these findings, Benjamini and Hochberg's False Discovery Rate (FDR) method was applied to adjust for the potential of falsely rejecting null hypotheses due to multiple comparisons. This effective method starts by arranging all *p*-values from the hypothesis tests in ascending order. Then, each p-value is compared to a calculated threshold based on its rank, allowing us to confidently reject the null hypothesis for those tests meeting the criteria.

4 Quality Assurance and Quality Control and Statistical Analysis Using Correlation, and Hierarchical Clustering

4.1 Quality Assurance and Quality Control of Analytical Data for Statistical Analysis

Quality Assurance and Quality Control (QA/QC) in the chemical analysis of water samples for statistical research involves implementing procedures to ensure the accuracy and reliability of the collected data. This includes using standard reference materials, conducting replicate analyses, and incorporating blank and spiked samples throughout the sampling and analysis process. These practices help minimize errors and identify potential issues that could impact the validity of the statistical conclusions drawn from the data. The reliability of a statistical study depends on the quality of its data. Data reliability is a crucial part of QA/QC, which is needed to draw meaningful conclusions and make decisions confidently.

The QA/QC data analysis and validation include evaluating data for outliers, precision, and accuracy using statistical methods like standard deviation, confidence intervals, and percent recovery, as well as data flagging–identifying and documenting data points that may be suspect due to potential issues during sampling or analysis. The QA/QC analysis was performed using the methods outlined in the paper by Faybishenko et al. (2022).

The Appendix contains an Excel file with the chemical analysis results of 83 water samples used for calculations. Table 11.1 and Table 11.6 in the Appendix provide summary statistics of the original dataset. The dataset includes missing data (NAs) in reported depth, DO, and SC data, and the outlier in temperature data. The reported depths are subject to the recall bias of the well owner or principal user; any results using "depth" should be interpreted cautiously. The R library "*vtable*" was used to conduct a statistical analysis of the chemical data from 83 water samples (Table 11.6 in the Appendix). The dataset's missing values (NAs) consist of one value each for SC and DO and four for Depth. The imputation of missing data (NAs) was conducted using a new imputation method from the R library "*FuzzyImputationTest*."

Due to limitations of chemical analysis procedures, small concentrations cannot be precisely measured. The limit of detection (LOD) for measuring arsenic using silver diethyldithiocarbamate was rounded to 5 μg/L. Nineteen samples showed arsenic concentrations below this detection limit (see the Appendix). These values were substituted with a constant value, the detection limit of 5 μg/L divided by the square root of 2 (Croghan and Egeghy 2003).

The QA/QC-ed dataset used for calculations in Sects. 4 and 5 is in the Appendix.

Boxplots of arsenic concentration, dissolved oxygen, oxidation-reduction potential, specific conductance, pH, temperature, and reported depth are shown in Fig. 11.3. The red line on the arsenic boxplot indicates the WHO 10 μg/L drinking water guideline (WHO 2022). The red line on the specific conductance boxplot indicates the United States Environmental Protection Agency (U.S. EPA) recommended 1000 μS/cm maximum for drinking water (New Mexico EPHT 2021).

Table 11.1 The summary statistics of the quality assured and quality controlled (QA/QC-ed) chemical analysis data, which were used in calculations in Sects. 4 and 5

Variable	Number of samples	Mean	Standard deviation	Minimum	Pctl. 25	Pctl. 75	Maximum
As (μg/L)	83	78	107	3.5	6.3	80	448
DO (mg/L)	83	4.3	2.5	0.16	2.6	5.4	13
ORP (mV)	83	19	109	−120	−71	103	276
SC (μS/cm)	83	642	743	40	164	794	5360
pH	83	6.6	0.7	3.9	6.2	6.9	8
T (°C)	83	27	1.5	24	26	27	33
Depth (m)	83	90	89	4.6	25	120	366

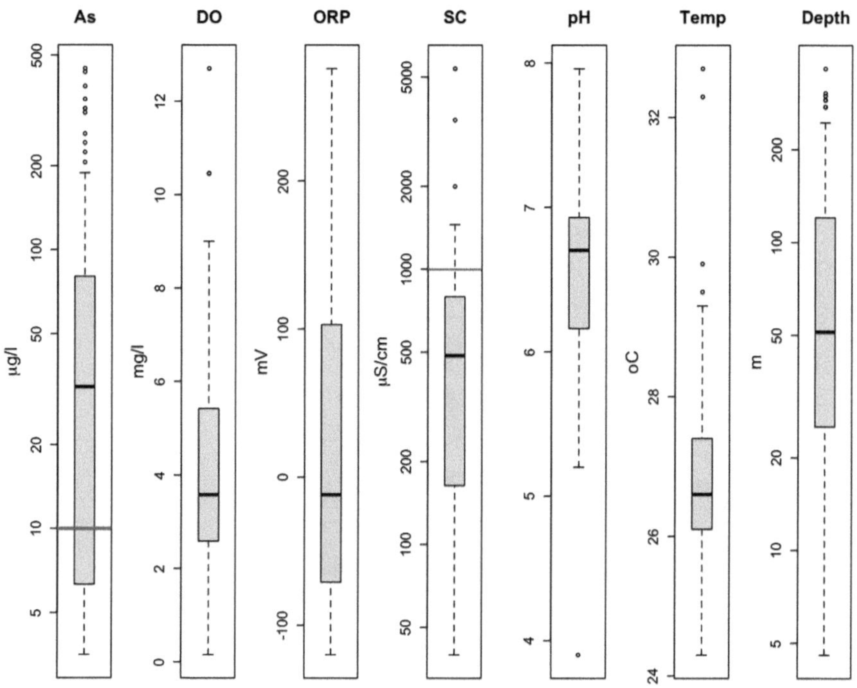

Fig. 11.3 Boxplots of arsenic (As) concentration in micrograms per liter (µg/L), dissolved oxygen (DO) concentration in milligrams per liter (mg/L), oxidation-reduction potential (ORP) in millivolts (mV), specific conductance (SC) in microsiemens per centimeter (µS/cm), pH, temperature in Celsius (°C), and reported depth in meters (m)

The statistical analysis described in the rest of Sect. 4 includes evaluating the pairwise correlation between variables, determining an optimal number of clusters, Principal Component Analysis (PCA) of variables and sampling locations, and Hierarchical Clustering on Principal Components (HCPC).

4.2 *Pairwise Correlation Analysis*

Figure 11.4 shows the histograms and correlation coefficients of variables calculated using the Pearson method. Dots represent the chemical composition of individual variables, red lines represent regression lines, and histograms are shown in light blue. The highest correlation coefficient is 0.65, and the rest are less than 0.5, indicating that there is no strong relationship between the variables in the analyzed dataset.

We applied a generalized linear model to assess the relationship between arsenic concentration and the independent variables. The R function *glm* of the R library *stats* was used to fit generalized linear models, specified by giving a symbolic

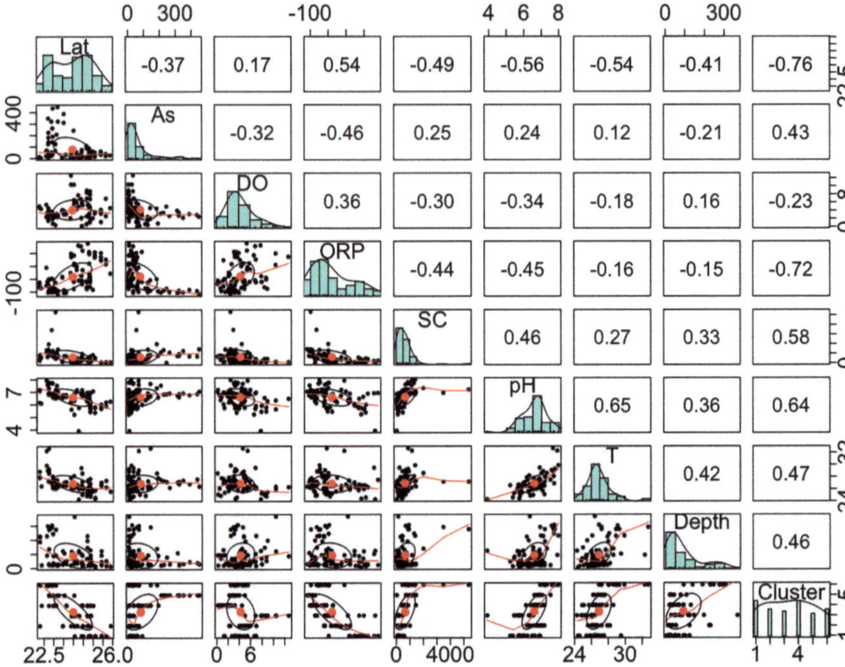

Fig. 11.4 Histograms and correlation coefficients of variables calculated using the Pearson method. Table 11.6 in the Appendix provides pairwise correlation coefficients of variables

Table 11.2 Coefficients of linear relationships between the dependent variable arsenic concentration and the independent variables (DO, ORP, SC, pH, T, and Depth), which are shown as the output of calculations using the generalized linear model function *glm*

Variable	Estimate	Standard error	*t*-value	Pr(>|t|)
(Intercept)	−230.163	210.862	−1.092	0.278
DO (mg/L)	−1.698	4.748	−0.358	0.722
ORP (mV)	−0.414	0.112	−3.699	0.000
SC (μS/cm)	0.020	0.016	1.215	0.228
pH	−1.690	22.105	−0.076	0.939
T (°C)	13.504	9.523	1.418	0.160
Depth (m)	−0.466	0.137	−3.411	0.001

description of the linear predictor and a description of the error distribution. Table 11.2 shows the coefficients of the linear relationships between arsenic concentration and the independent variable dissolved oxygen concentration, oxidation-reduction potential, specific conductance, pH, temperature, and reported depth.

The linear regressions' *p*-values are less than 0.05 only for the relationship between arsenic concentration and oxidation-reduction potential and between arsenic concentration and reported depth.

4.3 Determining an Optimal Number of Clusters

Determining the optimal number of clusters is a necessary step required for partitioning the dataset for further analysis. However, many methods exist to determine the number of clusters because the optimal number is generally purpose-dependent and subjective. We determined the optimal number of clusters employing the R libraries such as *factoextra* and *NbClust*, with the application of "silhouette" (for average silhouette width), "*wss*" (for total within the sum of the square), and "*gap_stat*" (for gap statistics) functions.

The silhouette value measures how similar an object is to its cluster compared to other clusters. It shows how well each object fits within its cluster. A high average silhouette value indicates good clustering. The plot of the silhouette method is shown in Fig. 11.15 in the Appendix.

The basic idea of partitioning methods is to define clusters with the total minimal intra-cluster variation, which can be expressed with the minimal total within-cluster sum of square (*wss*). This method is also called the elbow method. The total *wss* is determined as a function of the number of clusters. The plot of the *wss* (elbow) method is shown in Fig. 11.16 in the Appendix.

The gap statistic method compares the total within-cluster variation for different numbers of clusters (k) to their expected values under the data's null reference distribution (Tibshirani et al. 2001). The optimal number of clusters is identified by maximizing the gap statistic, representing the most significant difference between the observed clustering structure and a random uniform distribution of points. Figure 11.5 shows that the optimal number of clusters is six (6), which is used in our calculations for Hierarchical Clustering on Principal Components (HCPC) and fuzzy systems clustering.

After selecting the number of clusters, we can partition the dataset into clusters as described below.

4.4 Principal Component Analysis

Principal Component Analysis (PCA) is a statistical method used to summarize and partition data sets into smaller sets of variables. It is also used to reduce the dimensionality of a multi-variable dataset because it captures information from the original dataset and helps identify correlations between data points. PCA helps visualize relationships between variables.

The PCA calculations were conducted using the normalized variables. Figure 11.6 is a principal component analysis graph of our seven-variable dataset. Figure 11.7 demonstrates a two-dimensional (2D) principal component analysis graph of the variables used in this analysis. Table 11.8 in the Appendix is a matrix containing all the eigenvalues, the percentage of variance, and the cumulative percentage of variance. It shows that the first two components describe 62.5% of the variance of the

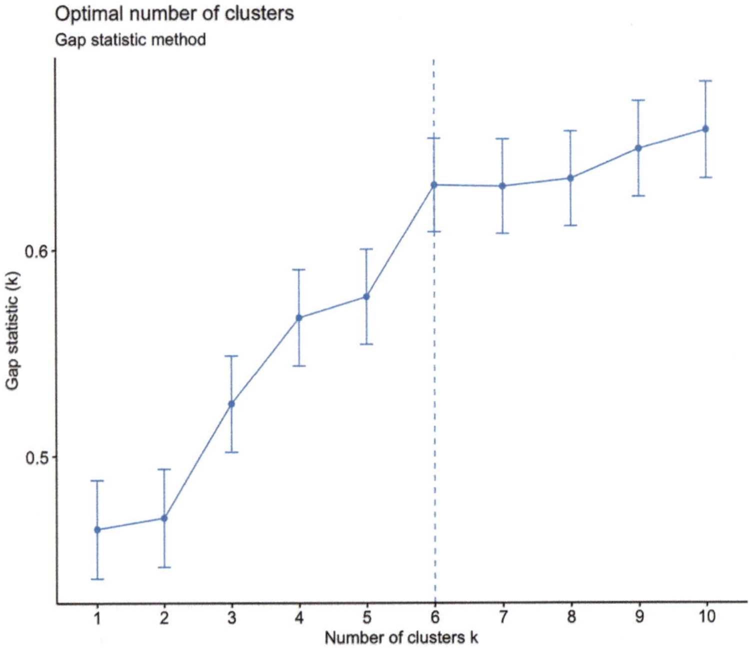

Fig. 11.5 The plot of the gap statistics method for evaluating an optimal number of clusters

studied multi-variable dataset. Table 11.9 in the Appendix is a matrix showing the contribution of different variables to the principal components of the analyzed dataset. The most significant contributions to Dimension 1 are from pH and oxidation-reduction potential, and Dimension 2 is from arsenic concentration and reported depth.

Figure 11.7 displays a two-dimensional Principal Component Analysis plot, and Fig. 11.8 displays a three-dimensional Principal Component Analysis plot illustrating six clusters of samples based on their similarity and the connections between the clusters.

4.5 Hierarchical Clustering on Principal Components

Hierarchical Clustering on Principal Components (HCPC) is an agglomerative hierarchical clustering method applied to the factor analysis results. This technique constructs a hierarchical tree, also known as a dendrogram, illustrated in Fig. 11.9. The analysis partitioned 83 water samples into six clusters, which were identified as the optimal number of clusters. The clustering of individual samples using the Hierarchical Clustering on Principal Components method is summarized in Table 11.8 in the Appendix.

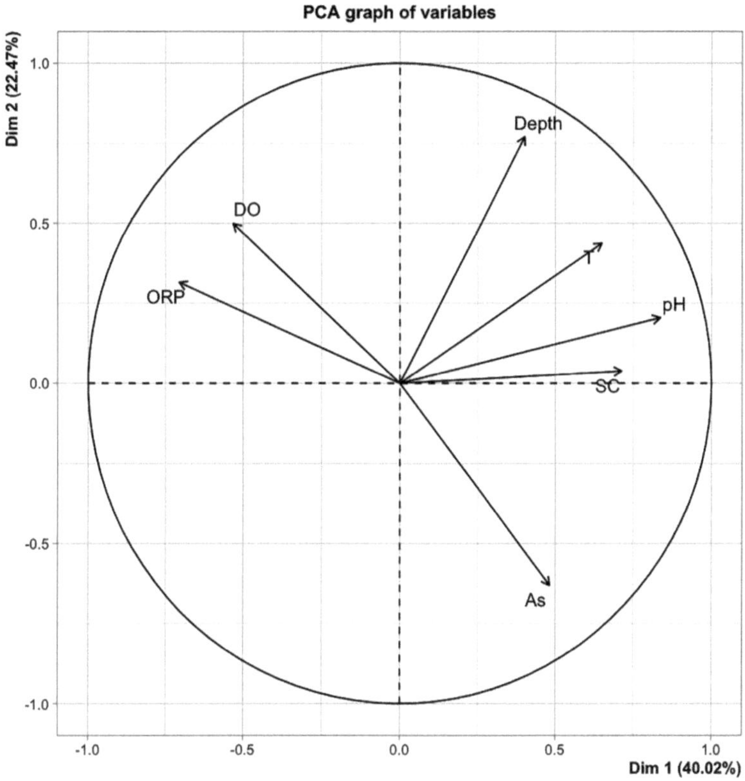

Fig. 11.6 A Principal Component Analysis graph of the seven-variable dataset (Figs. 11.6 through 11.8 were plotted using the R package "FactoMineR" dedicated to multivariate exploratory data analysis: https://cran.r-project.org/web/packages/FactoMineR/FactoMineR.pdf)

4.6 Summary of the Pros and Cons of Traditional Statistical Methods

Traditional statistical analysis of groundwater chemical composition can identify general trends and patterns in water quality, allowing for comparisons between different sites. However, these methods have limitations when the number of observations is small, and they may lead to misinterpretations due to the complex interactions between chemicals. Traditional statistical methods, such as correlation and regression analyses and basic descriptive statistics, can reveal significant trends in chemical concentrations for different sampling locations and times. These methods are well-established, readily available in statistical software packages, and cost-effective for initial data exploration. Nonetheless, traditional methods may not identify complex relationships between multiple chemical parameters, potentially leading to misinterpretations when dealing with intricate hydrogeochemical processes, spatial

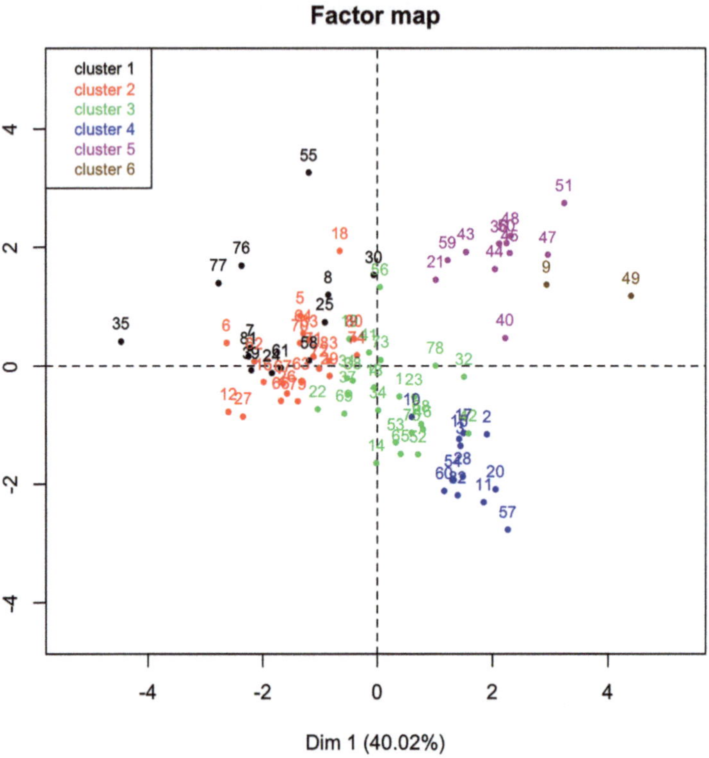

Fig. 11.7 A two-dimensional (2D) Principal Component Analysis graph of individual water samples partitioned into six clusters (see Fig. 11.5). The numbers on this graph go from 1 to 83 and are the same as the sample numbers in Table 11.7 in the Appendix

variability, and sampling errors. They may not accurately reflect the underlying processes when dealing with highly variable data and might not pinpoint the underlying relationship between the variables.

5 Fuzzy Systems Modeling Approach

5.1 Concepts of Fuzzy System Modeling

Fuzzy systems modeling is suggested as a viable alternative to conventional statistical analysis, particularly in situations involving limited and uncertainty (Zadeh 1965). Fuzzy logic-based systems methods have been applied to control water treatment processes, effectively reducing arsenic levels below WHO guidelines (Larroca et al. 2020). Additionally, fuzzy logic-based agricultural decision support systems have been proposed to help farmers make informed choices about crop production,

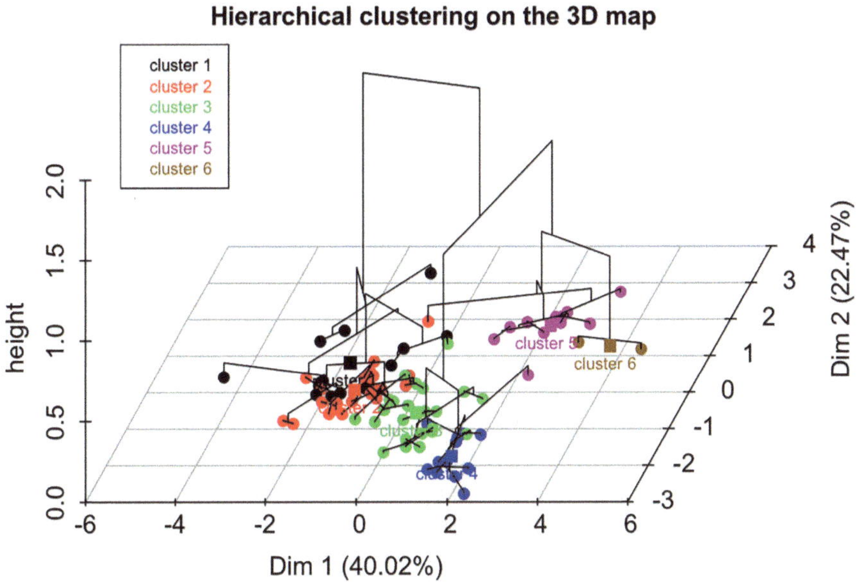

Fig. 11.8 A three-dimensional (3D) Principal Component Analysis map demonstrating the connections between the clusters

considering factors like shallow groundwater utilization (Rafiuzzaman and Cil 2016). Faybishenko et al. (2023) applied the fuzzy regression and fuzzy clustering approach to analyze the chemical composition of arsenic-contaminated acidic mine drainage downstream of a gold mine. While multiple hypotheses exist regarding arsenic contamination sources and mechanisms, no explanation fully accounts for the observed variability, suggesting that several biogeochemical processes contribute to arsenic mobilization in different geological settings (Khalequzzaman et al. 2005).

Various studies have applied fuzzy analysis techniques to evaluate water quality and chemical composition. Fuzzy regression and clustering were used to analyze arsenic-polluted water samples, revealing complex relationships between variables and identifying spatial patterns of contamination (Faybishenko et al. 2023). Fuzzy linear discriminant analysis successfully classified mineral waters based on their composition, achieving an 88% correct classification rate (Guidea et al. 2020). Fuzzy clustering analysis was developed as a complementary tool to water quality indices, providing valuable information for decision-making in water quality management (Kung et al. 1992). A fuzzy logic approach was employed to determine the Water Quality Index (WQI) of groundwater samples, resulting in a unitless number ranging from 1 to 10, with higher numbers indicating better water quality (Jinturkar et al. 2010). These studies demonstrated the effectiveness of fuzzy analysis techniques in evaluating and classifying water samples based on their chemical composition. Nasr et al. (2013) analyzed groundwater contamination using the Fuzzy Water Quality Index (FWQI). Vadiati et al. (2016) demonstrated that applying groundwater quality modeling plays a vital role in water resource management decisions. They concluded that artificial intelligence models, particularly Fuzzy Inference Systems (FIS), can effectively evaluate groundwater quality for complex aquifers.

A fuzzy logic system is a mapping of input and output data vectors. The method is based on the principle of Zadeh's fuzzy set theory, representing and reasoning with the ambiguous, missing, uncertain, or noisy input information (Zadeh 1965). Fuzzy theory is generally a mathematical framework that deals with uncertainty and imprecision, allowing for a more flexible and nuanced approach to modeling and decision-making. Fuzzy theory finds applications in various fields, such as control systems, artificial intelligence, decision analysis, and pattern recognition. It provides a means to handle complex and uncertain situations that traditional mathematical models cannot easily capture. Fuzzy logic allows for approximate reasoning and incorporates human-like reasoning into computational systems.

In a crisp set, an element's membership is strictly binary, either 0 or 1, leaving no room for ambiguity. Fuzzy set theory is an extension of what one might call classical set theory. Unlike classical binary logic, which operates with precise true or false values, fuzzy logic allows for degrees of truth or membership between 0 and 1.

5.2 Software for Calculations

Calculations presented in Sects. 5.3 and 5.4 were conducted using Rstudio libraries *cluster*, *dplyr*, *factoextra*, *FactoMineR*, *fclust*, *fuzzyreg*, *NbClust*, *ppclust*, *rnaturalearth*, *sf*, *SpatialEpi*, *SpatialEp*, *stats*, and *zoo*. The references to these libraries are cited in the list of references. These libraries were used for implementing advanced linear regression, fuzzy regression, and fuzzy clustering.

5.3 Fuzzy Clustering

Fuzzy clustering is an unsupervised method for data analysis and model construction that allows data points to belong to multiple clusters with varying degrees of membership (Rui 2008). Ruspini (1977) developed a rigorous mathematical theory for fuzzy clustering, formalizing the concept of clusters as fuzzy subsets satisfying specific conditions. This approach provides richer representations of taxonomical structures compared to conventional methods. The Fuzzy C-Means (FCM) algorithm is a widely used technique in fuzzy clustering due to its simplicity and broad applicability (Bezdek 1981; Rui 2008). Fuzzy C-Means involves dividing data points into homogeneous classes based on similarity, with each point having membership grades indicating its degree of belonging to different clusters (Makhalova 2013). While Fuzzy C-Means has advantages, it also has limitations that require further study (Rui 2008).

To analyze the dataset, we employed fuzzy multivariate clustering with the application of the FCM algorithm, aiming to identify clusters or groupings within the water samples based on their arsenic concentrations and other relevant chemical components. Fuzzy clustering assigns membership values to each sample to indicate the degree of belongingness to different clusters, allowing for overlapping clusters and providing a more comprehensive understanding of the dataset (Tokushige et al. 2007; Ferraro 2024). As in the Principal Component Analysis and Hierarchical Clustering on Principal Components calculations, Fuzzy C-Means calculations were conducted using the standardized data (based on the scale function). Fuzzy analysis clustering was conducted using the fanny function of the cluster and *ppclust* packages in R.

Table 11.9 in the Appendix presents the results of the fuzzy clustering for all 83 samples. Table 11.3 demonstrates an example (the upper six rows of Table 11.9 in the Appendix) of the variables' contribution of Fuzzy Membership Functions (FMFs) to each cluster. The cluster number for each sample corresponds to the largest value among the clusters. For instance, the largest value is 0.686 is in column

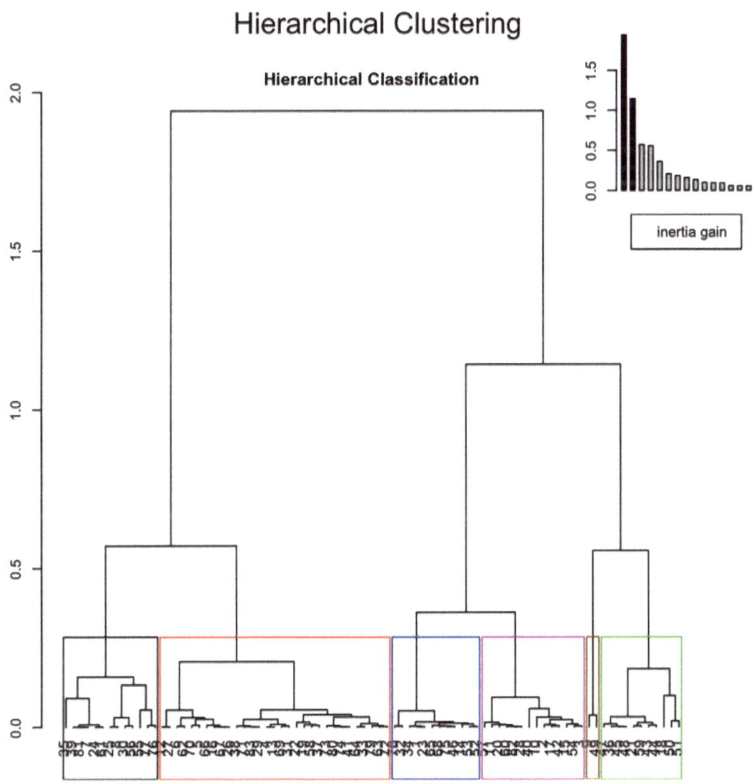

Fig. 11.9 Hierarchical clustering tree (dendrogram) shows water sample partitioning into six clusters

Table 11.3 The contribution of the variables' Fuzzy Membership Functions (FMFs) to each of the six clusters

Sample	Cluster 1	Cluster 2	Cluster 3	Cluster 4	Cluster 5	Cluster 6	Sum	Cluster
13	0.05060252	0.14008016	0.04704339	0.6815434	0.04616917	0.03456131	1	4
16	0.06520892	0.11604572	0.05456398	0.2368417	0.43729931	0.09004036	1	5
29	0.05219907	0.08498138	0.04411644	0.1804299	0.58846594	0.04980729	1	5
43	0.11080365	0.32893203	0.05048321	0.4311503	0.04010544	0.03852538	1	4
48	0.34823151	0.26651323	0.12537167	0.1201744	0.06309838	0.07661081	1	1
60	0.38789652	0.16082832	0.28647208	0.0803613	0.04418997	0.04025181	1	1

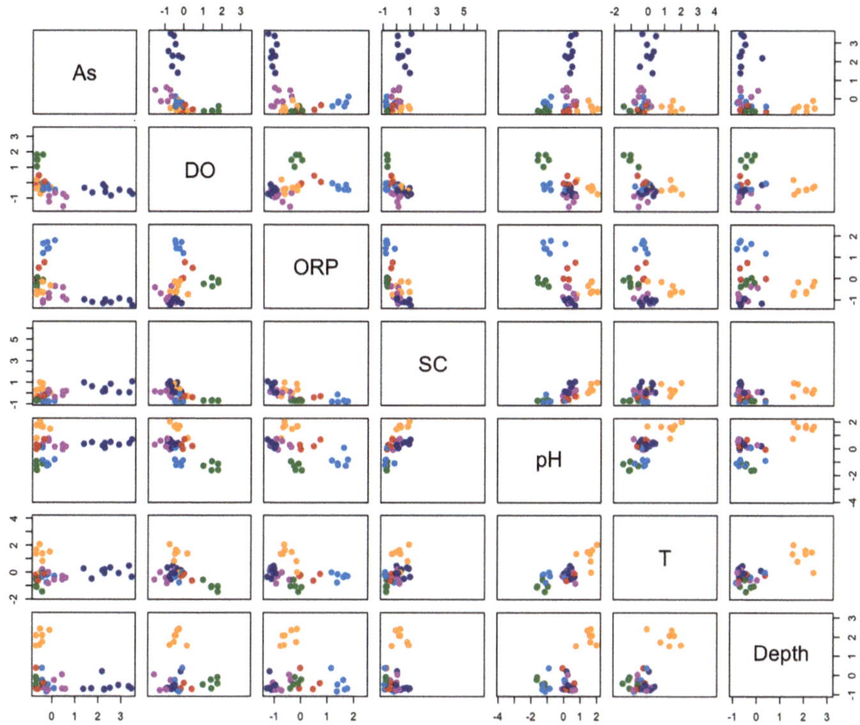

Fig. 11.10 Pair-wise correlation plots based on the Fuzzy C-Means clustering. Colors indicate individual clusters

"Cluster 2" for Sample 13. The largest value of 0.438 is in column "Cluster 3" for sample 16. The sum of all numbers in each row is one.

Figure 11.10 illustrates the pairwise correlation between the variables representing the six clusters identified through Fuzzy C-Means (FCM) clustering. The distinct separation among individual clusters is evident, accounting for the low correlation coefficients calculated using the traditional statistical method outlined in Sect. 4.2.

Figure 11.11 presents the results of Fuzzy C-Means clustering on the Principal Component Analysis map. It highlights the overlapping areas shared with Clusters 1, 4, and 6, those shared with Clusters 2 and 3, and those shared with Clusters 2 and 6. In contrast, Cluster 5 is clearly distinguished as a separate group of points. Cluster 5 is characterized by a more extensive range of arsenic concentrations depicted using the boxplots in Fig. 11.3.

Figure 11.19 in the Appendix shows the boxplots of the ranges of arsenic concentrations in Clusters 1–6 (these clusters are defined from the fuzzy clustering results). Figure 11.12 shows the Fuzzy Membership Functions for arsenic concentration, dissolved oxygen concentration and oxidation-reduction potential, specific conductance and pH, temperature, and reported depth. These plots show the Fuzzy

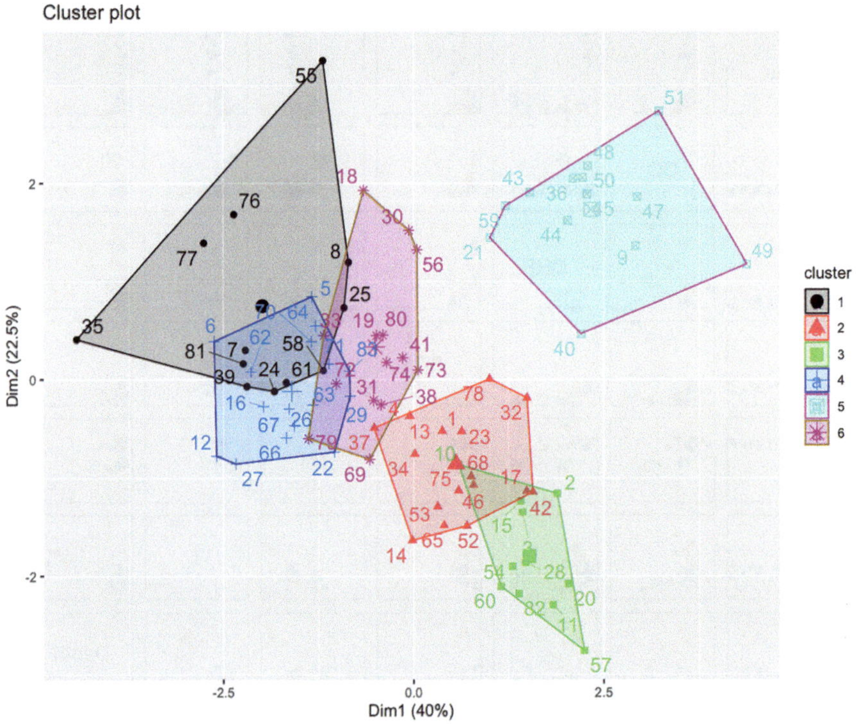

Fig. 11.11 Results of Fuzzy C-Means clustering showing the overlap between clusters. Numbers are from 1 to 83, and the correspondence to sample numbers is given in Table 11.7 in the Appendix

Membership Functions for the different clusters that overlay each other in Fig. 11.11. Figure 11.12 shows that the Fuzzy Membership Functions of arsenic concentration in Cluster 5 are greater than those in the other clusters.

5.4 *Fuzzy Regression*

Fuzzy regression analysis augments traditional regression analysis that accommodates vague or imprecise data using fuzzy set theory (Kacprzyk and Fedrizzi 1992). It addresses linguistic uncertainties and complex predictions where statistical regression falls short (Dongale et al. 2012). Two main approaches to fuzzy regression are possibilistic and least squares methods (Diamond and Tanaka 1998). The possibilistic approach optimizes a measure of the spreads of estimated outputs, while the least squares method minimizes the distance between observed and estimated outputs. Choi and Yoon (2010) propose a general fuzzy regression model that separates the response function on mode and spreads an α-level (i.e., the Fuzzy Membership Functions) set for observed fuzzy numbers. This model uses least

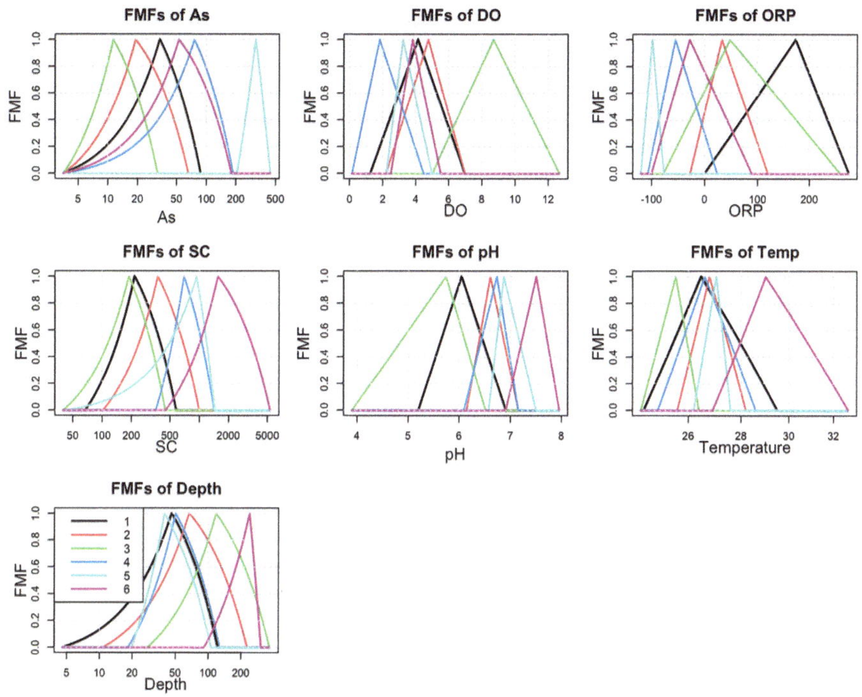

Fig. 11.12 The Fuzzy Membership Functions of arsenic concentration, dissolved oxygen concentration and oxidation-reduction potential, specific conductance and pH, temperature, and reported depth for each cluster

squares estimation and can predict fuzzy numbers' crisp mean and variance. Fuzzy regression has wide applications in areas involving human judgments and can be implemented algorithmically with available software (Kacprzyk and Fedrizzi 1992).

Fuzzy regression analysis was conducted to explore the relationships between the measured variables within the water samples. Fuzzy sets were used to represent the variables, assigning membership values to each data point based on their degree of belongingness to different clusters. The fuzzy regression model was then applied to estimate the relationships and dependencies among the variables, considering their fuzzy memberships. Notably, Fuzzy Linear Regression, introduced by Tanaka et al. (1982), varies from traditional regression by utilizing possibility distributions based on membership values of the dependent variable instead of conventional probability distributions. This approach defines the relationship between dependent and independent variables using fuzzy concepts instead of statistical ones. One of the central objectives of a regression model is to approximate the value of the dependent variable as closely as possible to the observed data. In the context of a fuzzy linear regression model, the level of alignment between the estimated fuzzy linear model and the given data is determined by the "h-level" within the possibility distribution. Given that a membership function characterizes the dependent variable, the estimated fuzzy output, also depicted by a membership function, should closely match the membership function of the observed data (Škrabánek and Martínková 2021). The disparity in membership values is utilized to assess the degree of concordance between the estimated fuzzy linear model and the observed data (Lee and Tanaka 1999; Tansu and Naeem 2022).

A fuzzy regression model can be represented by Eq. (11.1):

$$\tilde{F}_i = \tilde{B}_0 \bullet x_{i0} + \tilde{B} \bullet x_{i1} + \cdots + \tilde{B}_n \bullet x_{in} \tag{11.1}$$

where $\tilde{F}_i\left(\forall i = 1,2,\cdots,m\right)$ represents the predicted output as a fuzzy number; m is the number of variables; $x_{ij}(\forall j = 1,2,\cdots,n)$ denotes the jth independent variables, and $1+n$ represents the number of terms in the fuzzy regression model; and $\tilde{B}_j$ is the fuzzy coefficient of the jth independent variable (Eq. 11.1).

The fuzzy coefficients are given by Eq. (11.2) (Jiang et al. 2019):

$$\tilde{B}_j = \left(b_j^c, b_j^s\right) : \tilde{B}_0 = \left(b_0^c, b_0^s\right), \tilde{B}_1 = \left(b_1^c, b_1^s\right), \tilde{B}_2 = \left(b_2^c, b_2^s\right), \tilde{B}_n = \left(b_n^c, b_n^s\right) \tag{11.2}$$

where b_j^c and b_j^s represent the central and spread values of the jth fuzzy coefficients, respectively (Eq. 11.2).

The fuzzy linear regression intercept refers to the dependent variable's predicted value when the independent variable is set to zero (i.e., the x-axis value is zero, like a standard linear regression). In contrast, the intercept of a fuzzy regression represents a fuzzy number, meaning it has a central value and a range of uncertainty due to the fuzzy nature of the data involved. The intercept value is calculated together with the fuzzy slope coefficients. The intercept and coefficients of the fuzzy regression are typically represented as triangular fuzzy numbers with a central value and lower and upper bounds. The range of the fuzzy number indicates the degree of uncertainty associated with the regression prediction. Calculations were provided using the R package *fuzzyreg* developed based on the possible linear regression combined with the least squares method, specifically adapted to operate within the framework of fuzzy sets.

Figure 11.13 demonstrates the fuzzy linear regression between the arsenic concentration given as ln(As) and the specific conductance given as ln(SC). The clusters in the upper plate of Fig. 11.13 are shown in colors. The middle and lower plates of Fig. 11.13 depict the intercept and the regression slope's asymmetric triangular Fuzzy Membership Functions (FMFs). The red horizontal dashed line indicates the WHO 10 μg/L drinking water guideline (WHO 2022). The red vertical dashed line on the specific conductance indicates the United States Environmental Protection Agency recommended 1000 μS/cm maximum for drinking water (New Mexico EPHT 2021). The points in the region labeled "I" are those with satisfactory drinking water quality. The points in the region labeled "II" have satisfactory specific conductance, but the arsenic concentration is above the WHO 10 μg/L drinking water guideline. The points in region "III" do not meet drinking water criteria regarding specific conductance and arsenic concentration.

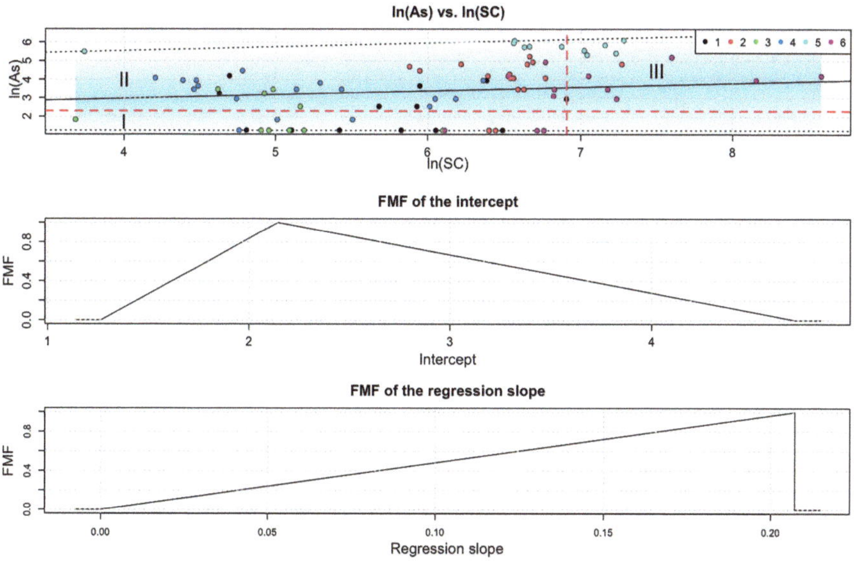

Fig. 11.13 Fuzzy regression between ln(As) and ln(SC) showing the asymmetric triangular membership functions for the intercept and the regression slope of the fuzzy regression ln(As) vs. ln(SC)

Table 11.4 The fuzzy regression of ln(As) vs. ln(SC) model coefficients of the fuzzy regression shown in Fig. 11.13

Coefficients	Center	Left spread	Right spread
Intercept	2.14106567	0.87820135	2.57363128
Slope	0.20723902	0.20723902	−1.73E-09

In the upper plate in Fig. 11.13, a thick black line indicates the central tendency, and dotted lines indicate the left and right spreads of the membership. This defines a support interval for the potential values of the relationships shown as Fuzzy Membership Functions. The Fuzzy Membership Functions' intervals encompass values with non-zero degrees of membership for the model predictions. The regression graph is plotted using a blue color gradient, transitioning from intense near the central tendency to white near the dotted lines, visually representing the degree of membership of the fuzzy regression. Table 11.4 includes all the model coefficients related to the fuzzy regression shown in Fig. 11.13.

Figure 11.14 illustrates the fuzzy linear regression between arsenic concentration, represented as ln(As), and reported depth, expressed as ln(Depth). Regions with arsenic concentrations below and above the WHO 10 μg/L drinking water guideline are indicated by symbols "I" and "II." Figure 11.14 shows that the highest arsenic concentrations (Cluster 5) occurred in water samples at reported depths of approximately 10–40 m. Table 11.5 includes all the model coefficients related to the fuzzy regression shown in Fig. 11.14.

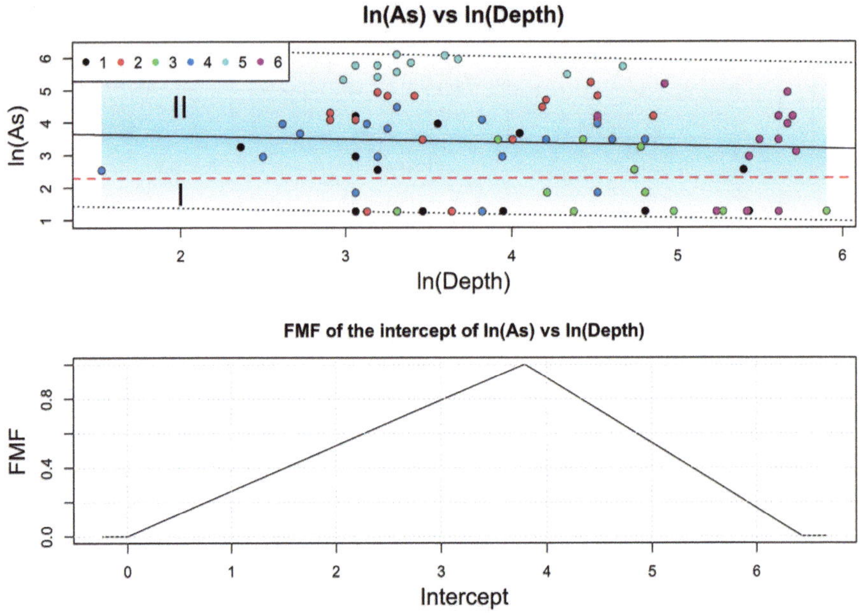

Fig. 11.14 Fuzzy regression between ln(As) and ln(Depth) showing the asymmetric triangular Fuzzy Membership Functions (FMFs) for the intercept of the fuzzy regression ln(As) vs. ln(Depth)

Table 11.5 The fuzzy regression of ln(As) vs. ln(Depth) model coefficients of the fuzzy regression shown in Fig. 11.14

Coefficients	Center	Left spread	Right spread
Intercept	3.78008835	2.21152896	2.655001884
Slope	−0.099866	1.76e-10	1.76e-10

Figures 11.17 and 11.18 in the Appendix show the fuzzy regression of ln(As) vs. pH and OPR vs. pH, respectively, together with their Fuzzy Membership Functions of regression model coefficients.

6 Conclusions

Groundwater arsenic contamination in Bangladesh has had a devastating impact on public health, affecting millions of people. This chapter delves into how a combination of traditional statistical and fuzzy systems analysis can be leveraged to better understand and address this critical issue. By combining statistical and fuzzy systems analysis strengths, researchers can develop more robust and accurate models for predicting arsenic contamination. These models can then identify areas at high risk of contamination, enabling targeted interventions and preventive measures.

Fuzzy logic can be integrated into real-time monitoring systems to provide immediate assessments of water quality, aiding in timely decision-making for water treatment and management (Bokingkito and Caparida 2018). Fuzzy systems analysis offers a robust framework for analyzing the chemical composition of water samples by addressing uncertainties and complexities inherent in water quality data. Its adaptability and ability to combine disparate parameters make it a valuable tool in research and practical applications.

The chapter also explores the application of clustering zonation techniques to the studied area in Bangladesh. Clustering zonation involves dividing the region into distinct zones based on similarities in water quality parameters. This approach can help identify areas with similar arsenic contamination patterns, facilitating more effective and efficient management strategies. Furthermore, the chapter demonstrates how fuzzy regression and clustering models can be used to investigate the link between sea level rise and arsenic contamination in Bangladesh. Sea level rise can lead to saltwater intrusion into coastal aquifers, altering the chemical composition of the groundwater and potentially mobilizing arsenic from sediments. By applying fuzzy modeling techniques, researchers can gain valuable insights into the complex interplay between sea level rise, groundwater chemistry, and arsenic release, informing adaptation and mitigation strategies.

Overall, this chapter highlights the potential of combining statistical and fuzzy systems analysis to address the pressing issue of groundwater arsenic contamination in Bangladesh. By incorporating uncertainty and imprecision, these advanced modeling techniques can provide a more comprehensive understanding of the factors driving arsenic contamination and enable more effective and targeted interventions to protect public health.

7 Discussion Questions

1. What are the pros and cons of applying the traditional statistical analysis of the chemical composition of groundwater?
2. What are the pros and cons of applying fuzzy systems analysis to the analysis of groundwater?
3. What are the advantages and disadvantages of applying a combined traditional statistical and fuzzy systems analysis approach?
4. What are other potential applications of fuzzy systems analysis for water treatment?

Acknowledgment BF's scientific collaboration on the data analysis and an application of the fuzzy systems modeling approach was partially supported by the U.S. Department of Energy under Award Number DE-AC02-05CH11231.

Competing Interests This study was funded by [X] [grant number X].

Appendix

Table 11.6 Summary statistics of the original dataset analyzed in this chapter

Variable	N	Mean	Std. dev.	Min	Pctl. 25	Pctl. 75	Max
Sample	83	385	213	13	232	560	787
Lon	83	90	1.1	88	89	91	92
Lat	83	24	0.97	22	23	25	26
As (μg/L)	83	77	107	0	6.3	80	448
DO (mg/L)	82	4.2	2.5	0.16	2.5	5.4	13
ORP (mV)	83	19	109	−120	−71	103	276
SC (μS/cm)	82	644	747	40	163	798	5360
pH	83	6.6	0.7	3.9	6.2	6.9	8
T (°C)	83	28	10	24	26	27	120
Depth (m)	79	92	91	4.6	24	122	366

Table 11.7 Pairwise correlation coefficients of variables. Arsenic concentration (As in μg/L) is the dependent variable and dissolved oxygen (DO in mg/L), oxidation-reduction potential (ORP in mV), specific conductance (SC in μS/cm), pH, temperature (T in °C), and reported depth (Depth in m) are the independent variables

	As	pH	ORP	SC	T	DO	Depth
As	1	0.235	−0.462	0.251	0.124	−0.325	−0.215
pH	0.235	1	−0.455	0.458	0.648	−0.34	0.361
ORP	−0.462	−0.455	1	−0.44	−0.157	0.362	−0.146
SC	0.251	0.458	−0.44	1	0.274	−0.299	0.332
T	0.124	0.648	−0.157	0.274	1	−0.176	0.419
DO	−0.325	−0.34	0.362	−0.299	−0.176	1	0.163
Depth	−0.215	0.361	−0.146	0.332	0.419	0.163	1

Table 11.8 A matrix containing all the eigenvalues, the percentage of variance, and the cumulative percentage of variance from the Principal Component Analysis (PAC) calculations

	Eigenvalue	Percentage of variance	Cumulative percentage of variance
comp 1	2.80332908	40.0475583	40.04755827
comp 2	1.56861411	22.408773	62.45633122
comp 3	0.82482928	11.7832754	74.2396066
comp 4	0.6852166	9.7888086	84.0284152
comp 5	0.50421057	7.20300808	91.23142328
comp 6	0.36280766	5.18296659	96.41438987
comp 7	0.25099271	3.58561013	100

Table 11.9 A matrix of different variables' contribution to the analyzed dataset's principal components

	Dim.1	Dim.2	Dim.3	Dim.4	Dim.5
As (μg/L)	8.26766664	25.3507525	0.14247382	37.5254209	10.5952817
DO (mg/L)	10.0558564	15.9387159	14.0264657	41.2190991	1.65604438
ORP (mV)	17.7038938	6.30279071	18.0760968	1.6665052	40.105165
SC (μg/cm)	18.1214537	0.09916302	21.0613022	12.8004382	40.889398
pH	24.9589201	2.60079151	6.83409818	0.29577286	2.46496765
T (°C)	15.0593747	12.0824745	29.8674194	6.49212784	2.07224578
Depth (m)	5.83283466	37.6253119	9.99214397	0.00063588	2.21689745

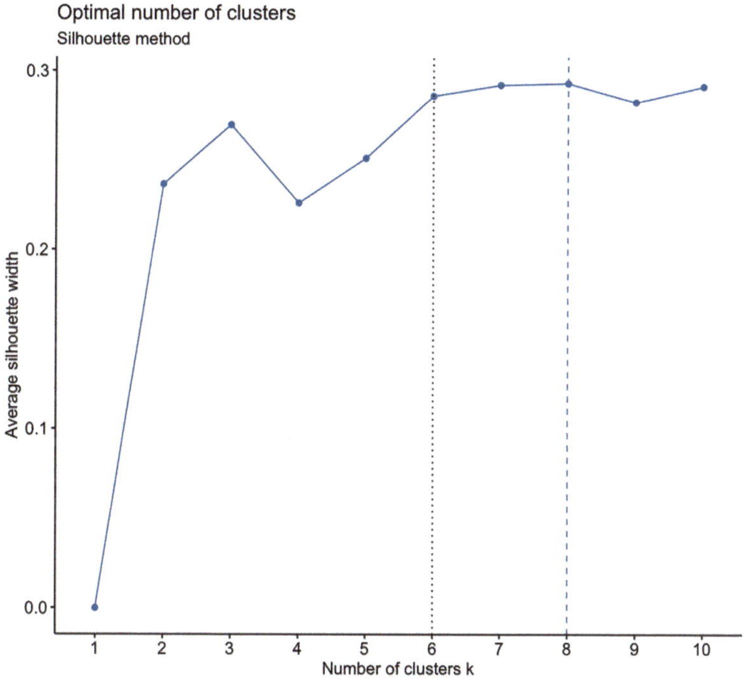

Fig. 11.15 The optimal number of clusters using the silhouette method is 6 to 8

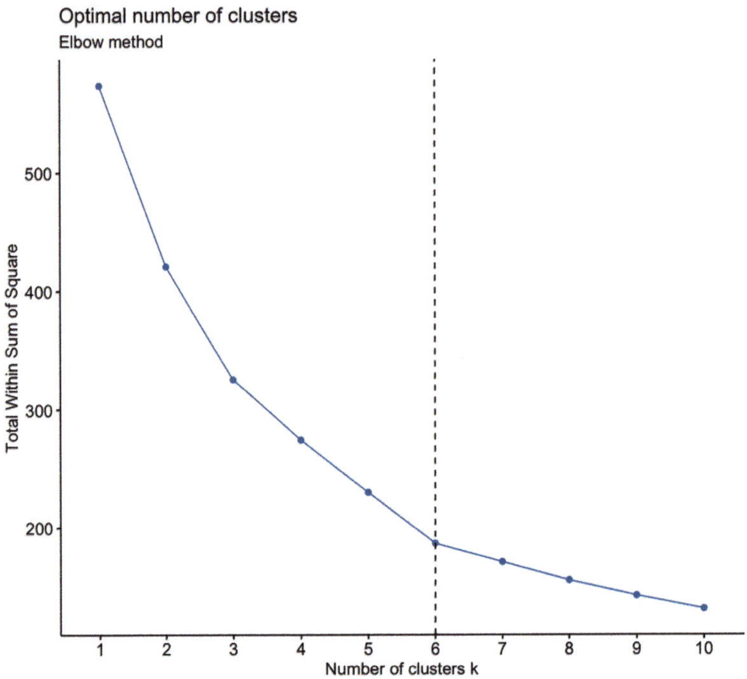

Fig. 11.16 The optimal number of clusters using the *wss* (elbow) method is 7

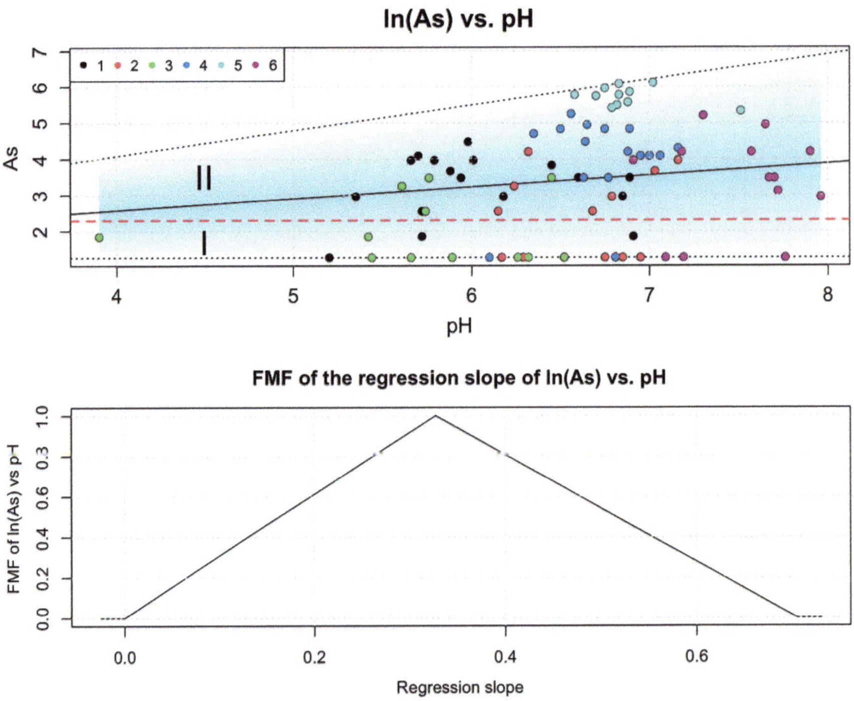

Fig. 11.17 Fuzzy regression of ln(As) vs. pH and the Fuzzy Membership Functions (FMF) of the regression slope

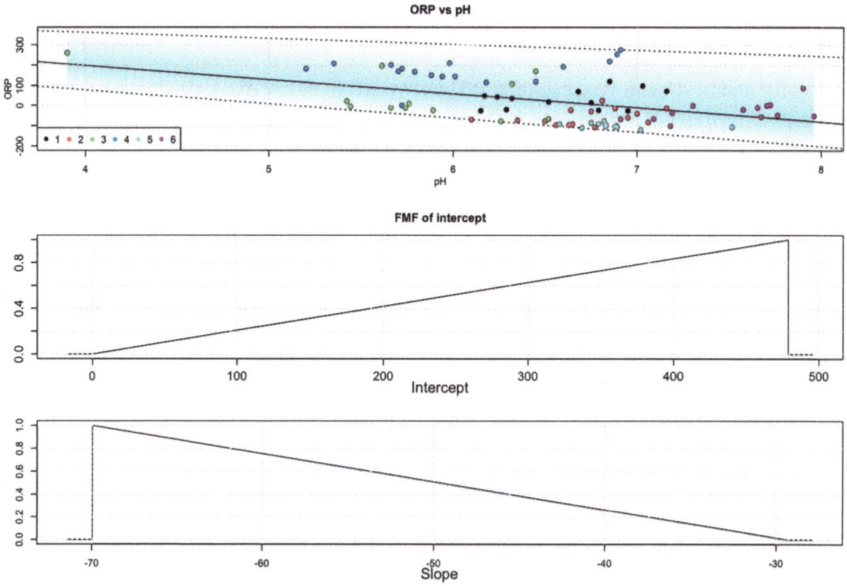

Fig. 11.18 The fuzzy regression of oxidation-reduction potential (ORP) vs. pH. Oxidation-reduction potentials are in millivolts (mV)

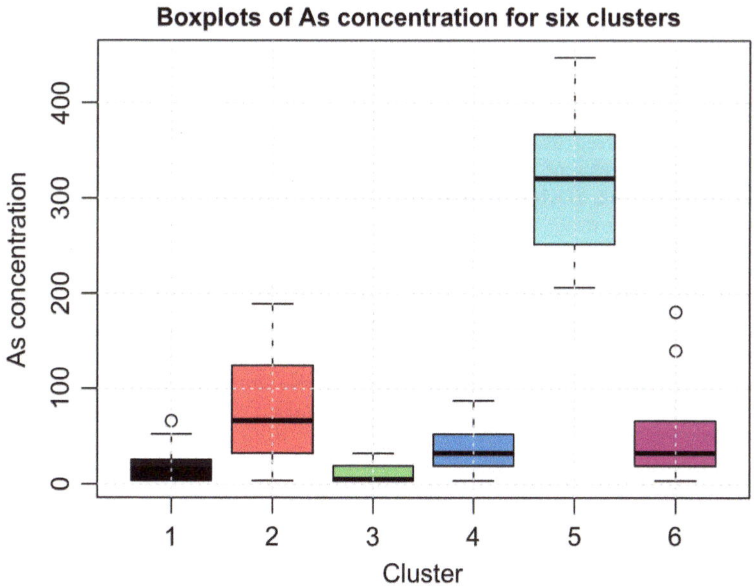

Fig. 11.19 Boxplots of arsenic (As) concentration in Clusters 1–6 from fuzzy clustering results. The arsenic concentrations are in micrograms per liter (μg/L)

References

American Public Health Association (APHA), American Water Works Association (AWWA), Water Environment Federation (WEF) (1995) Standard methods for the examination of water and wastewater, 19th edn. APHA, Washington, DC, parts 2000, 3000, 4000, 5000

Beck HE, Zimmermann NE, McVicar TR, Vergopolan N, Berg A, Wood EF (2018) Present and future Köppen-Geiger climate classification maps at 1-km resolution. Sci Data 5(1):180214. https://doi.org/10.1038/sdata.2018.214

Bezdek JC (1981) Pattern recognition with fuzzy objective function algorithms. Plenum Press, New York

Bokingkito P, Caparida L (2018) Using fuzzy logic for real-time water quality assessment monitoring system. International Conference on Automation, Control and Robots (ICACR). https://doi.org/10.1145/3293688.3293695

Choi SH, Yoon JH (2010) General fuzzy regression using least squares method. Int J Syst Sci 41(5):477–485. https://doi.org/10.1080/00207720902774813

Croghan C, Egeghy PP (2003) Methods of dealing with values below the limit of detection using SAS. Presented at Southeastern SAS User Group, St. Petersburg, FL, September 22–24, 2003. https://cfpub.epa.gov/si/si_public_record_report.cfm?Lab=NERL&dirEntryId=64046. Accessed 1 Mar 2025

Diamond P, Tanaka H (1998) Fuzzy regression analysis. In: Słowiński R (ed) Fuzzy sets in decision analysis, operations research and statistics. The handbooks of fuzzy sets series, vol 1. Springer, Boston, MA. https://doi.org/10.1007/978-1-4615-5645-9_11

Dongale TD, Kulkarni TG, Mudholkar RR (2012) Fuzzy modelling of voltage standing wave ratio using fuzzy regression method. Int J Emerg Technol Adv Eng 2(6):21–26

Fahim AKF, Maksud Kamal ASM, Shahid S (2023) Spatiotemporal change in groundwater sustainability of Bangladesh and its major causes. Stoch Environ Res Risk Assess 37:665–680. https://doi.org/10.1007/s00477-022-02294-z

Faybishenko B, Bakhtavar E, Hewage KN, Sadiq R (2023) Chemical composition of arsenic-based acid mine drainage in the downstream of a gold mine: fuzzy regression and clustering analysis. J Hazard Mater 465(4):133250. https://doi.org/10.1016/j.jhazmat.2023.133250

Faybishenko B, Versteeg R, Pastorello G, Dwivedi D, Varadharajan C, Agarwal D (2022) Challenging problems of quality assurance and quality control (QA/QC) of meteorological time series data. Stoch Environ Res Risk Assess 36:1049–1062. https://doi.org/10.1007/s00477-021-02106-w

Ferraro MB (2024) Fuzzy k-means: history and applications. Econ Stat 30:110–123. https://doi.org/10.1016/j.ecosta.2021.11.008

Frisbie SH, Maynard DM, Hoque BA (1999) The nature and extent of arsenic-affected drinking water in Bangladesh. In: Sarkar B (ed) Metals and genetics. Kluwer Academic/Plenum Publishers, New York, pp 67–85

Frisbie SH, Mitchell EJ, Molla AR (2024) Sea level rise from climate change is expected to increase the release of arsenic into Bangladesh's drinking well water by reduction and by the salt effect. PLoS ONE 19(1):e0295172. https://doi.org/10.1371/journal.pone.0295172

Guidea A, Găceanu R, Pop H, Sârbu C (2020) Mineral waters classification using fuzzy linear discriminant analysis. Studia Universitatis Babeş-Bolyai Chemia, LXV 2:45–56. https://doi.org/10.24193/subbchem.2020.2.04

Jiang H, Kwong CK, Chan CY, Yung KL (2019) A multi-objective evolutionary approach for fuzzy regression analysis. Expert Syst Appl 130:225–235. https://doi.org/10.1016/j.eswa.2019.04.033

Jinturkar AM, Deshmukh SS, Agarkar SV, Chavhan GR (2010) Determination of water quality index by fuzzy logic approach: a case of ground water in an Indian town. Water Sci Technol 61(8):1987–1994. https://doi.org/10.2166/wst.2010.095

Kacprzyk J, Fedrizzi H (1992) Fuzzy regression analysis. Omnitech Press, Physica-Verlag, Warsaw, Heidelberg

Khalequzzaman M, Faruque FS, Mitra AK (2005) Assessment of arsenic contamination of groundwater and health problems in Bangladesh. Int J Environ Res Pub Health 2(2):204–213. https://doi.org/10.3390/ijerph2005020002
Kung H-t, Ying L-g, Liu C (1992) A complementary tool to water quality index: fuzzy clustering analysis. JAWRA 28(3):525–533. https://doi.org/10.1111/j.1752-1688.1992.tb03174.x
Larroca F, Olschewski E, Quino-Favero J, Rosales-Huamaní J, Castillo-Sequera J (2020) Water treatment plant prototype with pH control modeled on fuzzy logic for removing arsenic using Fe(VI) and Fe(III). Water 12(10):2834. https://doi.org/10.3390/w12102834
Lee H, Tanaka H (1999) Fuzzy approximations with non-symmetric fuzzy parameters in fuzzy regression analysis. J Oper Res Soc Jpn 42(1):98–112. https://doi.org/10.15807/jorsj.42.98
Makhalova E (2013) Fuzzy C means clustering in MATLAB. The 7th International Days of Statistics and Economics, Prague, September 19–21, 2013, Prague, p 905–914. https://msed.vse.cz/files/2013/209-Makhalova-Elena-paper.pdf. Accessed 5 Mar 2025
Nasr AS, Rezaei M, Barmaki MD (2013) Groundwater contamination analysis using Fuzzy Water Quality index (FWQI): Yazd province, Iran. GeoPersia 3(1):47–55. https://doi.org/10.22059/jgeope.2013.31931
New Mexico Environmental Public Health Tracking Program (New Mexico EPHT) (2021) pH and conductivity and drinking water. https://nmtracking.doh.nm.gov/environment/water/PHConductivity.html. Accessed 1 Mar 2025
Rafiuzzaman M, Cil I (2016) A fuzzy logic based agricultural decision support system for assessment of crop yield potential using shallow ground water table. Int J Comput Appl 149:20–31. https://doi.org/10.5120/ijca2016911526
Rui L (2008) Fuzzy C-Means clustering algorithm. J Chongqing Inst Technol. https://api.semanticscholar.org/CorpusID:203672932
Ruspini EH (1977) A theory of fuzzy clustering. 1977 IEEE Conference on Decision and Control including the 16th Symposium on Adaptive Processes and A Special Symposium on Fuzzy Set Theory and Applications, New Orleans, LA, p 1378–1383. https://doi.org/10.1109/CDC.1977.271519
Škrabánek P, Martínková N (2021) Algorithm 1017: fuzzyreg: an R package for fitting fuzzy regression models. ACM Trans Math Softw 47(3):1–18. https://doi.org/10.1145/3451389
Stumm W, Morgan JJ (1981) Aquatic chemistry. Wiley, New York, pp 465–469, 569
Tanaka H, Uejima S, Asai K (1982) Linear regression analysis with fuzzy model. IEEE Trans Syst Man Cybern 12(6):903–907. https://doi.org/10.1109/TSMC.1982.4308925
Tansu A, Naeem Z (2022) Fuzzy regression analysis with a proposed model. Technium: Romanian J Appl Sci Technol 4(10):250–273. https://doi.org/10.47577/technium.v4i10.8121
Tibshirani R, Walther G, Hastie T (2001) Estimating the number of clusters in a data set via the gap statistic. J R Stat Soc Ser B: Stat Meth 63(2):411–423. https://doi.org/10.1111/1467-9868.00293
Tokushige S, Yadohisa H, Inada K (2007) Crisp and fuzzy k-means clustering algorithms for multivariate functional data. Comput Stat 22:1–16. https://doi.org/10.1007/s00180-006-0013-0
Vadiati M, Vadiati M, Asghari-Moghaddam A, Nakhaei M, Adamowski J, Akbarzadeh A (2016) A fuzzy-logic based decision-making approach for identification of groundwater quality based on groundwater quality indices. J Environ Manag 184(Pt 2):255–270. https://doi.org/10.1016/j.jenvman.2016.09.082
World Health Organization (WHO) (2022) Guidelines for drinking-water quality, 4th edn. WHO, Geneva, p 125–166, 249–309, 361. https://apps.who.int/iris/rest/bitstreams/1414381/retrieve. Accessed 13 Mar 2023
Zadeh LA (1965) Fuzzy sets. Inform Control 8(3):338–353. https://doi.org/10.1016/S0019-9958(65)90241-X

Chapter 12
The Burden of Disease: The Economics of Drinking Water

Tara S. Kulkarni, Jacqueline Strenio, Kahwa Douoguih, and William F. Lyons Jr.

1 Learning Objectives

1. Identify the impacts of climate change on global water resources.
2. Define key terms to explore the economics of drinking water.
3. Apply economic evaluations, including cost-effectiveness, cost-utility, and cost–benefit analysis, to identify and compare the costs and benefits of drinking water interventions.
4. Evaluate the disparate effects of contaminated water at the macro level (i.e., by region) and micro level (i.e., by gender).
5. Explain the burden of disease and secure water access through an economic lens.
6. Employ systems thinking to evaluate interrelated components of drinking water economics.

T. S. Kulkarni (✉)
David Crawford School of Engineering, Norwich University, Northfield, VT, USA
e-mail: tkulkarn@norwich.edu

J. Strenio · K. Douoguih
William Lasky School of Business, Norwich University, Northfield, VT, USA
e-mail: jstrenio@norwich.edu; kdouogui@norwich.edu

W. F. Lyons Jr.
College of Graduate & Continuing Studies, Norwich University, Northfield, VT, USA
e-mail: wlyons@norwich.edu

A. R. Molla et al. (eds.), *Climate Change, Drinking Water Security, and Public Health*, https://doi.org/10.1007/978-3-032-12074-8_12

2 Introduction: The Economic Impact of Climate Change via Global Water Resources

Water is indispensable to human health and security. Civilization is dependent on a sufficient supply of clean water for human existence and health. We need clean water for the cultivation and preparation of food. We use water to generate electricity, accelerate industrial production to meet the demands of a growing population, and support local economies.

The World Bank asserts that "Climate change expresses itself through water" (World Bank n.d.). Virtually every aspect of human civilization involves water resources that are at risk of degradation due to climate change. Ninety percent of all climate change induced natural disasters involve some nexus with water resources—either an overabundance of water or insufficient water. This climate change–water nexus directly includes floods, storm damage, droughts, and wildfires. In each case—floods, droughts, or contamination, the result is unusable water that challenges the health and well-being and the economics of individuals and nations.

2.1 *Costs of Climate Disasters*

The National Oceanic and Atmospheric Administration's National Center for Environmental Information (NCEI) reports that the United States has spent $2.77 trillion on 293 weather and climate disasters since 1980. In 2024, 19 disasters that cost $1 billion each occurred through August 8, 2024. Only 1 of these 19 was a wildfire. The remaining were storm events, tropical cyclones, and winter storm events (NCEI 2024). While the scale of spending may not compare, other regions in the world, including Asia, the Caribbean, and Central America, have also experienced disproportionate impacts from flooding, displacement, and loss of lives and livelihoods during extreme weather events (Ebi et al. 2021).

Limited water resources also play a role in contributing to other natural and man-made disasters, such as famine, desertification, and human insecurity. Poor WASH (water, sanitation, and hygiene) services, including unreliable and nonfunctional water supply and underused or deteriorating toilets, impact health, livelihoods, school attendance, and dignity (Satriani et al. 2022; Haque and Freeman 2021). For instance, poor water sanitation is the primary cause of diarrhea in children under 5 years in Bangladesh. It is estimated that, in 2018, the economic cost of treating diarrhea in Bangladesh was $79 million, or roughly 4% of gross domestic product (GDP) (Hasan et al. 2021).

The World Bank has noted that economic growth is a "thirsty business." Water scarcity could impact as much as 6% of the global GDP. This impact figure may be conservative, given a 2023 World Wildlife Fund-Dalberg report that estimates the annual value of water and freshwater ecosystems to be $58 trillion, which is equivalent to 60% of global GDP. At least one prominent economist has suggested that water will be the most important physical commodity asset class, exceeding the importance of most other extractives (Stuckenberg 2018).

2.2 *Production Disruption Due to Climate Change*

The ability of water scarcity to disrupt entire sectors and, therefore, societies is not new. Tzanakakis et al. (2020) document the role of megadroughts experienced by ancient kingdoms, including those in Egypt and Khmer, impacting the agricultural sectors most severely and leading to the collapse of entire civilizations. The compounding factors of the past several decades, from the dust bowls of the 1930s through forest fires, deforestation, soaring temperatures, and growing conflicts, have catapulted water scarcity into one of the top global risks.

Petrakopoulou (2021) reports that in 2021 the energy sector relied on 41% and 42% of freshwater withdrawals in the United States and Europe, specifically for the cooling process in thermoelectric power generation. China used one-tenth of their freshwater sources for power plants. The cost of water impacts can be optimized by controlling for some of the operational conditions such as efficiency, temperature rise and the minimum temperature difference in the condenser. Revollo-Fernández and Rodrigues-Tapia (2022) provide an example of the Valley of the Mexican Basin (VMB) that houses 20% of the nation's population and contributes to over a third of its Gross National Product (GNP). A fifth of that GNP is a result of the manufacturing sector in this region through employment opportunities and the offer of goods and services. A Relative Water Stress Index (ratio of total water volume use to mean freshwater availability) of 140% also puts this region in a high-water stress category.

A further consideration in the economics of water is its relationship to food. As one example, water is now being traded throughout the world in the form of grain. Arid countries, such as those in the Middle East and North Africa (MENA), must import grain. It takes 1000 cubic meters of water to produce 1 ton of grain. By importing grain, countries are, in effect, importing water. Water needed to produce food for one person is 2000 L/day for a plant-based diet and 4000 L/day for a livestock-based diet (Brown 2009).

Therefore, in this chapter, we introduce an economic perspective to examine drinking water and introduce tools used by economists and policymakers to evaluate the burden of disease. First, we provide an overview of using cost–benefit analysis to provide insight into the economic causes, costs, and consequences associated with contaminated water, along with related insights for public policy. We also provide an overview of recent analyses detailing these economic costs and consequences associated with contaminated drinking water. This chapter employs a gender-aware lens, highlighting that the costs and consequences of contaminated water are highly gendered, with women and girls often bearing a greater brunt of the burden.

3 Economic Evaluation of Drinking Water

Economists often use tools of economic evaluation to calculate the costs associated with a particular disease, illness, social ill, or intervention (a partial economic evaluation that just includes analysis of costs, i.e. cost identification). Often, these evaluations also include analysis of effects and/or benefits in addition to the costs. Analyses may highlight specific effects in terms of natural units (i.e., cost-effectiveness analysis), multiple effects in terms of some health-adjusted quality of life measurement (i.e., cost-utility analysis using quality-adjusted life years (QALYs), or Disability Adjusted Life Years (DALYs), or with multiple effects measured in the same monetary units as the costs to allow for explicit comparison of costs and outcomes (i.e., cost–benefit analysis).

In the early 90s, the Harvard School of Public Health and the World Health Organization (WHO) collaborated on the first-ever Global Burden of Disease (GBD) study. This effort quantified the health effects of over 100 diseases and injuries in eight global regions. The report also introduced a new metric—the Disability Adjusted Life Year (DALY)—the number of years lost due to ill-health, disability, or a premature loss of life (Mathers 2020). The WHO and the Institute for Health Metrics and Evaluation (IHME) have also specifically focused on the disease burden posed by unsafe drinking water as well as inadequate hygiene and sanitation (WASH) conditions. These studies estimated the burden specifically for diarrhea, acute respiratory infections, as well as undernutrition, malaria, soil-transmitted helminthiasis, trachoma, and schistosomiasis (Wolf et al. 2023). Researchers looked at the infrastructure, such as access to piped treated water and sewage systems, handwashing, etc. However, the availability and accuracy of the infrastructure and related epidemiological data are limited. Economists and health researchers, therefore, continue using various tools to evaluate the burden of disease. For example, Wolf et al. (2023) used a comparative risk assessment methodology to develop their estimates of the disease burden attributable to unsafe drinking water and sanitation conditions. The economic methodologies are described in the following sections.

3.1 Cost-Effectiveness Analysis

Cost-effectiveness analysis is a way to examine both the costs and health outcomes of one or more interventions. It compares an intervention to another intervention (or the status quo) by estimating how much it costs to gain a unit of a health outcome, like a life year gained or a death prevented.

3.2 Cost-Utility Analysis

Cost-utility analysis is a type of cost-effectiveness analysis that compares different procedures and outcomes relative to a person's health-related quality of life.

When analyzing any good or service, economists first determine whether that product is a private good—a product or service produced by a privately owned business and purchased to increase the utility, or satisfaction, of the buyer, or a public good—a good that is both non-excludable and non-rivalrous. Most goods and services traded in a market economy are private goods. Water is unique in this regard in that while it is a commodity traded in markets, it can be described as both a public and a private good. This determination depends on how it is accessed and managed.

Water is a public good when it cannot easily be restricted to paying customers only (non-excludable). In its natural state, like rivers, lakes, and oceans, water is often considered non-excludable because access to it can't be easily restricted. For example, a river that flows through a community is naturally available to everyone in the area. Public goods are also non-rivalrous, meaning one person's use does not diminish the availability of the resource for others. Water in natural settings, to some extent, can be non-rivalrous. For example, the ocean can provide water to many people without one person's use directly affecting another's access.

Water becomes a private good when access can be controlled and limited to paying consumers. For instance, when water is treated, purified, and delivered to homes through a municipal water system or bottled by private companies, it becomes excludable. People have to pay for access to this water, and those who don't pay can be denied service. Water is also rivalrous when it is treated as a private good because one person's consumption of the water can reduce the quantity available for others. For example, when water is distributed through a piped system to households, the use by one household means less water is available in the system for others, especially in areas where water is scarce.

In the private sector, water can be owned, traded, and managed by corporations or individuals. Bottled water companies, for example, treat water as a commodity that is extracted, processed, packaged, and sold for profit. Here, water is subject to market forces like any other private good.

In the case of water, its hybrid status as a public and private good means that the costs and benefits of its quality and scarcity may be borne by many different stakeholders that may or may not be involved directly in the market activities that determine how the water is allocated and whether it is safe to drink. This makes it difficult, if not impossible, to rely exclusively on markets to ensure its safety and accessibility and poses a challenge to policymakers, especially as climate change makes the dynamics of our water systems more volatile. If a factory pollutes a drinking water source, in the absence of any intervention by the government, the cost of that pollution is borne by the downstream communities. This is an example of an externality—when a production or consumption decision by one party imposes a cost or benefit on another party that was not involved in the initial decision.

The following sections will introduce evidence from around the world of various direct and indirect costs associated with the burden of disease created by unsafe drinking water and the economic benefits that can be realized from clean and accessible water supplies.

4 Breaking Down the Economics of Drinking Water

This section provides a breakdown of the economic costs and benefits of providing clean drinking water, and the economic costs of contaminated drinking water. A Dig Deep report found that 2.2 million people in the United States have insufficient access to running water and adequate sanitation in their residences. The report quantified the "water access gap" with a price tag for time lost at $846 million, physical health at $762 million, mental health at $218 million, water purchase costs at $291 million, and additional GDP impacts from lost productivity at $924 million (Thatte 2022).

It is not only expensive to access, treat, and distribute clean drinking water to homes, businesses, industries, and farms; it is that much harder when the original source is contaminated by natural or manmade sources of pollution. Costs also present themselves indirectly in the form of lost income, poor health, time away from educational and recreational pursuits, and loss of dignity. The benefits of clean drinking water go far beyond good health to improved well-being that helps societies flourish through improved economies and restoration of peace and dignity.

4.1 Economic Costs of Providing Clean Drinking Water

Drinking water sources may span from surface water (rivers, lakes, ponds) to aquifers holding groundwater reserves. It has cost the United States over $1.9 trillion, or $140 per person per year, since 1960 to reduce the contamination of natural drinking water sources (Frazer 2018). Additional sources also include wastewater effluent that may be further treated to serve as a drinking water source or ocean water that has undergone a desalinization process. Potable water, considered safe for human consumption, is governed by various water quality regulations depending on the standards provided by the WHO, and that may vary by country. It is estimated that by 2030, water and sanitation infrastructure will involve an estimated investment of over 0.9–1.5 trillion United States Dollars (USD) per year, with 70% of these dollars for urban areas in the global south (UNESCO 2023).

4.2 Economic Costs of Contaminated Drinking Water

The evidence on the economic consequences of contaminated drinking water includes, for example, the direct costs associated with increased healthcare utilization resulting from water-borne disease as well as the costs associated with treating water. There are also significant indirect costs, including reduced productivity and loss of paid or unpaid work resulting from morbidities and premature mortality. Other indirect costs include time costs associated with fetching and treating water. These costs span both the short- and long-run, affecting individuals and caregivers, and accruing to the individual household, educational institutions, employers, and society. Costs can be estimated at the individual/unit or macro level. This section provides an overview of the types and recent estimates of the economic costs associated with contaminated drinking water.

Direct costs include those resulting from adverse health effects, and increased healthcare use resulting from water-borne diseases. For example, arsenic-contaminated drinking water exposure in Bangladesh during pregnancy is associated with lower birth weight among pregnant women (Kile et al. 2016) and negative health effects among children (Asadullah and Chaudhury 2011). DeFlorio-Barker et al. (2018) studied the annual economic burden of waterborne illnesses from recreational activities in the United States. The activities included swimming, fishing, paddling, or motor boating, and found that the over 90 million cases of gastrointestinal, ear, eye, respiratory, or skin infections and illnesses each year cost the United States approximately $2.9 billion.

Indirect costs include those related to the time costs of water treatment (i.e., fetching and boiling water) as well as those associated with illness from waterborne diseases that can result in adverse consequences in terms of cognitive function, educational attainment, and labor market outcomes. Pitt et al. (2021) found long-term negative effects of arsenic contamination in well water in rural Bangladesh on cognition and lower educational attainment. In a systematic review of arsenic exposure in drinking water and neurobehavioral effects, Boffetta et al. (2024) note the limitation in this area of inquiry: limited 'high quality' causal studies. Two articles that Boffetta et al. (2024) categorize as high-quality find consistent reporting that exposure to arsenic-contaminated drinking water is associated with declines in cognitive functioning. Nevertheless, the authors' conclusion is that there is a lack of evidence for a causal association between exposure to arsenic in drinking water and neurobehavioral effects.

Regardless of cognitive effects, research has documented adverse educational effects. For example, among public school children in Assam, India, exposure to contaminated groundwater is associated with increased school absenteeism, grade retention, and decreased test scores and grade point averages (Aggarwal et al. 2024). In the same study, but using nationally representative data, the authors find that exposure to arsenic-contaminated drinking water is also negatively associated with school attendance. Asadullah and Chaudhury (2011) found similar results in Bangladesh: using nationally representative data, arsenic-contaminated drinking tube wells are negatively and significantly associated with children's educational achievement, proxied by mathematics scores.

There are also significant indirect costs associated with water-fetching. For example, Dhital et al. (2022) find that time spent water-fetching in remote hilly and mountainous areas of Nepal is negatively and significantly associated with the completion of primary school for girls. While boys' completion rates are less affected, increased time in water fetching is associated with an increased likelihood of boys repeating a grade.

Differential costs by gender aren't surprising given that globally, women and girls are responsible for water collection in 80% of households without water on premises (WHO, World Bank, UNICEF 2022). Such responsibilities may also expose women and girls to further harm, including the direct and indirect costs of adverse health effects of carrying water (Das 2017) as well as sexual assault and resulting intangible costs (WHO, World Bank, UNICEF 2022).

Indirect costs stemming from reduced educational attainment affect occupational trajectories. Pitt et al. (2021) found that the young men in their study with higher concentrations of arsenic were less likely to have a skilled occupation or become entrepreneurs. For women, they found that arsenic poisoning is associated with reduced domestic productivity. These findings emphasize that lost productivity costs should include paid and unpaid labor analyses. This lost productivity spans missed workdays (absenteeism) and reduced productivity while at work (presenteeism). Further, such illnesses also result in the need for a caregiver, meaning that not only are there productivity losses associated with the sick individual, but also those that result from a caregiver having to skip paid or unpaid work.

Women and girls are much more likely to be responsible for water treatment and caregiving of sick individuals within a household, meaning that the indirect costs of contaminated drinking water are highly gendered (Das 2017). There is also some argument that the provision of clean water could free up unpaid work time so that women could instead engage in paid work, which may contribute to increasing economic growth. Zhou and Turvey (2018) found that drinking water access and quality affect the off-farm labor supply in China, and the impacts are greater for women than men. They also found that there are significant within-gender effects, that is, while daughters are not sensitive to either water access or quality, daughters-in-law are sensitive to water quality. The authors hypothesize two channels previously discussed in this chapter: obtaining safe drinking water is time-consuming, and water accessibility and quality may affect labor participation through health effects (either directly or indirectly through reallocating time to care for sick family members).

4.3 Economic Benefits of Safe Drinking Water

In cost–benefit analysis (CBA), economic benefits are measured in monetary units, which allows for explicit comparison of benefits to costs in the same units. Because the benefits are being expressed in monetary units, CBA can include single or multiple effects. Although there is no one way to aggregate economic benefits, a common approach includes considering resources saved (including morbidities and mortalities averted), peoples' willingness to pay for the intervention, and any other value created.

Often then, when considering the economic benefits of access to safe drinking water, many of the benefits are the inverse of the costs associated with contaminated drinking water. A World Health Organization report found that India's Jal Jeevan Mission could prevent 400,000 diarrheal deaths and 14 million DALYs for a cost saving of $101 billion (Pitt et al. 2021). Other examples include saved resources, including time savings from not having to fetch and treat water and/or care for the sick, reduced morbidities and mortalities from waterborne illnesses, and the associated increased productivity in paid and unpaid work. Resources saved also include those associated with reduced healthcare costs. In a global CBA of WASH services, Hutton (2018) found that the economic returns to access to drinking water and sanitation exceed the costs in all regions and in all interventions examined, with benefit–cost ratios of 3.4 and 6.8 in urban and rural areas, respectively. That is, for every dollar of costs, water access results in 3 to nearly 7 dollars in benefits.

As the variation in benefit–cost ratios hint at, access to safe drinking water can reduce disparities, including those between rural and urban individuals and rich and poor households (WHO, World Bank, UNICEF 2022), and especially by gender. In a joint report on the state of drinking water globally, the WHO, UNICEF, and World Bank describe safe drinking water as "a service that provides immeasurable economic and health benefits, and essential gender equality outcomes" (World Bank 2007). And indeed, gender equality, in addition to being intrinsically important, is also an economic benefit; as the World Bank described, gender equality is "smart economics" (World Bank 2007), and Sustainable Development (SDG) Goal 5. In fact, even though SDG Goal 5 is specifically called out here, progress in many of the 17 SDGs of the United Nations can directly impact and are impacted by progress or lack thereof in SDG 6, Clean Water and Sanitation.

5 Safe, Secure, and Affordable Drinking Water

One specific aspect that connects the burden of disease and the economics of drinking water is not typically addressed in scientific and technical textbooks. It is catalyzed by growing conflict due to political unrest or environmental degradation, exacerbated by climate change. Often embedded in policy and strategy literature, this chapter calls out these issues to underscore that the discussion on disease and water economics is incomplete without highlighting these. For example, the 2011 Syrian civil war was triggered by an extended drought (Stuckenberg 2018).

The Republic of Yemen, located on the Arabian Peninsula, is experiencing one of the world's most severe water supply and sanitation crises in the world. As a result of extreme poverty, water scarcity, and persistent conflict, the percentage of Yemen's population without access to clean water increased from 48% in 2015 to 69% in

2017. As of 2018, approximately 19.4 million people lack clean drinking water and adequate sanitation. The lack of adequate sanitation resulted in almost 249,900 suspected cases of cholera in 2024 through December 1, 2024 (Abu-Lohom et al. 2018).

In 2018, Cape Town was experiencing its most severe drought in over a century (Taing et al. 2019). The city embarked on a countdown to "Day Zero," the day when Cape Town would become the first major city in the world to run out of municipal water. The government implemented severe restrictions on municipal water use to avoid Day Zero. These restrictions had widespread economic and societal impacts, requiring immediate and drastic mitigation measures.

An example of the impacts of climate change-induced natural disasters is ongoing in South Sudan. The impact of climate change on migration and conflict in South Sudan is well documented (Tiitmamer 2020). The flood seasons have increased in intensity and duration and caused food insecurity throughout the Nile River valley and the communities in the Sudd (Mayen et al. 2023). In South Sudan, there are 2 million climate refugees, and 76% of the population requires food assistance because historic farmlands and pastures have been underwater for several years. As a result of this food insecurity, the United States Agency for International Development (USAID) committed $288 million in humanitarian aid to South Sudan in 2023, largely in conjunction with their partner, the United Nations World Food Programme (WFP; USAID 2023).

The cost of overabundant water extends well beyond the monetary costs of feeding the food-insecure population of South Sudan. The nexus between climate change, water resource migration, and reduced human security is well established in South Sudan. The resulting degradation in human security has resulted in intertribal violence and political strife, from cattle rustling to extrajudicial killings (Mednick 2020). The costs also included a lack of societal stability and corresponding loss of productivity as well as the cost of stationing a United Nations mission in the country. Estimates of the cost of the unrest in South Sudan exceed $200 billion (Breckenridge 2015).

6 Systems Thinking

One approach to examining the relationship between such diverse considerations as water, public health, economics, and human security is to use systems thinking. The use of systems thinking is a critical concept in evaluating the role of water resources in a complex context of societal considerations. A causal loop diagram is a useful way to conceptualize the inter-relationship between water resources and public health, economic vitality, and human security. A causal loop diagram is not necessarily supported by facts or citations but represents a hypothesis of causation to provide a framework for systematically solving problems while avoiding unintended second and third-order effects.

A causal loop diagram conceptualizing the systems approach to the relationship between water resources, public health, economics, and human security is illustrated in Fig. 12.1 (Lyons 2024). By placing water resources in the hypothesized causal loop and relating the impacts of other factors on water resources, and then relating the impact of water resources on still other factors, the interrelationships can all be considered in the full context of the environment being analyzed.

For instance, in the causal loop diagram, migration causes urbanization, which causes a loss of agriculture, leading to water resource depletion. Simultaneously, climate change is causing floods, storms, desertification, famine, and other effects that also impact water resources. Urbanization causes concentrated poverty, which results in water resource depletion. Urbanization leads to water pollution, which leads to water resource depletion. Balancing all these factors, international development aid improves physical and social infrastructure, addressing water pollution and improving water resources generally, thereby improving human security and reducing the likelihood of wider conflict. And so on, expanding the range of impacted systems to consider the entire system of societal factors.

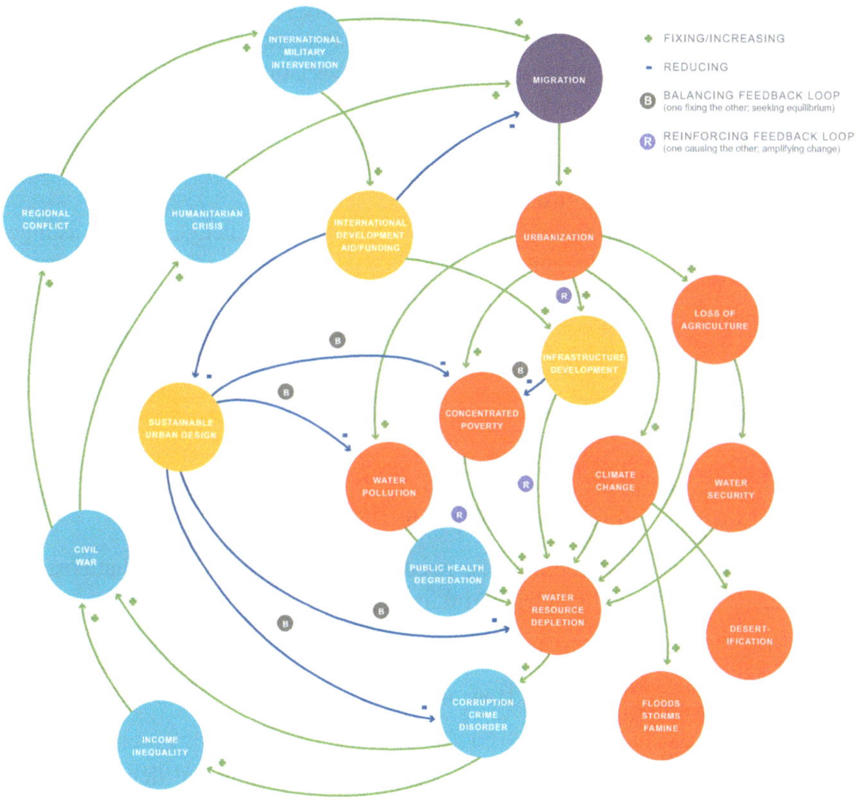

Fig. 12.1 Causal loop diagram: a systems approach to explore the relationship between water resources, public health, economics, and human security (Lyons 2024) [Reproduced from Lyons 2024, Figure 1. https://doi.org/10.5604/01.3001.0054.7399, licensed under the terms of the Creative Commons Attribution 4.0 International License (http://creativecommons.org/licenses/by/4.0/)]

Considering the relationship between any two of these factors would prevent consideration of the wider system within which these factors are operating. Creating a causal loop diagram that is sufficiently inclusive of most factors allows us to consider the second and third-order effects of our actions before testing them and potentially suffering unintended consequences. The use of systems thinking is an important construct in assessing societal problems and ideating solutions.

7 Conclusions

In this chapter, the impacts of climate change on global water resources were identified with a closer look at the economics of drinking water. Water is tagged as a public or private good, depending on the context. Economists use tools, including cost–benefit analyses, to assign value to this resource. However, given its indispensable nature and role in building and sustaining thriving societies, it is imperative that any discussions on the economics of drinking water include additional variables. The burden of disease is an example of one such variable that calls out the impact of decreasing or unsafe supplies of clean drinking water. This chapter highlights other considerations such as migration, urbanization, gender disparity, social and political unrest, agricultural losses, and poverty that accelerate the impacts of climate change and barricade sustainable development.

A systems thinking approach that accounts for all the challenges as well as ingenuity to innovate with and against constraints will be critical in finding long-lasting and resilient solutions to the wicked problem of providing clean drinking water to all.

8 Discussion Questions

1. The World Bank has noted that economic growth is a "thirsty business." Do you agree? Justify your response with specific examples.
2. When is water considered a public good versus a private good? What costs are associated with providing clean drinking water?
3. List and explain three benefits of providing clean drinking water.
4. Define the burden of disease. How are the burden of disease and the economics of drinking water correlated?
5. What are some ways in which climate-driven environmental degradation has intersected with civil and political unrest? What are the consequences?
6. What is a causal loop diagram? How is it used as a tool in a systems thinking approach?

References

Abu-Lohom NM, Konishi Y, Mumssen Y, Zabara B, Moore SM (2018) Water supply in a war zone: a preliminary analysis of two urban water tanker supply systems in the Republic of Yemen. World Bank Group, Washington, DC. https://doi.org/10.1596/30107

Aggarwal K, Barua R, Vidal-Fernandez M (2024) Still waters run deep: groundwater contamination and education outcomes in India. Econ Educ Rev 100:102525. https://doi.org/10.1016/j.econedurev.2024.102525

Asadullah MN, Chaudhury N (2011) Poisoning the mind: arsenic contamination of drinking water wells and children's educational achievement in rural Bangladesh. Econ Educ Rev 30(5):873–888. https://doi.org/10.1016/j.econedurev.2011.05.001

Boffetta P, Sambati L, Sassano M (2024) Systematic review of studies on exposure to arsenic in drinking water and cognitive and neurobehavioral effects. Crit Rev Toxicol 54(3):174–193. https://doi.org/10.1080/10408444.2023.2297751

Breckenridge A (2015) South Sudan: the cost of war: an estimation of the economic and financial costs of ongoing conflict. Frontier Econ. https://www.frontier-economics.com/uk/en/news-and-insights/articles/article-i4976-south-sudan-the-cost-of-war/. Accessed 1 Oct 2024

Brown LR (2009) Plan B 4.0: mobilizing to save civilization. W.W. Norton, New York

Das, Maitreyi Bordia. (2017). "The Rising Tide: A New Look at Water and Gender." World Bank, Washington, DC.

DeFlorio-Barker S, Wing C, Jones RM, Dorevitch S (2018) Estimate of incidence and cost of recreational waterborne illness on United States surface waters. Environ Health 17:3. https://doi.org/10.1186/s12940-017-0347-9

Dhital RP, Ito T, Kaneko S, Komatsu S, Yoshida Y (2022) Household access to water and education for girls: the case of villages in hilly and mountainous areas of Nepal. Oxf Dev Stud 50(2):142–157. https://doi.org/10.1080/13600818.2021.1965978

Ebi KL, Vanos J, Baldwin JW, Bell JE, Hondula DM, Errett NA, Hayes K, Reid CE, Saha S, Spector J, Berry P (2021) Extreme weather and climate change: population health and health system implications. Annu Rev Public Health 42:293–315. https://doi.org/10.1146/annurev-publhealth-012420-105026

Frazer K (2018) Study examines costs, benefits of clean water measures. Cornell Chronicle, Ithaca, NY. https://news.cornell.edu/stories/2018/11/study-examines-costs-benefits-clean-water-measures. Accessed 1 Oct 2024

Haque SS, Freeman MC (2021) The applications of implementation science in water, sanitation, and hygiene (WASH) research and practice. Environ Health Perspect 129(6):065002. https://doi.org/10.1289/EHP7762

Hasan MZ, Mehdi GG, De Broucker G, Ahmed S, Ali MW, Martin Del Campo J, Constenla D, Patenaude B, Uddin MJ (2021) The economic burden of diarrhea in children under 5 years in Bangladesh. Int J Infect Dis 107:37–46. https://doi.org/10.1016/j.ijid.2021.04.038

Hutton G (2018) Global benefits and costs of achieving universal coverage of basic water and sanitation services as part of the 2030 Agenda for Sustainable Development. In: Lomborg B (ed) Prioritizing development: a cost benefit analysis of the United Nations' Sustainable Development Goals. Cambridge University Press, Cambridge, pp 422–445. https://doi.org/10.1017/9781108233767.025

Kile ML, Cardenas A, Rodrigues E, Mazumdar M, Dobson C, Golam M, Quamruzzaman Q, Rahman M, Christiani DC (2016) Estimating effects of arsenic exposure during pregnancy on perinatal outcomes in a Bangladeshi cohort. Epidemiol 27(2):173–181. https://doi.org/10.1097/EDE.0000000000000416

Lyons WF (2024) Climate action as a strategic deterrent. Sci J Military Univ Land Forces 56(3):25–39. https://doi.org/10.5604/01.3001.0054.7399

Mathers CD (2020) History of global burden of disease assessment at the World Health Organization. Arch Public Health 78(77). https://doi.org/10.1186/s13690-020-00458-3

Mayen JV, Wood E, Frazier T (2023) Practical flood risk reduction strategies in South Sudan. J Emerg Manag 20(8):123–136. https://doi.org/10.5055/jem.0669

Mednick S (2020) Burkinabe civilians caught in jihadist in-fighting. The New Humanitarian. https://www.thenewhumanitarian.org/2020/08/17/burkina-faso-s-new-conflict-front-jihadists-against-jihadists. Accessed 1 Oct 2024

National Center for Environmental Information (NCEI) (2024) U.S. billion-dollar weather and climate disasters, 1980—Present. https://www.ncei.noaa.gov/access/metadata/landing-page/bin/iso?id=gov.noaa.nodc:0209268. Accessed 1 Oct 2024

Petrakopoulou F (2021) Defining the cost of water impact for thermoelectric power generation. Energy Rep 7:2101–2112. https://doi.org/10.1016/j.egyr.2021.04.001

Pitt MM, Rosenzweig MR, Hassan MN (2021) Identifying the costs of a public health success: arsenic well water contamination and productivity in Bangladesh. REStu 88(5):2479–2526. https://doi.org/10.1093/restud/rdaa087

Revollo-Fernández, D. A., & Rodríguez-Tapia, L. (2022). Water scarcity reduces the efficiency of the manufacturing industry in the valley of Mexico Basin: DEA-based two-stage efficiency analysis. Applied Economics Letters, 29(13), 1193–1198. https://doi.org/10.1080/13504851.2021.1918623

Satriani S, Ilma IS, Daniel D (2022) Trends of water, sanitation, and hygiene (WASH) research in Indonesia: a systematic review. Int J Environ Res Public Health 19(3):Article 3. https://doi.org/10.3390/ijerph19031617

Stuckenberg MDJ (2018) Water scarcity: the most understated global security risk. Harv Nat'l Sec J. https://harvardnsj.org/2018/05/18/water-scarcity-the-most-understated-global-security-risk/. Accessed 1 Oct 2024

Taing L, Chang CC, Pan S, Armitage NP (2019) Towards a water secure future: reflections on Cape Town's Day Zero crisis. Urban Water J 16(7):530–536. https://doi.org/10.1080/1573062X.2019.1669190

Thatte K (2022) How America's hidden water poverty is costing the U.S. economy over $8.5 billion each year. Spotlight on Poverty and Opportunity. https://spotlightonpoverty.org/spotlight-exclusives/how-americas-hidden-water-poverty-is-costing-the-u-s-economy-over-8-5-billion-each-year/. Accessed 1 Oct 2024

Tiitmamer N (2020) S. Sudan: climate change, migration and conflict—HLRN. https://mail.hlrn.org/activitydetails.php?title=S.-Sudan:-Climate-Change,-Migration-and-Conflict&id=pm9oag==. Accessed 1 Oct 2024

Tzanakakis VA, Paranychianakis NV, Angelakis AN (2020) Water supply and water scarcity. Water 12(9):Article 9. https://doi.org/10.3390/w12092347

United Nations Educational, Scientific and Cultural Organization (UNESCO) (2023) Valuing hydraulic infrastructure. 2021 World water development report. https://www.unesco.org/reports/wwdr/2021/en/valuing-hydraulic-infrastructure. Accessed 1 Oct 2024

United States Agency of International Development (USAID) (2023) United States provides more than $288 million in additional humanitarian assistance to South Sudan. https://www.usaid.gov/news-information/press-releases/feb-16-2023-united-states-provides-more-288-million-additional-humanitarian-assistance-south-sudan. Accessed 1 Oct 2024

World Health Organization (WHO), World Bank, United Nations International Children's Emergency Fund (UNICEF) (2022) State of the world's drinking water: an urgent call to action to accelerate progress on ensuring safe drinking water for all. https://www.who.int/publications/i/item/9789240060807. Accessed 1 Oct 2024

Wolf J, Johnston RB, Ambelu A, Arnold BF, Bain R, Brauer M, Brown J, Caruso BA, Clasen T, Colford JM, Mills JE, Evans B, Freeman MC, Gordon B, Kang G, Lanata CF, Medlicott KO, Prüss-Ustün A, Troeger C et al (2023) Burden of disease attributable to unsafe drinking water, sanitation, and hygiene in domestic settings: a global analysis for selected adverse health outcomes. The Lancet 401(10393):2060–2071. https://doi.org/10.1016/S0140-6736(23)00458-0

World Bank (n.d.) Overview. World Bank. https://www.worldbank.org/en/topic/water/overview. Accessed 8 Aug 2024

World Bank (2007) Global monitoring report 2007. Millennium development goals: confronting the challenges of gender equality and fragile states. World Bank. https://documents.worldbank.org/en/publication/documents-reports/documentdetail/309311468324011934/Global-Monitoring-Report-2007-Millennium-Development-Goals-confronting-the-challenges-of-gender-equality-and-fragile-states. Accessed 1 Oct 2024

Zhou L, Turvey CG (2018) Drinking water and off-farm labour supply: between-gender and within-gender bias. Aust J Ag Res Econ 62(1):103–120. https://doi.org/10.1111/1467-8489.12239

Chapter 13
Climate Change and the Economics of Accessing Safe Drinking Water: Contexts of the Physiographic and Climatic Vulnerable Areas of the Bengal Basin

Anwar Zahid and Ibrahim R. Shams

1 Learning Objectives

1. Explain how people in different countries, regions, and development levels pay for drinking water (in units of a standard currency and percent of median yearly income per capita).
2. Explain how much people pay for water treatment.
3. Describe the interaction between the economic burden of safe drinking water supply and human health.
4. Explain how this will change with climate change.

2 Introduction

Ensuring the availability and sustainable management of water for all is the prime target of the 2030 Agenda for Sustainable Development Goal Target 6.1. In 2020, safely managed drinking water was available to about 74% of the world's population, which was 62% in 2000 (WHO-UNICEF-World Bank 2022). However, about 2 billion people still lack safe drinking water. People in developing countries are most at risk in places with high population density and demand for water resources

A. Zahid (✉)
Department of Disaster Science and Climate Resilience, University of Dhaka, Dhaka, Bangladesh

Bangladesh Water Development Board, Dhaka, Bangladesh

I. R. Shams
WEB-AIS Project, University of Applied Sciences, Koeln, Germany

A. R. Molla et al. (eds.), *Climate Change, Drinking Water Security, and Public Health*, https://doi.org/10.1007/978-3-032-12074-8_13

(Luijten et al. 2001). The problem is particularly frequent in rural areas, where water scarcity impacts the health and livelihood of most people (FAO 2007). Hence, the world is not even near meeting the SDG's goal of safe water by 2030. Continued support for SDG Target 6.1 and the additional acceleration to gain universal access is threatened by the enhancing impacts and challenges of competing agricultural and ecological water demands, climate change, competing financial priorities, and the threats of existing and emerging degradation and depletion of water resources. Diarrheal diseases, endemic in 69 countries, are largely attributable to unsafe drinking water and poor sanitation, causing severe disease pressure. Bangladesh largely relies on groundwater to meet its domestic demand for potable water. The nation reached near-complete coverage in providing drinking water until its progress was endangered by high levels of arsenic in the shallow groundwater and saltwater intrusion into the coastal groundwater basins. The World Health Organization (WHO suggests that in Bangladesh, over 97% of the population has access to drinking water, 40% have sufficient access to sanitation facilities, and more than 60% consume water contaminated with arsenic; Saima 2022). In Bangladesh, urban water supply coverage is >85%, and piped service is 31%. This is about 82% for the rural population and only 1%, respectively. Nevertheless, further progress in ensuring better water and sanitation service delivery in rural areas, particularly in physiographic and climatic vulnerable Hard-to-Reach (HtR) areas, remains a significant challenge in the country. In areas designated as HtR, the availability of clean water is just 24%. When looking at the various indicators and their specific criteria, it was found that 1144 HtR unions, representing 21% of these areas, fall into 11 different geographical types, such as chars (sand bars), haors (large swamps), floodplains, hill tracts, coastal regions, and more. This spans across 257 upazilas (sub-districts) and 50 districts in Bangladesh. A study by the World Bank in 2019 on the combined risk of groundwater contamination from arsenic and salinity, along with the depletion of groundwater storage, categorized a significant portion of Bangladesh's land area, ranging from 5 to 24%, as being at an extremely high to high risk. This puts between 6.5 million and 24.4 million people at risk from a triple threat of high arsenic levels, salinity, and the depletion of groundwater storage. The research indicates that the most vulnerable populations, particularly the poor, face groundwater risks to varying extents. Currently, around 5 million of the poor are at risk, with the potential for another 6 million to be affected in the future. Additionally, 11 million poor individuals are at risk from shallow groundwater contaminated with arsenic and salinity. Consequently, numerous government agencies, international organizations, and non-governmental organizations (NGOs) are working tirelessly to address, survey, and resolve water and sanitation issues to ensure these essential services are accessible to everyone. However, there is a pressing need for more efforts and activities to improve water and sanitation facilities, especially in remote and HtR areas. It is also crucial to enhance and tailor water and sanitation services to meet the basic needs of these communities. The demand for these services is particularly high among marginalized groups and vulnerable regions.

The main goals of the Bangladesh safe water supply and sanitation policies focus on making sure everyone can easily access basic water and sanitation services, encouraging people to change how they use water and sanitation, reducing water-borne diseases by helping local governments and communities deal with water-related issues, and promoting long-lasting water and sanitation services. The Department of Public Health Engineering (DPHE) is the main government organization responsible for providing water supply and sanitation facilities in rural Bangladesh. NGOs like NGO Forum for Public Health, Bangladesh Rural Advancement Committee (BRAC), and WaterAid also play a significant role in this area. NGO Forum works with partner organizations to study and address problems, assess demand, evaluate the impact of existing facilities, and choose appropriate technologies for different areas, including HtR areas. For urban water supplies, local municipalities and water supply and sewerage authorities are linked to these responsibilities.

2.1 The Major Sources and Technologies of Water Supply in Bangladesh

In rural Bangladesh, the shallow hand tubewell (STW) represents the most prevalent low-cost tube-well technology, functioning in suction mode. Typically, an STW comprises a number 6 pump head, along with well pipes and filters that have a diameter of 38.0 mm (millimeters; 1.5 in.). This system is straightforward to install using local hand percussion techniques and is suitable for shallow water table areas. The primary benefits of STWs include their relatively low installation costs and the absence of maintenance expenses. However, these tubewells often become nonfunctional during dry irrigation when the groundwater table falls below the suction limit of 7.5 m from the tubewell head.

Conversely, deep hand tube wells (DTW) also operate in suction mode, extracting groundwater from deeper aquifers where the potentiometric surface of the groundwater table is located within 7.5 m below the ground surface. The term "deep aquifer" can vary based on its intended use. Both STWs and DTWs can feature multiple outlets. Generally, deep groundwater is of high quality, with no arsenic detected in DTWs that reach depths of over 250–300 m in Bangladesh. Hand pumps are applicable in many regions, as the groundwater table typically remains within the suction limit of 7.5 m from the surface. While these systems can be costly for individual households, they are often provided for community water supply, potentially reducing costs through multiple heads.

In flood-prone areas, double platform tubewells (STW and DTW) are employed to utilize water resources during dry and flood seasons, ensuring a reliable supply even during heavy flooding. One platform is constructed at a height of approximately 1.0–1.5 m, generally above the highest flood level, thus remaining unaffected by floods and usable during flood seasons. The second platform is situated at the level of homestead land, serving its purpose during the dry season. An elbow pipe connects the main boring pipe of the lower platform to the higher platform.

Piped water supply systems in rural Bangladesh are developed using both STW and DTW, as well as treated pond water, to serve clustered rural settlements. In terms of convenience for collection and usage, piped water systems are the only ones that can effectively compete with existing water supply methods. A significant number of stand-posts equipped with water delivery taps are linked to groups of households within distances of up to 1 km. The benefits of a rural piped water supply include the delivery of water in close proximity to users, safeguarding well-maintained piped water from external contamination, and regulating the volume of water supplied. Nevertheless, this system is susceptible to pipeline damage and surface pollution during natural disasters such as storm surges and floods.

The decline of the water table and its seasonal variations between wet and dry periods pose significant challenges for rural water supply systems reliant on tubewells. Given that number 6 suction mode pumps are ineffective in extracting groundwater when the water table falls below 7.5 m from the surface, there is a pressing need to innovate tubewell technology suitable for low water table regions such as Barind. Consequently, force-mode pumps have been introduced. In instances where the water table depth is marginal and experiences a reduction during the dry irrigation season, Tara tube wells are designed to be preferred, as they can extract water from depths of 20–25 m below the ground surface. The deep-set hand tubewell (DSHTW), which operates in force mode, can draw water from depths ranging from 30 to 37 m, contingent upon the specific type of DSHTW utilized.

Arsenic-iron removal plants (AIRPs) represent a cost-effective solution for diminishing arsenic levels in groundwater, facilitating access to safer drinking water. Various options have received provisional certification from the Bangladesh Council of Scientific and Industrial Research (BCSIR), most designed for household use and one for community application. These systems are particularly advantageous when tube wells and dug wells are impractical. It is essential to replace the filter media once it becomes clogged. The performance of AIRPs can be affected by several chemical processes, including the oxidation of iron and arsenic, co-precipitation of arsenic with iron, multiple iron additions, interference from organic materials, and iron crystallization. AIRPs demonstrate significant potential for deployment in areas with elevated natural iron levels, where concerns regarding arsenic and iron in drinking water are prevalent. However, the efficacy of AIRPs is contingent upon factors such as arsenic speciation, the type of oxidant used, and the specific point of oxidant application. AIRPs provide a relatively low-cost method for arsenic removal from groundwater, ensuring safer drinking water access. The development of reverse osmosis (RO) technology in the late 1960s was aimed at desalting saline water supplies in coastal regions. This method relies on the principle of osmosis and necessitates a membrane barrier to segregate salts from water. RO technology is deemed more energy-efficient than distillation, making it a more viable option for desalination in regions facing water scarcity. The process utilizes an osmotic membrane that permits water to flow through significantly higher rates than the dissolved salts. Saline water can be sourced from surface water bodies or tube wells. Groundwater is generally favored due to its lower turbidity, which reduces the need for extensive pretreatment. Desalination plants often utilize shallow coastal groundwater with moderate salinity levels (EC 5000 to 6000 micro-siemens per centimeter, μS/cm) as raw water. Although the initial setup costs for RO systems are substantial, they require diligent maintenance and careful management of the remaining untreated saline water.

Dug wells represent one of the oldest methods for extracting groundwater for water supply, but they are still utilized in certain areas where alternative sources are deemed unsafe. The flow from dug wells is influenced by the lowering of the water table due to water extraction. Notably, water from dug wells is free from dissolved arsenic and iron, even in regions where shallow tube wells are contaminated. These wells are viable in locations with a stable upper soil layer, preferably clay, and a sandy layer within 9–12 m. While dug well water is generally free from dissolved arsenic and iron, it remains susceptible to bacterial contamination. A promising and cost-effective approach for developing a surface water-based supply system is the establishment of community-scale Slow Sand Filters, commonly called PSF. This technology is both economical and highly effective in removing turbidity and bacteria. PSF represents an economical technology that efficiently removes turbidity and bacteria. While PSF may not eliminate pathogens from heavily contaminated surface water, it exhibits significant bacterial removal capabilities. Chlorination is effective in eradicating 100% of pathogens.

In Bangladesh, where the annual rainfall averages approximately 2300 mm, rainwater harvesting is a viable option throughout the country. If hygienic practices are observed, rainwater can be considered safe for consumption. The primary drawback of this method is the need for rainwater availability throughout the year; however, it can serve as a valuable supplementary source. Adequate catchment areas are necessary for effective rainwater harvesting, with rooftops of buildings typically utilized for this purpose. The construction of storage containers can be accomplished using relatively inexpensive materials, and the maintenance costs are low. If collected from a clean surface, rainwater can be consumed without treatment.

Nonetheless, there is a risk of contamination from bird or animal droppings on catchment surfaces and gutter systems, necessitating cleaning or flushing prior to use. An additional filtration system can effectively eliminate such bacterial contamination. This approach ensures a reliable supply of safe water in proximity to homes, schools, or communities, promoting increased consumption and reducing the time women and children spend gathering water.

3 How Much Do People in Different Countries, Regions, and Development Levels Pay for Drinking Water Supply?

The WHO-UNICEF (WHO-United Nations Children's Fund 2023) report indicates that the inhabitants of the earth rose to 8 billion from 6.1 billion between 2000 and 2022. During this period, individuals without access to the required supply of potable water reduced to 703 million from 1.2 billion. Notably, about 66% of the population who has the availability of safe water supply resided in urban areas, which saw an increase in population to 4.5 billion from 2.9 billion. In contrast, the rural population experienced minimal change, increasing to 3.4 billion from 3.3 billion. During this timeframe, the number of populations who have availability of safe water supply was 704 million, and the number decreased to 549 million from 1 billion of those without minimum safe water supply.

Nevertheless, in 2022, about 80% of the rural population was still deprived of essential safe potable water supply. Availability of safe and fresh potable water rose to 73% in 2022 from 69% in 2015. Rural water supply improved to 62% from 56%, and coverage in urban areas increased by 1% from 80%. Although the urban situation improves higher in most regions, rural water supply has also improved significantly. The disparity between urban and rural coverage is most pronounced in sub-Saharan Africa (38%) and Caribbean and Latin America (27%). Conversely, this gap is narrowed in Southern and Central Asia because of a more rapid change in rural areas (1.39%) compared to urban areas (0.21%).

Furthermore, safe water supply in rural areas has consistently increased in Eastern and South-Eastern Asia, Latin America, and the Caribbean. In sub-Saharan Africa, urban areas have experienced a faster annual rate of change (0.56%) than rural areas (0.42%). The disparities in access to the minimum required water for drinking are less pronounced than those for safely managed water. In Oceania, there exists a significant 42% difference between urban (93%) and rural (51%) areas (WHO-UNICEF 2023). Since 2015, the proportion of the population with access to safely managed drinking water has risen from 69% to 73%, with rural access increasing from 56% to 62% and urban access from 80% to 81% (WHO-UNICEF 2023). Individuals utilizing piped water supplies grew from 3.5 billion to 5.3 billion, while those relying on non-piped sources rose to 2.3 billion from 1.7 billion between 2000 and 2022.

Following the establishment of global Sustainable Development Goal (SDG) baselines in 2017, there has been a threefold increase in the number of countries providing estimates for drinking water supply services in rural areas and a more than twofold increase for urban areas. From 2000 to 2020, the population relying on limited services nearly doubled, escalating from 154 million to 282 million, whereas the number of individuals using unimproved sources was halved, decreasing from 721 million to 367 million. The percentage of people using surface water declined from 4% to 2%, yet this still left 122 million individuals in 2020 with the lowest level of service, who have yet to benefit from any investments (WHO-UNICEF-World Bank 2022).

Water supply technologies such as Pond Sand Filter (PSF), Community-Based Rainwater Harvesting (RWH), Household-Based RWH, Deep Set Pump, Dug Well, Deep Tube Well with Deep Set Pump, Tube Well (shallow layer), Arsenic-Iron Removal Plant (AIRP), and Tube Well (double platform) have been evaluated to determine their efficiency, accessibility, cost, acceptance, and affordability in High-Risk Transition areas (Zahid et al. 2016). The study employed participatory techniques, including document reviews, field visits, Focus Group Discussions (FGD), and Key Informant Interviews (KII) with drillers, mechanics, NGO staff, and local government officials and representatives. The technology review addressed scientific and environmental factors, hydrogeology, subsurface lithology, groundwater levels, water quality, installation and maintenance costs, user acceptability, durability, seasonality, and climate change and disaster management strategies. Key issues affecting water sources were taken into account when selecting various options, including declining water tables, arsenic contamination in shallow groundwater, the presence of iron, manganese, and salinity in groundwater, the lack of suitable aquifers, and the absence of protected surface water sources throughout the year.

Installation costs of shallow and deep hand tube wells range from about US$ (United States Dollars) 100–150 and US$1000–1500, considering the depths between 15–40 meters (m) and 200–300 m, respectively, and generally used by 20–30 and 30–40 people. The lifetime of such tubewells is 15–20 years, with negligible operation and maintenance costs if natural disasters do not suffer them. For the construction of a dual platform with shallow or deep tubewells in the floodplain or flood-prone areas, an additional cost of US$80–100 must be added. The installation cost of a shallow and deep production tubewell for piped water supply in rural areas with about 40 stand posts (water delivery points) are approximately US$20,000 and 30,000, respectively; those can serve about 1000 people each. The pipeline network mainly incurs operation and maintenance costs. To provide water supply for about 150 and 1000 people, the construction cost of pond sand filters may reach up to US$1000 and 1300, respectively, with minor maintenance effort. Though rainwater is a seasonal options and available for 5–6 months, household-level rainwater harvesting systems are widespread and valuable in many places, including saline prone coastal areas. The construction cost of such technologies suitable for 8–10 and 20–25 people is US$300 and 800, respectively. The construction cost of an arsenic iron removal plant is between US$300 and 350. Those may serve 30–40 people. Maintenance, that is, cleaning of filter materials, is required every 3–4 months. About US$1500–1800 is needed to install a reverse osmosis system to supply potable water to 700–800 people. Regular operation and maintenance costs are required to run this technology. The construction cost of a 10–12 m deep dug well is between US$500 and 600 and is suitable for about 80–100 people.

4 How Much Do People Pay for Water Treatment?

It must meet microbiological and relevant chemical water quality standards to qualify as safe drinking water. Only for 138 of the 234 countries, data on drinking water that is "free from contamination" are available nationally, representing only 45% of the global population (WHO-UNICEF-World Bank 2022). Approximately 75% (1.3 billion) of individuals receiving piped water supplies and two-thirds (373 million) of those with non-piped supplies reside in urban areas. From 2000 to 2022, the urban population utilizing non-piped improved water sources has more than doubled, increasing from 333 million to 706 million. In 2020, 75% of the global population accessed improved drinking water sources free from contamination.

Nonetheless, there are considerable disparities between urban and rural environments and among different countries and regions. For instance, one-third of the population in Central and Southern Asia and two-thirds in sub-Saharan Africa continue to depend on unsafe drinking water.

Chemical contamination of drinking water is another significant health concern, with the WHO providing guideline values or "safe maximum concentrations" for over 100 chemicals that pose public health risks through water consumption (WHO 2022). While numerous chemicals can be identified in drinking water, only a limited number have been confirmed to have substantial effects on public health, and comprehensive data regarding the prevalence of chemical contamination still need to be made available. In most instances, microbiological contamination presents a far greater risk than chemical contamination; however, certain chemicals, notably arsenic and fluoride, are known to cause serious health issues and are relatively common in drinking water sources. Arsenic contamination of shallow groundwater in the Bengal Basin since the early 1990s (Zahid et al. 2020) and saltwater encroachment in the coastal belt (Zahid et al. 2016) increased financial involvement to provide alternate options for safe drinking water instead of very cheap hand tubewells in the affected areas (Fig. 13.1). Significant chemicals found in drinking water include lead and iron, while there has been growing concern regarding substances such as pesticides, pharmaceuticals, and microplastics in recent years (WHO-UNICEF-World Bank 2022). Unlike the immediate effects of waterborne microbial diseases, most chemical contaminants tend to exert their influence only after prolonged

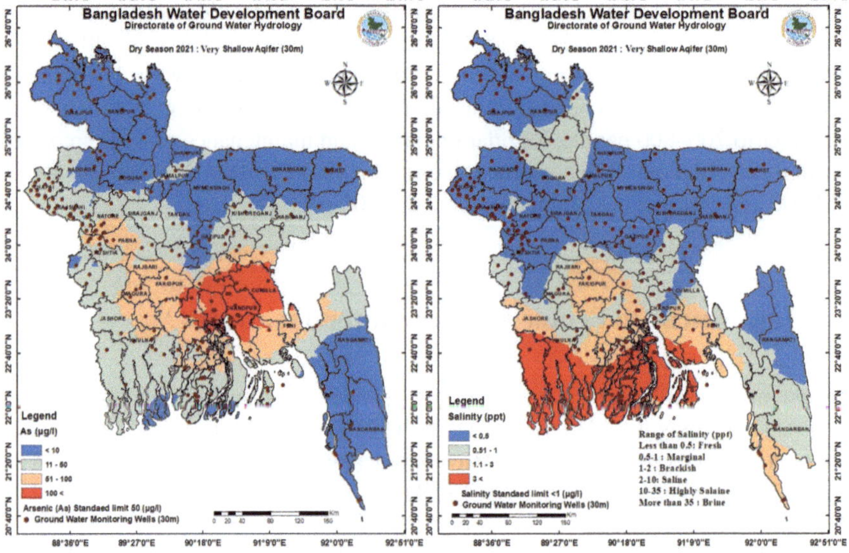

Fig. 13.1 Arsenic contamination of shallow groundwater and saline water encroachment in the coastal belt of the Bengal Basin posed more financial involvement to provide alternate options for safe drinking water (BWDB 2022)

exposure. For example, long-term consumption of drinking water with high arsenic levels, as well as food irrigated with such contaminated water, can lead to skin lesions and cancer and has also been linked to cardiovascular diseases and diabetes (WHO 2018). It is estimated that between 94 million and 220 million individuals may be exposed to drinking water with elevated arsenic concentrations (Podgorski and Berg 2020). Emerging contaminants in drinking water include pharmaceuticals, pesticides, per- and poly-fluoroalkyl substances (PFAS), and microplastics. Additionally, many pesticides are known to disrupt endocrine functions, leading to health implications that remain inadequately understood (WHO 2016).

Interventions aimed at improving drinking water quality have significantly decreased the incidence of childhood diarrhea. The provision of higher quality water supply at the household level has been shown to reduce diarrhea rates by 52% compared to unimproved sources (Wolf et al. 2022). Water treatment methods such as filtration, solar treatment, or chlorination have been found to reduce diarrhea by as much as 50% compared to untreated water from unimproved sources. A study conducted in 2016 estimated the financial requirements for achieving SDG Target 6.1, which pertains to safely managed water supply. The investment needed for universal availability was projected to be US$37.6 billion annually from 2015 to 2030, culminating in a total of US$564 billion if all users were to transition directly to safely managed potable water supplies. The operational and maintenance expenses are additional to the costs above, estimated at US$42 billion per year, totaling approximately US$635 billion from 2015 to 2030. The capital expenditures required to provide safe water for WASH uses—encompassing water supply, sanitation, and hygiene—are projected to about 0.39% of the GDP (gross domestic product) across the 140 states analyzed, which is roughly three times the historical investment in extending services to underserved populations. By applying the ratio of water service costs to the overall WASH service costs to this percentage, the average expenditure for gaining global availability to safe water supplies in the studied states is estimated at 0.14% of gross domestic product. Most aid allocations for water are directed toward what is classified as "large systems," with "basic systems" receiving an insignificant share. Big technologies typically refer to piped water supply in urban settings, whereas traditional technologies are generally common solutions in rural settings. Official Development Assistance (ODA) for basic water service technologies has been found to be consistent, covering between US$450 million and US$550 million annually, while funding for ample water supply systems has seen an upward trend and significant fluctuations. In 2020, repayable finance for basic technologies constituted about 29% of the total aid even though most impoverished and unserved populations reside in rural and peri-urban areas, which are more likely to benefit from essential systems. The considerable variability in disbursement amounts and the disparity between commitments and actual disbursements highlight ODA's unpredictability as a sector funding source.

The effective treatment of drinking water is capable of eliminating or inactivating over 20 waterborne bacteria, which include pathogens such as pathogenic *E. coli*, Shigella, and Campylobacter, as well as viruses like those responsible for hepatitis A and E, norovirus, and rotavirus, and parasites such as Giardia and

Cryptosporidium (WHO-UNICEF-World Bank 2022). Some pathogens have developed antibiotic resistance and are listed on the WHO global priority pathogen list. While specific chemical contaminants may not pose significant health risks at typical levels found in potable water, they can affect taste and color, potentially leading consumers to deny better sources of water in favor of more aesthetically pleasing but unsafe alternatives, such as surface water. Public health benefits arise not only from enhanced water quality but also from increased water usage. When water is accessible within the home through multiple taps, it sufficiently meets the needs for drinking, cooking, personal hygiene, and food safety while ensuring adequate quantities for improved hygiene practices during disease outbreaks (WHO 2020). There exist considerable disparities in the availability of microbiologically safe drinking water between urban and rural communities and among different wealth quintiles. Although Bangladesh's national water and sanitation coverage has significantly improved in recent years, certain areas remain underserved due to geophysical, socio-cultural, and economic factors. In regions with minimal infrastructural development, particularly regarding road connectivity, water and sanitation coverage continues to fall well below the bare minimum standards.

5 What Is the Interaction Between the Economic Burden of Safe Drinking Water Supply and Human Health?

The economic advantages of properly managed drinking water are evident in the immediate and extended future. For example, when water is sourced from more accessible locations, individuals expend less time and energy in its collection, allowing them to engage in more productive activities. The time saved in collecting, storing, and treating water disproportionately benefits women and girls, as they often bear the primary responsibility for water collection. Additional economic advantages include lower healthcare expenses for individuals and society, benefits from preventing premature deaths, and increased workforce participation. The water sector also contributes positively to employment by facilitating various income-generating activities, supporting other productive industries, and serving as an employer. For children, access to safely managed water services can improve health, resulting in fewer disruptions to school attendance, which has favorable long-term economic implications. A 2018 study estimated that the benefit-cost ratios are 3.4 in urban settings and 6.8 in rural areas for achieving primary water access, defined as a protected community borehole or dug well. These estimated ratios varied by region, with a peak of 15.9 for rural water supply in Eastern Asia and a minimum of 2.2 for urban water in South Asia.

At the present pace of advancement, 32 nations are projected to attain universal access (>99%) to safe drinking water by 2030, according to WHO-UNICEF 2023. Conversely, 78 nations are experiencing insufficient progress, and in 16 countries, the coverage is declining. Bhutan has demonstrated the most rapid annual progress since 2000, with an increase of 2.32 percentage points per year; however, its current coverage stands at 73%, indicating that more is needed to achieve the >99% target by 2030. From 2015 to 2022, global access to at least essential drinking water improved from 88% to 91%. By 2022, 83 countries had successfully reached universal coverage (>99%) of at least essential drinking water. Nonetheless, in 32 countries, coverage remained below 75%, including four nations in sub-Saharan Africa, where less than half of the population had access to essential drinking water in 2022. At the current rate of progress, 93 countries are on course to achieve universal coverage (>99%) by 2030, which includes the 83 countries that had already attained >99% by 2022, as reported by WHO-UNICEF (2023).

6 How Will This Change with Climate Change?

Water scarcity, which is increasingly intensified by climate change, serves both as a catalyst and a consequence of conflict as communities and entire populations vie for diminishing water resources. Over the past decade, incidents of violence related to water have more than doubled, and it is anticipated that water scarcity will continue to be a significant factor in future displacement.

The risks posed by climate change to water sources are escalating, manifesting in altered rainfall patterns, rising sea levels, elevated water temperatures, severe droughts, floods, and extreme storms. These effects have jeopardized the investments in safe drinking water since 1990 and threatened the progress toward achieving Sustainable Development Goal 6 (UNICEF 2021a; IPCC 2018, 2021). To ensure the provision of safely managed drinking water, it is essential to focus on adaptation and resilience-building, which includes making water infrastructure and services sustainable, secure, and resilient to climate-related threats while enhancing water systems to bolster community resilience against the impacts of climate change.

The risks associated with physical water scarcity are increasingly intensified by climate change, which includes water stress, fluctuations from year to year, seasonal changes, declines in groundwater levels, and drought occurrences. The accompanying map distinctly illustrates the primary regions experiencing high to extremely high water vulnerability, characterized by severe physical water scarcity and inadequate access to drinking water services. In 2020, it was estimated that approximately 1.4 billion individuals, including 450 million children, resided in these vulnerable areas (UNICEF 2021b). As global urbanization progresses, the demand for water in urban centers is anticipated to rise by 50–80% over the next 30 years (The World Bank 2017; He et al. 2021). This trend will remarkably heighten the vulnerability of urban residents, especially those from impoverished backgrounds, to the adverse effects of climate change.

Climate change compounds cities' demographic and supply-chain challenges, raising concerns about potential widespread water shortages exacerbated by intermittent climate-related disasters that can lead to significant social and economic upheaval. The repercussions of climate shocks on urban areas are expected to be especially severe in low- and middle-income nations (The World Bank 2016). In rural regions, inadequate management capabilities and the absence of professionalized service delivery further jeopardize drinking water services against climate change impacts.

6.1 Water Supply Practices in the Climatic and Physiographic Vulnerable Areas of Bangladesh

Bangladesh encompasses a significant portion of the Bengal Delta, shaped by the Ganges-Brahmaputra-Meghna (GBM) river system. Approximately 80% of the land is characterized by flat and low-lying terrain, crisscrossed by numerous rivers and tributaries. From a physiographic perspective, Bangladesh can be categorized into seven primary divisions: Hilly Regions, Pleistocene Uplands, Tippera Surface, Tista Fan, Flood Plains, the Delta Plain of the Ganges-Brahmaputra-Meghna Delta Complex, and the Sylhet Depression and Inland Marshes (Alam et al. 1990; GWTF 2002; Fig. 13.2). Each division can be further subdivided based on its unique characteristics. In light of climatic hazards and various vulnerabilities, including economic, physiographic, and social factors, six distinct Hazard and Risk (HtR) areas have been identified: (a) Drought-prone, (b) Flood-prone, (c) Char (sand bar), (d) Coastal, (e) Haor (swamp), and (f) Hilly areas. The socio-economic conditions in these regions indicate that most households fall within the poor category (BCAS-NGOF 2012). Around 70% of households live below the upper poverty line (UPL). Furthermore, a notable percentage of respondents demonstrated awareness regarding safe water, hygienic latrines, and natural disaster preparedness.

Barind is recognized for its susceptibility to drought, which is characterized by lower average annual precipitation than other regions of Bangladesh. Additionally, a thick upper layer of hard, nearly impermeable clay, ranging from 8 to over 25 m in thickness, hinders the vertical percolation of rainfall and diminishes recharge rates. For several decades, the Bangladesh Minor Irrigation Development Authority (BMDA) has been implementing extensive groundwater irrigation projects, which have led to a decline in the groundwater table below the suction limit of 7.5 m for hand tube wells in numerous areas. In regions where the groundwater table has decreased, the suction mode no. 6 Hand Tubewell is ineffective; therefore, force mode Deep Set Hand Tubewells (DSTW) and groundwater-based Piped Water Supply systems have been adopted. DSTWs, with depths of approximately 40–45 m, represent the region's most efficient water supply technology. Although the initial investment for Piped Water Supply utilizing Production Wells is substantial, it can lower overall costs by serving a larger community.

Furthermore, constructing additional Rain Water Harvesting Systems (RWHS) and Pond Sand Filters (PSF) could alleviate the pressure on the dwindling groundwater resources. Wherever feasible, existing ponds may also be utilized for Piped Water Supply. The lower aquitard, composed of clay, exceeds 100 m thickness,

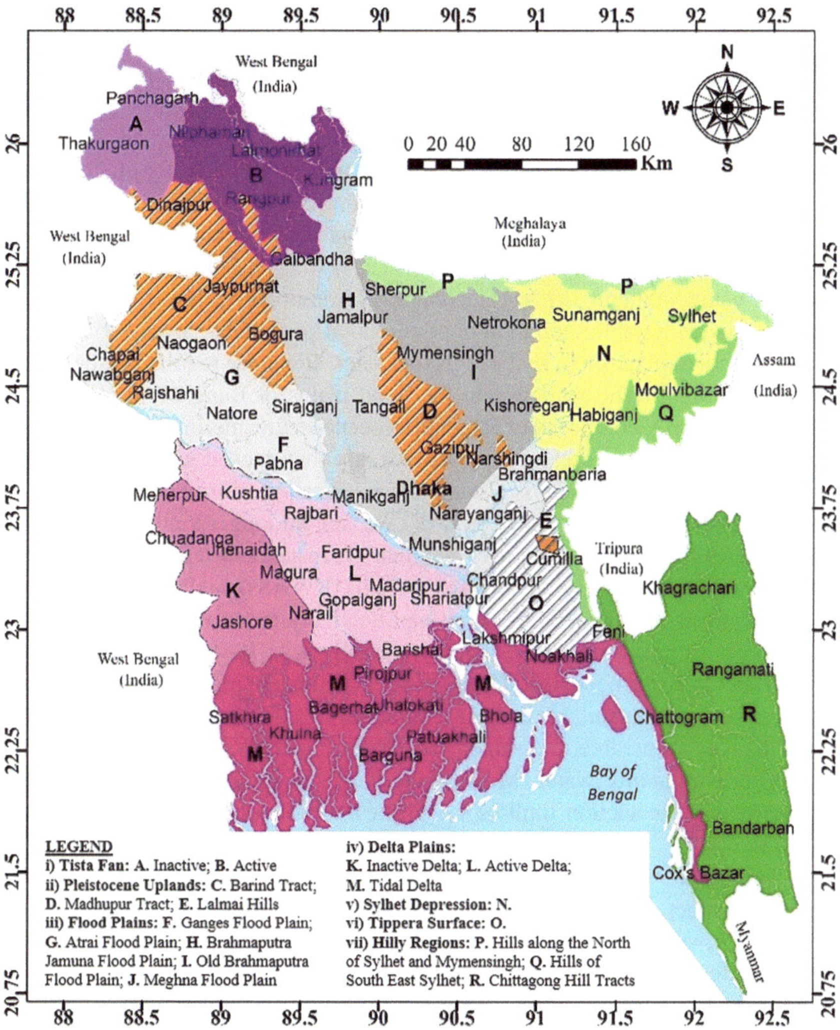

Fig. 13.2 Physiographic division and sub-divisions of Bangladesh (Alam et al. 1990)

making local drilling methods nearly impossible. Consequently, the deep aquifer in the area needs to be better understood and requires exploration to alleviate the stress on shallow groundwater resources.

Regular seasonal flooding and riverbank erosion present significant geophysical challenges for the Char Lands, characterized by sand bars. In these recently formed Char regions, shallow groundwater is notably free from arsenic contamination, making Shallow Hand Tubewells (STWs) a widely favored and economical source of safe drinking water. The groundwater table typically remains within a suction limit of 7.5 m throughout the year. Dual-platform tube wells have been implemented in areas prone to seasonal flooding to ensure access to fresh water during inundation by shifting the tubewell head from a lower platform to a raised one. Maintaining a safe distance between STWs and nearby pit latrines is essential to mitigate the risk of bacterial contamination, as the surface clay is insufficiently thick and prone to leakage. In regions where arsenic and iron pose challenges, the users have established Dug Wells and Arsenic Iron Removal Plants (AIRPs). Water quality assessments are conducted post-installation and require regular monitoring for both Dug Wells and AIRPs. In areas where shallow groundwater is contaminated with arsenic, Deep Tubewells (DTWs) and Rainwater Harvesting Systems (RWHS) can be installed using suitable drilling techniques. Additionally, low-cost Raised Pit Latrines provide a stable sanitation option in flood-prone Char areas.

Arsenic contamination in shallow groundwater, alongside regular seasonal flooding, poses a significant challenge in the flood plains of Bangladesh. These flood plains are primarily composed of Recent Holocene sediments, and in many regions, the shallow groundwater exhibits arsenic levels that exceed safe drinking water standards. Additionally, there is a notable concentration of iron in the upper aquifers. Consequently, shallow tube wells (STWs) are no longer viable for safe water supply in areas affected by arsenic contamination. Instead, installed dug wells, pond sand filters (PSFs), rainwater harvesting systems (RWHSs), and groundwater-based arsenic removal plants (AIRPs) serve as effective alternatives for ensuring safe water access. A gravel or pebble-rich layer at depths of 70–100 m complicates the installation of deep tube wells (DTWs) by local drillers in many floodplain regions.

Nevertheless, with suitable drilling techniques, DTWs can be successfully established in areas vulnerable to arsenic. Both PSFs and DTWs can also be utilized to supply the community water. Most of the year, the groundwater table typically remains within the suction limit of 7.5 m for traditional hand tube wells in many locations.

In the coastal regions of Bangladesh, natural disasters such as cyclones, storm surges, tornadoes, seawater encroachment, and saline water intrusion into groundwater are prevalent occurrences. Coastal areas are particularly susceptible to the expected effects of climate change, notably due to rising sea levels. The intrusion of saline water into upper aquifers, along with arsenic contamination in certain shallow groundwater areas, significantly restricts the availability of fresh water in these locales. Additionally, the surface water of tidal rivers remains saline or brackish throughout both wet and dry seasons. Various systems have been established to ensure a safe water supply in the coastal belt, including rainwater harvesting systems (RWHS) with and without filtration, community-based RWHS, pond water-based piped water supply, and desalination plants (Fig. 13.3). In locations where preserved ponds exist, using a slow sand filter (PSF) is the most economical and effective solution for coastal water supply, given the salinity of groundwater in upper aquifers. RWHS equipped with filtration units can also be implemented, albeit with limited functionality during the dry season. In areas lacking preserved ponds, desalination plants can be constructed, ensuring careful management of the disposal of any remaining untreated saline water. Furthermore, deep aquifers (greater than 300 m) may contain fresh water, making it feasible to install deep tube

Fig. 13.3 Photographs of different water supply technologies used in the climatic vulnerable areas as well as rural Bangladesh (Photo by: Anwar Zahid)

wells (DTWs) for community water supply. Managed Aquifer Recharge (MAR) techniques, which involve the infiltration of fresh rainwater, can enhance the quality of shallow groundwater above arsenic-affected zones by lowering salinity levels, thus enabling the use of Very Shallow Shrouded Tubewells for drinking water.

Haor regions are distinguished by abundant lakes (bells) and extensive swamps (haors), with waterlogging persisting for 6–8 months annually, posing significant challenges. The recurrent flooding and arsenic contamination in shallow groundwater represents critical threats to providing safe water and sanitation services in these areas. In Haor regions, Deep Tube Wells (DTWs) and Rainwater Harvesting Systems (RWHSs) are the most effective methods for ensuring safe water supply. In contrast, regular monitoring of water quality is essential for RWHSs. Additionally, a Piped Water Supply system is necessary. Given the persistent availability of surface water, Pond Sand Filters (PSFs) can also be utilized. Due to the presence of arsenic in shallow groundwater, DTWs emerge as the optimal solution for water supply. Constructing multiple outlets and a Piped Water Supply system using DTWs can enhance the overall benefits of water usage.

The Chattogram Hill Tracts represent a significant area of concern regarding climate change in Bangladesh. The challenging topography, considerable distances, and isolated nature of these regions often result in insufficient attention from government extension agencies, leading to limited physical and social infrastructure access. Bandarban, Rangamati, and Khagrachari districts are particularly vulnerable due to the reduction of hill streams attributed to global warming (UNDP 2011). In times of water scarcity, residents in the hill tracts resort to collecting and using untreated water, which contributes to various waterborne illnesses, including diarrhea, cholera, and typhoid. Without clean drinking water, individuals are compelled to consume contaminated sources. Practical solutions for ensuring safe water supply include the use of shallow tube wells (STW) and deep tube wells (DTW) in valley areas, as well as systems such as Gravity Flow System (GFS), Infiltration Gallery (IFG), Spring Water Capping, dug wells/ring wells, and rainwater harvesting systems.

7 Way Forward

Increased activities and efforts are essential to ensure that water supply facilities are accessible to all. Expanding and enhancing context-specific services is imperative to meet the fundamental needs of areas vulnerable to climate change. The demand is particularly acute for underprivileged groups and at-risk regions. The promotion of diverse technological options will support sustainable water supply solutions that address the needs of specific areas and socio-economic groups while considering broader environmental factors. Also, maintaining and monitoring existing technologies are crucial for their long-term sustainable use. Measures must be implemented to safeguard these technologies from damage during natural disasters. To fully realize the benefits of water services, greater emphasis is required on improving the design and implementation of technology. There is a notable decline in efficiency and the benefits received by households and communities due to the selection of inappropriate technologies, inadequate delivery of solutions, and the failure to sustain individual and community practices. Comprehensive awareness programs are necessary to familiarize users with less common technologies, such as dual-platform tubewells, AIRP, and desalination plants.

To foster climate resilience, various factors must be considered, including historical patterns of weather variability and extreme events like floods and droughts, as well as their impact on water quality and source reliability. Additionally, projected climatic conditions and water resource demands, including catchment development, land use, and population growth, must be considered.

8 Discussion Questions

1. What are the major sources and technologies of water supply in Bangladesh?
2. How much do people in different countries, regions, and development levels pay for drinking water?

3. How much do people pay for water treatment?
4. What is the interaction between the economic burden of safe drinking water supply and human health?
5. How will this change with climate change?

References

Alam MK, Hasan AKM, Khan MR, Whitney JW (1990) Geological map of Bangladesh. Ministry of Energy and Mineral Resources, Geological Survey of Bangladesh

Bangladesh Centre for Advanced Studies-Non-Government Organization Forum (BCAS-NGOF) (2012) Baseline study on water supply, sanitation, and hygiene in hard to reach areas

BWDB (2022) Report on Groundwater quality assessment of the multi-layered aquifer system of Bangladesh down to the depth of 335 meters. Bangladesh Water Development Board, 361p

Food and Agriculture Organization of the United Nations (FAO) (2007) Water news, hot issues: water scarcity. http://www.fao.org/nr/water/issues/scarcity. Accessed 19 Oct 2024

Ground Water Task Force (GWTF) (2002) Report of the Ground water task force, Ministry of Local Govt. Rural Dev. and Co-operatives, Govt. of the People's Republic of Bangladesh

He C, Liu Z, Wu J, Pan X, Fang Z, Li J et al (2021) Future global urban water scarcity and potential solutions. Nat Commun 12(1):4667. https://doi.org/10.1038/s41467-021-25026-3

Intergovernmental Panel on Climate Change (IPCC) (2021) Climate change 2022: Impacts, adaptation, and vulnerability. Contribution of Working Group II to the Sixth Assessment Report of the Intergovernmental Panel on Climate Change. Intergovernmental Panel on Climate Change; 2021. https://www.ipcc.ch/report/sixth-assessment-report-working-group-ii/ Accessed 19 Oct 2024

Intergovernmental Panel on Climate Change (IPCC) (2018) Global warming of 1.5°C. https://www.ipcc.ch/sr15/. Accessed 19 Oct 2024

Luijten JC, Knapp EB, Jones JW (2001) A tool for community-based assessment of the implications of development on water security in hillside watersheds. Agric Syst 70:603–622. https://doi.org/10.1016/S0308-521X(01)00061-0

Podgorski J, Berg M (2020) Global threat of arsenic in groundwater. Sci 368(6493):845–850. https://doi.org/10.1126/science.aba1510

Saima H (2022) Water in crisis-spotlight on Bangladesh. https://thewaterproject.org/water-crisis/water-in-crisis-bangladesh. Accessed 19 March 2022

The World Bank (2016) High and dry: climate change, water and the economy. World Bank, Washington. https://www.worldbank.org/en/topic/water/publication/high-and-dry-climate-change-water-and-the-economy. Accessed 19 Oct 2024

The World Bank (2017) Transforming water-scarce cities into water-secure cities through collaboration. World Bank, Washington. https://www.worldbank.org/en/news/feature/2017/05/15/water-scarce-cities-initiative. Accessed 19 Oct 2024

United Nations Development Programme (UNDP) (2011) Promotion of development and confidence building in the Chittagong Hill Tracts, annual report, Chittagong Hill Tracts Development Facility, Bangladesh.

United Nations Children's Fund (UNICEF) (2021a) Why water and sanitation must be top of your climate agenda. UNICEF, New York. https://www.unicef.org/documents/why-water-sanitation-and-hygiene-must-be-top-your-climate-agenda. Accessed 19 Oct 2024

United Nations Children's Fund (UNICEF) (2021b) Extreme water vulnerability index (EWVI): Methodology paper, WASH technical paper, TP/14/21. UNICEF, New York

World Health Organization-United Nations Children's Fund-World Bank (WHO-UNICEF-World Bank) (2022) State of the world's drinking water: an urgent call to action to accelerate progress on ensuring safe drinking water for all. WHO, Geneva, Licence: CC BY-NC-SA 3.0 IGO

World Health Organization-United Nations Children's Fund (WHO-UNICEF) (2023) Progress on household drinking water, sanitation and hygiene 2000–2022: special focus on gender. ISBN (WHO) 978-92-4-007692-1, ISBN (UNICEF) 978-92-806-5476-9

World Health Organization (WHO) (2016) Protecting surface water for health: Identifying, assessing and managing drinking-water quality risks in surface-water catchments. WHO, Geneva. https://www.who.int/publications/i/item/9789241510554. Accessed 19 Oct 2024

World Health Organization (WHO) (2018) Arsenic: Fact sheet. WHO, Geneva. https://www.who.int/news-room/fact-sheets/detail/arsenic. Accessed 19 Oct 2024

World Health Organization (WHO) (2020) Domestic water quantity, service level and health, 2nd edn. WHO, Geneva. https://www.who.int/publications/i/item/9789240015241. Accessed 19 Oct 2024

World Health Organization (WHO) (2022) Guidelines for drinking-water quality, 4th edn, incorporating the 1st and 2nd addenda. WHO, Geneva. https://www.who.int/publications/i/item/9789240045064. Accessed 19 Oct 2024

Wolf J, Hubbard S, Brauer M, Ambelu A, Arnold BF, Bain R et al (2022) Effectiveness of interventions to improve drinking water, sanitation, and handwashing with soap on risk of diarrhoeal disease in children in low-income and middle-income settings: a systematic review and meta-analysis. Lancet 400(10345):48–59. https://doi.org/10.1016/S0140-6736(22)00937-0

Zahid A, Rahman A, Ali MH, Hassan MR (2016) Determining sources of groundwater salinity in the multi-layered aquifer system of the Bengal Delta, Bangladesh. BRAC University Journal (Science and Engineering) XI(2):37–52

Zahid A, Roy SK, Shams IR, Shams R, Bashar F (2020) Review of arsenic contamination in the shallow groundwater of the Bengal Basin, Bangladesh. J Water Sci Eng 1(5-2029):1–7

Chapter 14
Climate Justice: Access to Safe Drinking Water and Health—A Legal Discourse

Anuli U. Njoku and Martin Agbalenyo

1 Learning Objectives

1. Describe the impacts of climate change on access to safe drinking water.
2. Emphasize the effects of climate change on drinking water disparities.
3. Highlight policies to mitigate the impact of climate change.
4. Discuss recommendations to address climate change and disparities in access to water.

2 Introduction

Water, a precious commodity that has been nicknamed the oil of the twenty-first century, has been a central focus of several interventions geared toward sustainable development (Joshi and Kapadia 2010). Adil et al. (2021) revealed that tackling the issues of water insecurities and unequal access to water has been a priority of the Sustainable Development Goals. History has taught us that access to potable water is the most effective way to prevent diseases such as typhoid, cholera, and polio (Jain 2012). Even though medical equipment and facilities are essential for the well-being of individuals, Jain (2012) revealed that they cannot be substitutes for safe drinking water. A study conducted by Patunru (2015) in Indonesia showed that

A. U. Njoku (✉)
Department of Public Health, College of Health and Human Services, Southern Connecticut State University, New Haven, CT, USA
e-mail: njokua3@southernct.edu

M. Agbalenyo
Southwestern AHEC, Inc., Shelton, CT, USA
e-mail: martin.agbalenyo@ynhh.org

A. R. Molla et al. (eds.), *Climate Change, Drinking Water Security, and Public Health*, https://doi.org/10.1007/978-3-032-12074-8_14

households without access to safe drinking water are 12% more susceptible to diarrhea than households with access to safe drinking water. The significance of using safe drinking water was evident in reducing waterborne diseases such as fever and typhoid in the twentieth century (Sosbey 2006).

Water contamination is not always attributed to the source of water. The safety of water may also depend on how the water is handled. Water can become unsafe due to the unhygienic handling of water from its source to storage (Osiemo et al. 2019). The challenges of water insecurities are not just about the quality but also the quantity. According to Leclerc et al. (2002), lack of water for domestic and hygienic purposes has resulted in more health problems than unsafe drinking water. A lot of progress made toward public health and sustainable development is being jeopardized by the adverse health impacts of climate change (Watts et al. 2017).

The CDC recorded 214 water-related outbreaks in the United States between 2015 and 2020 (Kunz 2024). Because of climate change, the world is expected to experience an increase in hydrologic stress, increasing the proportion of the global population facing water-stressed conditions (Mehran et al. 2017; Estrela et al. 2012).

3 Climate Change, Water and Health

3.1 History

Climate change is one of the most important topics in public health and one of public health's greatest challenges (American Public Health Association 2024). Climate change refers to long-term shifts in temperatures and weather patterns. Such shifts can be natural, due to changes in the sun's activity or large volcanic eruptions. But since the 1800s, human activities have been the main driver of climate change, primarily due to the burning of fossil fuels like coal, oil, and gas. Burning fossil fuels generate greenhouse gas emissions that act as a blanket wrapped around the Earth, trapping the sun's heat and raising temperatures. The main greenhouse gasses that cause climate change include carbon dioxide and methane. These gasses come from using gasoline for driving a car or coal for heating a building, for example. Clearing land and cutting down forests can also release carbon dioxide. Agriculture, oil, and gas operations are major sources of methane emissions. Energy, industry, transport, buildings, agriculture and land use are among the main sectors causing greenhouse gasses (United Nations n.d.).

Climate change awareness is not new, as it predates the First World Climate Conference in 1979. In 1827, Jean-Baptiste Fourier, a French scientist, raised a concern about the greenhouse effect, and by the 1930s, the public held the perception of global warming (Evans and Steven 2007). Although the issue of climate change has been indicated since the nineteenth century, it was not until 1979 that the issue catapulted to prominence at the global level for the first time (Gupta 2010). There are several definitions of climate change. However, the definitions by the

Framework Convention on Climate Change (FCCC) and the International Panel on Climate Change (IPCC) are very prominent. The FCCC defined climate change as "a change of climate that is attributed directly or indirectly to human activity, that alters the composition of the global atmosphere, and that is in addition to natural climate variability over comparable time periods" (Pielke 2004, p.1). The International Panel on Climate Change (IPCCC), on the other hand, defined climate change as "any change in climate over time whether due to natural variability or as a result of human activity" (Pielke 2004, p.1). Wilson et al. (2010, p.1) also defined climate as "any long-term significant change in the expected patterns of average weather (i.e., temperature, wind speed, humidity, precipitation, tornadic activity) of a specific region or the Earth due to natural or anthropogenic activities." Although there are variations in these definitions, it is evident that human activities have profound impacts on the climate. Population growth, uneven population distribution and climate change are significant reasons for water insecurity (Jain 2012). Addressing climate variability will not only improve access to water for the current generation but will also benefit posterity. According to Olmstead (2014), climate change affects the quantity of water in the long term and variability of water in the short term.

3.2 Increase in Extreme Weather Events

We are already witnessing an increase in extreme weather events, altered ecosystems that affect food security and the spread of infectious diseases, the uprooting of communities due to natural disasters, and a rise in sea level as a result of greenhouse gas emissions warming the earth's atmosphere. Climate change is the great multiplier when it comes to health disparities. People who have been less likely to have the means to protect their homes from damage, adjust, or recover from extreme weather events are those who have historically been marginalized (Njoku and Sampson 2023).

For example, among the top causes of death from disasters are deadly heat events that occur in an environment of extreme heat. In 2003, almost 70,000 people in Western Europe died from heat-related causes during 3 weeks of unrelentingly high temperatures. Heat waves in 1993 in Philadelphia and in 1995 in Chicago were blamed for about 120 and 600 deaths, respectively. Studies have indicated that differences in heat-related morbidity and mortality were observed between racial and socioeconomic groups. People who already have health issues, are homeless, don't have access to air conditioning, transportation, medical care, or social support have higher rates of sickness and death during heat waves (Njoku and Sampson 2023).

Within the United States of America (USA), climate impacts like raging storms and floods, increasing wildfires, severe heat, poor air quality, access to food and water, and disappearing shorelines can put Black, Brown, and Indigenous communities, low-income groups, people with disabilities, women, children, older people, and others at greater risk and make them more susceptible to the negative health effects of climate change (American Public Health Association 2024).

Globally, these disparities are evident since the United States and China account for almost 40% of greenhouse gas emissions, despite making up only 20–25% of the world's population. In the meantime, island states that are facing significant disruptions from rising sea levels, an increase in natural disasters, and droughts that are reducing important cash crops include Antigua and Barbuda, the Bahamas, the Marshall Islands, and the Maldives (Njoku and Sampson 2023). Furthermore, according to the World Health Organization, approximately 3.6 billion people live in areas that are very susceptible to climate change. Moreover, climate change is expected to cause approximately 250,000 additional deaths per year between 2030 and 2050 (WHO 2023).

3.3 Problems Associated with Climate Change

Problems associated with water quality, water availability, draughts and floods will be exacerbated due to climate change (Estrela et al. 2012). According to Delpla et al. (2009), significant effects of climate change on water quantity are mainly due to drought and flood. As a result of increased emission of greenhouse gases, the year 1998 recorded the highest temperature over the last millennium (Dominelli 2011). As the average temperature continues to rise, heat-related deaths are expected to increase (Wilson et al. 2010). In 2003, about 30,000 deaths in Europe were attributed to heat waves (Dominelli 2011). The number of heat-related deaths in Europe exceeded 61,000 in 2022 (Ballester et al. 2023).

Climate change has huge impacts on people around the world. One report revealed that about 4 million people are vulnerable, 325 million people are severely impacted, and more than 300,000 deaths are attributed to climate change per year (Douglas et al. 2012). Vairavamoorthy et al. (2008) revealed that human factors such as industrialization, population explosion, land use modification, and an increase in energy use are some causes of climate change. The temperature will increase by at least 1 °C by the next century (Vairavamoorthy et al. 2008). Climate change threatens water quality, availability, air quality, coastal zones, and wetlands (Dessler and Parson 2019). There has been a rise in average heat wave frequency in the United States over the past four decades (Gaffen and Ross 1998). A study done by Wilson et al. (2010) established a positive correlation between temperature variability and rates of morbidity and mortality. The United States has recorded 240 yearly heat-related deaths (McGeehan and Mirabelli 2001) and more than 14,000 heat-related deaths since 1979 according to the U.S. Environmental Protection Agency (2024). As the average temperature continues to rise, heat-related deaths are expected to increase (Wilson et al. 2010). The effects of climate change on groundwater quantity and quality cannot be overemphasized because groundwater serves 1.5 million to 3 million people worldwide (Kundzewicz and Döll 2009).

IPCC 2001 revealed "that the atmospheric concentration of carbon dioxide has increased from 280 ppm (parts per million) from 1750 to 367 ppm in 1999," representing an increase of 31% (Khandekar et al. 2005, p.1563). For the past two and a half centuries, greenhouse gases such as methane and nitrous oxide have increased by 145% and 15%, respectively (Khandekar et al. 2005). According to Douglas et al. (2012), the estimated cost of climate change per year is $125 million.

3.4 Impacts of Climate Change on Water

Climate change causes variations in precipitation and elevation of sea levels. According to Bates et al. (2008), climate change has contributed to a rise in sea levels, unpredictability of rainfall, and an increase in droughts and floods. A rise in sea level and storm surges increases the salinity of freshwater due to the intrusion of the seawater (Frederick and Major 1997). Increased salinity reduces the amount of freshwater available for hygiene and domestic use. According to Bates et al. (2008), climate change is a significant contributor to water stress in the world. The link between water security and anthropogenic factors is so sensitive that a slight change in these factors can induce more than proportional effects in water availability and quality. According to Vairavamoorthy et al. (2008) a slight increase in temperature can have an enormous impact on the water cycle.

According to the World Health Organization, in 2012, about 12.6 deaths representing 23% of global mortality rate can be linked to environmental factors, with climate change being the major cause among these environmental factors (Prüss-Üstün et al. 2016). Addressing environmental risk issues could have reduced the global mortality rate among children under 5 years by 26% (Prüss-Üstün et al. 2016). Climate change is a major cause of waterborne diseases such as diarrhea, as it influences the distribution and incidence of water (Levy et al. 2018). Increased temperature can promote the survival and replication of some pathogens that cause waterborne diseases (Levy et al. 2016). Heavy rainfall can also have impacts on water and sanitation infrastructure, and this can increase the incidence of waterborne diseases (Levy et al. 2016).

It is difficult to quantify the impacts of climate change. There has been disagreement among scholars and researchers about the magnitude of the impacts of climate change. While some perceive a minimal social cost of carbon (SCC) of about $10 per carbon released, others estimate a higher amount of impacts of about $200 per carbon (Pindyck 2013).

4 Climate Change, Water Quality, and Environmental Justice

4.1 United States Examples

Climate change is one of public health's greatest challenges and can produce disparate health outcomes among population subgroups. While many will experience the health impacts of climate change, certain groups, like children, the elderly, the underserved and minority communities, are less climate-resilient and, as a result, are more vulnerable to the adverse health effects of climate change (APHA 2024). For example, Hispanic neighborhoods in the USA are at increased risk for drinking contaminated water during droughts according to research from the American Journal of Public Health (Sum 2024). Heat waves can trigger high evaporation of water sources that are used by community systems, increasing the risk that toxins such as arsenic and nitrate will become concentrated. Neighborhoods of low income often lack modern equipment in water systems to properly filter out contaminants (The Nation's Health 2024). Moreover, a study examined records from 2007 to 2020 of over 2000 water service systems serving millions of California residents. The research collected demographic information from the American Community Survey and drought information from the Palmer Drought Severity Index. Results show that compared to other California communities, Hispanic communities, often in agricultural regions, had greater levels of nitrate and arsenic concentrations in their water when compared to other California communities. In many cases, nitrate concentrations were twice as high. The study illustrated that significant investments in adaptive or mitigation capacity are needed to address widening drinking water disparities that place already vulnerable communities at increased risk (Sum 2024).

Moreover, climate change has been shown to aggravate water insecurity in the United States. Much of the United States has been experiencing extreme streamflow events generated by climate change (Detheir et al. 2020). Numerous billion-dollar floods and droughts have occurred in the United States in the past decade, an increase in frequency compared to prior decades, raising concerns that extreme flow events are increasing in frequency (Smith 2022). Rising sea levels have led to increased risk of coastal flooding and saltwater intrusion into coastal aquifers. The Western part of the United States has experienced two decades of drought, and precipitation and snowpack have sharply decreased in some areas. Conflicts have escalated over the Colorado River's declining water flow, which supplies water to over 40 million people in seven states and several tribal nations, and many reservoirs are at historic lows as a result. Wildfires and catastrophic flooding brought on by parched soils unable to absorb excess rainfall are two examples of the cascading effects of drought, which have an impact on the quantity and quality of water. Food prices rise and agricultural crops are lost as a result of reduced river flows. Due to severe drought brought on by climate change, the Southeast also experiences reduced access to water and economic losses. Greater dependence on groundwater aquifers occurs as a result of diminished surface water resources (Cushing et al. 2023).

The causes of poor drinking water quality include natural elements (like arsenic), human activities (like nitrates from septic tanks and agriculture), industrial wastewater discharges, legacy contamination sites, oil and gas development, water infrastructure (like lead pipes), and water treatment (like disinfection byproducts). Severe weather events and service main breaks that cause depressurized distribution systems lead to pathogens contaminating drinking water. Poor water quality disproportionately affects communities of color. Furthermore, compared to their urban counterparts, small rural community water systems are more likely to experience contamination issues and frequently lack the financial, managerial, and technical resources necessary to address long-term drinking water quality issues (Cushing et al. 2023). Moreover, racial and ethnic disparities in climate change-related health effects can be illustrated by increases in the prevalence of chronic illnesses, new non-communicable diseases brought on by inadequate hurricane preparedness, interruptions in water and power supplies, and modifications to dietary and lifestyle choices (Berberian et al. 2022).

Climate change impacts and interventions are environmental injustices issues. Mohai (2009) stated that several studies have revealed that groups such as ethnic minorities, people of color and people of lower socio-economic status are more likely to face a water crisis. Strategies to reduce the emission of greenhouse gases do not pay much attention to equity (Tol 2001). Climate change is an environmental justice issue. According to Reddy and Assenzo (2009), the developing world is more susceptible to climate change because of inadequate adaptive strategies.

"Climate change reflects and increases social inequality in a series of ways, including who suffers most of its consequences, who caused the problem, who is expected to act, and who has the resources to do so" (Mohai et al. 2009, p. 420). Socially disadvantaged people in urban areas are disproportionately exposed to the impacts of climate change (Wilson et al. 2010). According to Doherty and Clayton (2011, p. 266), "climate change is as much a psychological and social phenomenon as a matter of biodiversity and geophysics and has impacts beyond the biophysical." Disasters such as Hurricane Katrina have proven racial and social income disparities in relation to extreme weather and climate change (Douglas et al. 2012).

4.2 Asia and Africa Examples

Climate change has caused food insecurity and inflation among the poor, especially in Asia and Africa, because it has affected water resources and agricultural activities (Misra 2014). Africa is one of the continents greatly affected by population climate change. With rapid population growth and an increase in factors such as drought, floods, and a rise in sea level, access to potable water has been cumbersome in some parts of the continents. Africa is faced with the problem of water scarcity as it is one of the arid continents in the world. Variations in precipitation will threaten food production in Africa. According to Bationo and Waswa (2011), 90% of farming in Sub-Saharan Africa is rain-fed and mostly subsistence.

The number of countries experiencing water stressed conditions has increased from 5 in 1955 to 20 in 1990, and by 2025, the number will range between 30 and 35 (Misra 2014). Misra (2014) stated that Africa may experience variations in drainage systems because of changes in precipitation and aridity (about 41% of the continent is dry).

A study conducted by Siegfried et al. (2012) revealed that while global warming is expected to deplete water resources in Asia, optimists argued that higher temperatures will increase water supply by melting ice and glaciers, and this will increase the amount of runoff.

North Africa is one of the regions in the world that needs serious attention. Diffenbaugh and Giorgi (2012) labeled North Africa as a "climate change hotspot." Some social scientists perceive the issues of climate change on water insecurity as a man-made problem in the Middle East and North Africa due to the impacts of high population growth (Schilling et al.2020). A study done by Droogers and Bouma (2014) in North Africa and the Middle East revealed that while water demand is expected to increase by 50%, the supply of water is expected to fall by 12% in the future. According to Milano et al. (2012), factors such as economic development and population growth will contribute to the water crisis in North Africa and the Middle East in the future.

In North Africa, human activities' impacts on aquifers are greater than climate variabilities (Schilling et al. 2020). If climate change adaptation strategies are not implemented, about 45% of the world's global gross domestic product and 52% of the global population will face acute water shortage by 2050 (Wang et al. 2016).

A decrease in precipitation in North Africa and the Middle East will threaten food and water security due to a reduction in water supplies to dams and reservoirs, which are the main sources of water for irrigation (Schilling et al. 2020).

4.3 Susceptible Groups

Women are disproportionately exposed to the risks of climate change and water insecurity. In his work "No Climate Justice without Gender Justice," Terry (2009) revealed that poor women are vulnerable to climate change because they lack coping and adaptation strategies. However, gender issues have been relegated as the climate debate has been masculinized (Terry 2009). Men and women may not encounter, perceive, and adapt to climate change similarly (Sultana 2018). Structural inequality and gender bias may negatively affect women's access, control, and ownership of resources (Sultana 2018). Thus, women are more likely to be affected by dwindling water resources. Although water insecurity is a global problem, some groups of people are disproportionately affected. Women bear the more significant burden of water scarcity because they usually provide water for their households (Sultana 2018).

Another group of people who are severely affected by water scarcity are children. Poor sanitation and inadequate access to drinking water are the second leading cause of death among children around the globe (Watkins 2006). According to the World Health Organization (WHO), approximately one out of five deaths in the world among children under 5 years can be attributed to diarrhea (Prüss-Üstün et al. 2016). People with weaker immune systems are more susceptible to Diarrhea (Herman and Zaslow 1996). Pal et al. (2018) revealed that diarrhea accounts for 11% of deaths among children (Pal et al. 2018). Other groups of people susceptible to unsafe drinking water are cancer patients and people living with HIV/AIDS (Pal et al. 2018).

5 Climate Change Discourse and Policies

The incalculable effects of climate change necessitate a moral responsibility to address this issue (Hedberg 2018; Tol 2001). For climate change policies to be effective, the focus should not just be quick fix but must also consider moral and ethical issues (Ney and Thompson 2000). However, recent data revealed that improvement in emission trends over the years has not been satisfactory (Nordhaus 2018). Rapid climate change is expected around the globe if attention is not paid to climate change policies (Nordhaus 2018). One of the reasons why the fight against climate change has achieved minimal results over the years is that emission reduction policies are not mandatory for all countries. According to Nordhaus (2018), efforts to curb emissions are voluntary in many countries. As the demand for fossil fuel surges over the years due to global population growth and industrialization, increased emissions of greenhouse gases have become inevitable. However, the long-term effects are far greater than the immediate economic costs. According to Pindyck (2013), the social cost of carbon (SCC) far outweighs the private cost. There is, therefore, a great need to address climate change for the betterment of humanity and the environment. A survey conducted by Dechezleprêtre et al. (2022) in 20 countries revealed that at least 75% of the respondents believed that climate change is a serious issue that needs to be addressed by their respective countries. Often, ambitious commitments to address climate change lack action plans for desirable results (Dechezleprêtre et al. 2022). Several organizations are trying to curb and mitigate the impacts of climate change both at local and international levels. However, some local organizations are not motivated to tackle climate change because they perceive that addressing the problem locally will not shield their immediate environment from the global risks of climate change (Brody et al. 2008).

Perceptions of policymakers, as well as existing and projected realities, can influence climate change and water interventions (Merton 1995). Inaccurate predictions of the impact of climate change on water can affect future generations, as this can exacerbate the water crisis and even result in conflict over access and control of water resources (Siegfried et al. 2012).

Although the impacts of climate change are ubiquitous, efforts to address the problem may exacerbate inequality if the principles of equity are not applied. According to Brody et al. (2008), policies geared toward the resiliency of humans and the environment to climate change may worsen disparities in societies where they are implemented. Societies of high carbon intensity may be disproportionately affected by the costs of carbon abatement and compliance (Brody et al. 2008).

There have been several debates on climate change over the years. Areas of debate include cost-benefit analysis, the social cost of carbon, and how the principles of equity should be applied (Gupta 2010). Addressing the impacts of climate change depends on the extent to which the current generation is willing to adjust their ways of life and consumption pattern for the speculative enhancement of the wellbeing of the future generation (Tompkins and Adger 2005).

Decision-making and policy implementation may be hampered by scientific uncertainties and the long duration of the manifestations of impacts (Tompkins and Adger 2005). Without certainty, policymakers will continue to make decisions based on their attitude toward risks (Tompkins and Adger 2005). According to Van Vliet et al. (2013), scientists have agreed that climate change will have an impact on water availability, but their analyses lack quantitative certainty, as it will be difficult to predict whether water resources will increase or dwindle in certain parts of the globe. According to Whitehead et al. (2009), although the 2007 United Nations Framework Convention on Climate Change established that there has been agreement among researchers that climate change is becoming a major global crisis, the degree of temperature variability impacts on precipitation and evaporation is still contentious among the members of scientific communities.

According to Schneider and Kuntz-Duriseti (2002), formulation and implementation of climate policies may be affected by uncertainties such as linguistic impression, errors in measurement, variability and disagreement among stakeholders. Uncertainties and disagreements can also result from paradigm shifts over the years. According to Dessler and Parson (2019), the perceptions of climate change have shifted as new knowledge is discovered. Lempert and Schlesinger (2000) revealed that future climate change prediction will be difficult because of factors such as uncertainty about economic growth in Africa, the inability to know the value the future generation will place on the environment, and changes in preference for goods consumed.

One of the main factors that prevent international organizations from making progress in formulating and implementing climate change policies is historical responsibility, as some countries keep blaming others for their environmentally unfriendly practices that contribute to climate change (Tompkins and Adger 2005).

Ney and Thompson (2000, p.71) revealed that "an analysis of global climate change policy debate in the mid-1990s revealed three policy stories. Each policy story provided a setting (in jargonese, the basic assumptions), villains (the policy problem), heroes (policy protagonists), and, of course, a moral (the policy solution)."

Despite advancements in technology, the depletion of the ozone layer in the 1970s went overlooked not because satellite instrumentation failed but because computers were programmed to ignore the changes that occurred (Schneider and

Kuntz-Duriseti 2002). Making objective decisions and arriving at a universal agreement have been very difficult. According to Schneider and Kuntz-Duriseti (2002), incomplete and ambiguous information on climate change and its impact may lead to subjective decision-making. According to Dessler and Parson (2019), climate change is one of the most controversial environmental issues, as there have been disagreements about how much the climate has been changing, how fast the change is occurring, whether man is the leading cause of climate change, and the cost involved to stop or curb climate change.

Furthermore, Reddy and Assenza (2009) also revealed that key reasons for the debate on climate change include causes of climate change, effects of climate change, and interventions to address climate change, with a focus on population growth, energy use, and economic growth. While the Global North has been blamed for increasing the emission of gases through industrialization, the Global South has been blamed for rapid population growth, which has adverse impacts on the climate (Ney and Thompson 2000).

One of the main obstacles to addressing climate change is politics. According to Ney and Thompson (2000), politics may manifest in two ways: "the spirit is willing, but the flesh is weak" (a situation where partisan implementation adversely affects good policies) or "healthy body blighted by madness" (the process of decision-making hampered by conflict of interest).

Addressing climate change requires enormous investment and commitment. According to Dessler and Parson (2019), although it is evident that tackling climate change requires strong actions, such actions can be cumbersome and expensive. While those who are skeptical of climate change may not want to avoid or postpone climate change interventions partly due to the economic cost involved, those who are advocating for measures to curb climate change may also be concerned about both the environmental and economic costs (Reddy and Assenza 2009). Although it has been estimated that reducing the emission of greenhouse gases by 7%, as directed by the Kyoto Protocol, will cost up to 3% of the United States GDP, others have predicted that an emission reduction will rather increase the GDP of the United States (Lempert and Schlesinger 2000).

Ludwig et al. (2014) stated that earlier research on climate change focused on consequences and interventions to reduce the emission of greenhouse gases. Having realized that climate change is inevitable, more attention has been paid to adaptation (Ludwig et al. 2014).

African-Americans are disproportionately affected by climate change in the United States. According to Schlosberg and Collins (2014), although African Americans are less responsible for greenhouse emissions, they are more susceptible to climate change-related diseases. Climate change interventions can disproportionately affect the Black community. For instance, Schlosberg and Collins (2014) stated climate mitigation measures such as increasing carbon prices impose a financial burden on disproportionately poor Black people. However, companies that focus on new energy technologies will not only lower pollution in Black communities but will also provide employment for them (Schlosberg and Collins 2014).

The United Nations organized several conferences to address climate change. Although Western countries admitted that their activities have contributed to climate change, they were not fully committed to reducing greenhouse gases (Dominelli 2011). For instance, the United States and Australia refused to ratify the Kyoto Protocol Agreement (Dominelli 2011).

One of the main policies used to control greenhouse gas emissions is a carbon tax. However, this may impose an extra financial burden on the vulnerable. According to Stiglitz (2019), this tax is mostly regressive, and this will be unfavorable to the poor.

6 Recommendations

Employment opportunities that focus on sustainable environmental practices should be created. This will not only address issues of climate change in the long run but also improve water infrastructure and natural water resources in vulnerable communities. Willson et al. (2010) revealed that focusing on the green economy and green-collar jobs would address infrastructure deficit and unemployment in socially vulnerable communities. Schlosberg and Collins (2014) revealed that sustainable technologies will create jobs and reduce pollution in vulnerable communities.

As efforts are being made to curb the emission of greenhouse gases, special attention must be paid to vulnerable local communities in order not to exacerbate the already existing disparities. Although vulnerable communities are less responsible for climate change, they may bear relatively more significant burdens of the impacts. Policymakers should use the polluter pays principle (PPP) to regulate the emission of greenhouse gases. To promote equity, part of abatement costs, such as carbon taxes, should be used to provide basic needs such as water infrastructure in disadvantaged communities. This measure will help mitigate the impact of climate change and improve access to water for the vulnerable.

Education and participation are needed to address the issues of climate change. Douglas et al. (2012) revealed that in order to advance from concepts to implementation, opportunities should be created for grassroots participation and education during the planning process. According to Gandure et al. (2013), policy makers need to consider the perceptions and coping strategies of the people directly exposed to climate change. In order to adapt to climate change, genetically modified crops were introduced to Guanajuato communities in Mexico (Bee 2014). However, this intervention was greeted with varied opinions due to inadequate information. This shows that in order to enhance the sustainability of adaptation strategies, grassroots consultation and education must precede the implementation of policies. Education is also essential for positive behavior and environmentally friendly strategies such as tree planting and sustainable farming. Sustained federal investment in community development programs is also necessary for ensuring a healthy, equitable, and climate change-ready economic recovery from the COVID-19 pandemic (Njoku and Joseph 2023).

While there is limited literature on the impacts of climate change on adaptation, there is a dearth of knowledge on adaptation (Olmstead 2014). Terry 2009 stated climate change policies are geared toward mitigating greenhouse gases or adaptation. While efforts are being made to curb the emission of greenhouse gases, more research on adaptation and resilience needs to be carried out to lighten the burden on those already affected by climate change. Most climate change research has been generalized. However, some groups of people are more vulnerable to the impacts of climate change. New studies need to examine climate change and its impact and interventions in relation to regions, race, gender, and socio-economic status.

7 Conclusions

One of the Sustainable Development Goals has been to address the problems of water insecurity and unequal access to water. The second most common cause of death for children worldwide is poor sanitation and limited access to drinking water. A basic human right is the availability of clean drinking water. One of the biggest problems facing public health is climate change, which also acts as a huge multiplier for health disparities. The World Health Organization estimates that 3.6 billion people reside in regions that are particularly vulnerable to climate change. Furthermore, it is anticipated that between 2030 and 2050, climate change will result in an additional 250,000 deaths annually.

Black, Brown, and Indigenous communities, low-income groups, people with disabilities, women, children, the elderly, and others may be more vulnerable to the adverse health effects of climate change in the United States due to climate impacts such as severe heat, raging storms and floods, an increase in wildfires, poor air quality, and disappearing shorelines. Addressing the negative effects of climate change requires the creation and application of climate policies. Continued efforts are needed to illuminate how climate change affects access to safe drinking water and contributes to drinking water disparities. Regulations are needed to alleviate the effects of climate change and disparities in access to drinking water among underrepresented groups and communities of color and low-income. Additional research is also needed to examine climate change and its impact and interventions in relation to geography, race, gender, and socio-economic status.

8 Discussion Questions

1. Based on the examples discussed in this chapter, explain at least five problems associated with climate change.
2. Describe at least five impacts of climate change on access to safe drinking water.
3. Discuss at least three population subgroups that are disproportionately exposed to the risks of climate change and water insecurity.

4. Based on the recommendations described in this chapter, what are at least three strategies to prevent or mitigate instances of water injustice, climate change, and unequal access to safe drinking water?

References

Adil S, Nadeem M, Malik I (2021) Exploring the important determinants of access to safe drinking water and improved sanitation in Punjab, Pakistan. Water Policy 23(4):970–984. https://doi.org/10.2166/wp.2021.001

American Public Health Association (2024) Climate changes health: Climate justice. https://www.apha.org/topics-and-issues/climate-health-and-equity/climate-justice. Accessed 24 Sept 2024

Ballester J, Quijal-Zamorano M, Méndez Turrubiates RF, Pegenaute F, Herrmann FR, Robine JM, Basagaña X, Tonne C, Antó JM, Achebak H (2023) Heat-related mortality in Europe during the summer of 2022. Nat Med 29(7):1857–1866. https://doi.org/10.1038/s41591-023-02419-z

Bates B, Kundzewicz Z, Wu S (2008) Climate change and water. Intergovernmental Panel on Climate Change Secretariat

Bationo A, Waswa BS (2011) New challenges and opportunities for integrated soil fertility management in Africa. Innovations as key to the green revolution in Africa: exploring the scientific facts, pp. 3–17. https://doi.org/10.1007/978-90-481-2543-2_1

Bee BA (2014) Si no comemos tortilla, no vivimos: women, climate change, and food security in central Mexico. Agric Hum Values 31:607–620. https://doi.org/10.1007/s10460-014-9503-9

Berberian AG, Gonzalez DJ, Cushing LJ (2022) Racial disparities in climate change-related health effects in the United States. Curr Environ Health Rep 9(3):451–464. https://doi-org.scsu.idm.oclc.org/10.1007/s40572-022-00360-w

Brody SD, Zahran S, Grover H, Vedlitz A (2008) A spatial analysis of local climate change policy in the United States: risk, stress, and opportunity. Landsc Urban Plan 87(1):33–41. https://doi.org/10.1016/j.landurbplan.2008.04.003

Cushing LJ, Babson Dobbin K, Osborne Jelks N, Liu X, Morello-Frosch R (2023) Water insecurity and population health: implications for health equity and policy. Health Affairs Health Policy Brief. https://www.healthaffairs.org/content/briefs/water-insecurity-and-population-health-implications-health-equity-and-policy. Accessed 24 Oct 2024

Dakhlaoui H, Ruelland D, Tramblay Y, Bargaoui Z (2017) Evaluating the robustness of conceptual rainfall-runoff models under climate variability in northern Tunisia. J Hydrol 550:201–217. https://doi.org/10.1016/j.jhydrol.2017.04.032

Dechezleprêtre A, Fabre A, Kruse T, Planterose B, Chico AS, Stantcheva S (2022) Fighting climate change: international attitudes toward climate policies (No. w30265). National Bureau of Economic Research

Delpla I, Jung AV, Baures E, Clement M, Thomas O (2009) Impacts of climate change on surface water quality in relation to drinking water production. Environ Int 35(8):1225–1233. https://doi.org/10.1016/j.envint.2009.07.001

Dessler AE, Parson EA (2019) The science and politics of global climate change: a guide to the debate. Cambridge University Press

Dethier EN, Sartain SL, Renshaw CE, Magilligan FJ (2020) Spatially coherent regional changes in seasonal extreme streamflow events in the United States and Canada since 1950. Sci Adv 6(49):eaba5939. https://www.science.org/doi/full/10.1126/sciadv.aba5939

Diffenbaugh NS, Giorgi F (2012) Climate change hotspots in the CMIP5 global climate model ensemble. Clim Chang 114:813–822. https://doi.org/10.1007/s10584-012-0570-x

Doherty TJ, Clayton S (2011) The psychological impacts of global climate change. Am Psychol 66(4):265–276. https://doi.org/10.1037/a0023141

Dominelli L (2011) Climate change: social workers' roles and contributions to policy debates and interventions. Int J Soc Welf 20(4):430–438. https://doi.org/10.1111/j.1468-2397.2011.00795.x
Douglas EM, Kirshen PH, Paolisso M, Watson C, Wiggin J, Enrici A, Ruth M (2012) Coastal flooding, climate change and environmental justice: identifying obstacles and incentives for adaptation in two metropolitan Boston Massachusetts communities. Mitig Adapt Strateg Glob Chang 17:537–562. https://doi.org/10.1007/s11027-011-9340-8
Droogers P, Bouma J (2014) Simulation modelling for water governance in basins. Int J Water Resour Dev 30(3):475–494
Estrela T, Pérez-Martin MA, Vargas E (2012) Impacts of climate change on water resources in Spain. Hydrol Sci J 57(6):1154–1167. https://doi.org/10.1080/02626667.2012.702213
Evans A, Steven D (2007) Climate change: the state of the debate. Center for International Cooperation, New York University
Frederick KD, Major DC (1997) Climate change and water resources. Clim Chang 37(1):7–23. https://doi.org/10.1023/A:1005336924908
Gaffen DJ, Ross RJ (1998) Increased summertime heat stress in the US. Nature 396(6711):529–530
Gandure S, Walker S, Botha JJ (2013) Farmers' perceptions of adaptation to climate change and water stress in a South African rural community. Environ Dev 5:39–53. https://doi.org/10.1016/j.envdev.2012.11.004
Gupta J (2010) A history of international climate change policy. Wiley Interdiscip Rev Clim Chang 1(5):636–653. https://doi-org.scsu.idm.oclc.org/10.1002/wcc.67
Hedberg T (2018) Climate change, moral integrity, and obligations to reduce individual greenhouse gas emissions. Ethics, Policy & Environment 21(1):64–80. https://doi.org/10.1080/21550085.2018.1448039
Herman G, Zaslow S (1996) Health effects of drinking water contaminants. North Carolina Cooperative Extension Service, Publication No HE-393
Jain R (2012) Providing safe drinking water: a challenge for humanity. Clean Techn Environ Policy 14:1–4. https://doi.org/10.1007/s10098-011-0446-1
Joshi MB, Kapadia VP (2010) Sharing water in the 21st century: rethinking the rationale? Irrigation and Drainage: The Journal of the International Commission on Irrigation and Drainage 59(1):92–101. https://doi-org.scsu.idm.oclc.org/10.1002/ird.563
Khandekar ML, Murty TS, Chittibabu P (2005) The global warming debate: a review of the state of science. Pure Appl Geophys 162:1557–1586. https://doi.org/10.1007/s00024-005-2683-x
Kundzewicz ZW, Doell P (2009) Will groundwater ease freshwater stress under climate change? Hydrol Sci J 54(4):665–675. https://doi.org/10.1623/hysj.54.4.665
Kunz JM (2024) Surveillance of waterborne disease outbreaks associated with drinking water—United States, 2015–2020. MMWR Surveill Summ 73. https://doi.org/10.15585/mmwr.ss7301a1
Leclerc H, Schwartzbrod L, Dei-Cas E (2002) Microbial agents associated with waterborne diseases. Crit Rev Microbiol 28(4):371–409. https://doi.org/10.1080/1040-840291046768
Lempert RJ, Schlesinger ME (2000) Robust strategies for abating climate change. Clim Chang 45(3–4):387–401. https://doi.org/10.1023/A:1005698407365
Levy K, Smith SM, Carlton EJ (2018) Climate change impacts on waterborne diseases: moving toward designing interventions. Curr Environ Health Rep 5:272–282. https://doi.org/10.1007/s40572-018-0199-7
Levy K, Woster AP, Goldstein RS, Carlton EJ (2016) Untangling the impacts of climate change on waterborne diseases: a systematic review of relationships between diarrheal diseases and temperature, rainfall, flooding, and drought. Environ Sci Technol 50(10):4905–4922. https://doi.org/10.1021/acs.est.5b06186
Ludwig F, van Slobbe E, Cofino W (2014) Climate change adaptation and Integrated Water Resource Management in the water sector. J Hydrol 518(2014):235–242. https://doi.org/10.1016/j.jhydrol.2013.08.010

McGeehin MA, Mirabelli M (2001) The potential impacts of climate variability and change on temperature-related morbidity and mortality in the United States. Environ Health Perspect 109(Suppl 2):185–189. https://doi.org/10.1289/ehp.109-1240665

Mehran A, AghaKouchak A, Nakhjiri N, Stewardson MJ, Peel MC, Phillips TJ, Wada Y, Ravalico JK (2017) Compounding impacts of human-induced water stress and climate change on water availability. Sci Rep 7(1):6282. https://doi-org.scsu.idm.oclc.org/10.1038/s41598-017-06765-0

Merton RK (1995) The Thomas theorem and the Matthews effect. Soc F 74:379. https://doi-org.scsu.idm.oclc.org/10.2307/2580486

Milano M, Ruelland D, Fernandez S, Dezetter A, Fabre J, Servat E (2012) Facing climatic and anthropogenic changes in the Mediterranean basin: what will be the medium-term impact on water stress? Comptes Rendus Géoscience 344(9):432–440. https://doi.org/10.1016/j.crte.2012.07.006

Misra AK (2014) Climate change and challenges of water and food security. Int J Sustain Built Environ 3(1):153–165. https://doi.org/10.1016/j.ijsbe.2014.04.006

Mohai P, Pellow D, Roberts JT (2009) Environmental justice. Annu Rev Environ Resour 34(1):405–430. https://doi-org.scsu.idm.oclc.org/10.1146/annurev-environ-082508-094348

Ney S, Thompson M (2000) Cultural discourses in the global climate change debate. In: Society, behaviour, and climate change mitigation. Springer Netherlands, Dordrecht, pp 65–92. https://doi.org/10.1007/0-306-48160-X_2

Njoku AU, Sampson NR (2023) Environment injustice and public health. In: Liamputtong P (ed) Handbook of social sciences and global public health. Springer International Publishing, Cham, pp 1987–2006. https://link-springer-com.scsu.idm.oclc.org/referenceworkentry/10.1007/978-3-030-96778-9_37-1. Accessed 24 Sept 2024

Njoku A, Joseph M (2023) COVID-19, housing, and environmental injustice. Hygie 3(4):396–405. https://doi.org/10.3390/hygiene3040029

Nordhaus W (2018) Projections and uncertainties about climate change in an era of minimal climate policies. Am Econ J Econ Pol 10(3):333–360. https://doi.org/10.1257/pol.20170046

Olmstead SM (2014) Climate change adaptation and water resource management: a review of the literature. Energy Econ 46:500–509. https://doi.org/10.1016/j.eneco.2013.09.005

Osiemo MM, Ogendi GM, M'Erimba C (2019) Microbial quality of drinking water and prevalence of water-related diseases in Marigat Urban Centre, Kenya. Environ Health Insights 13:1178630219836988. https://doi-org.scsu.idm.oclc.org/10.1177/1178630219836988

Pal M, Ayele Y, Hadush M, Panigrahi S, Jadhav VJ (2018) Public health hazards due to unsafe drinking water. Air Water Borne Dis 7(1000138):2. https://doi.org/10.4172/2167-7719.1000138

Patunru AA (2015) Access to safe drinking water and sanitation in Indonesia. Asia Pac Policy Stud 2(2):234–244. https://doi-org.scsu.idm.oclc.org/10.1002/app5.81

Pielke RA Jr (2004) What is climate change? Energy Environ 15(3):515–520. https://doi-org.scsu.idm.oclc.org/10.1260/0958305041494576

Pindyck Robert S (2013) Climate change policy: what do the models tell us? J Econ Lit 51(3):860–872. https://doi.org/10.1257/jel.51.3.860

Prüss-Üstün A, Wolf J, Corvalán C, Bos R, Neira M (2016) Preventing disease through healthy environments: a global assessment of the burden of disease from environmental risks. World Health Organization

Reddy BS, Assenza GB (2009) The great climate debate. Energy Policy 37(8):2997–3008. https://doi.org/10.1016/j.enpol.2009.03.064

Schilling J, Hertig E, Tramblay Y, Scheffran J (2020) Climate change vulnerability, water resources and social implications in North Africa. Reg Environ Chang 20:1–12. https://doi.org/10.1007/s10113-020-01597-7

Schlosberg D, Collins LB (2014) From environmental to climate justice: climate change and the discourse of environmental justice. Wiley Interdiscip Rev Clim Chang 5(3):359–374. https://doi-org.scsu.idm.oclc.org/10.1002/wcc.275

Schneider SH, Kuntz-Duriseti K (2002) Uncertainty and climate change policy. Climate Change Policy: A Survey:53–87. http://stephenschneider.stanford.edu/Publications/PDF_Papers/Ch02ClimatePolicy.pdf

Siegfried T, Bernauer T, Guiennet R, Sellars S, Robertson AW, Mankin J, Bauer-Gottwein P, Yakovlev A (2012) Will climate change exacerbate water stress in Central Asia? Clim Chang 112:881–899. https://doi.org/10.1007/s10584-011-0253-z

Smith AB (2022) US Billion-dollar weather and climate disasters, 1980-present. NOAA National Centers for Environmental Information. https://doi.org/10.25921/stkw-7w73. Accessed 14 Oct 2024

Sobsey MD (2006) Drinking water and health research: a look to the future in the United States and globally. J Water Health 4(S1):17–21

Stiglitz JE (2019) Addressing climate change through price and non-price interventions. Eur Econ Rev 119:594–612. https://doi.org/10.2166/wh.2006.0039

Sultana F (2018) Gender and water in a changing climate: challenges and opportunities. Water security across the gender divide:17–33. https://doi.org/10.1007/978-3-319-64046-4_2

Sum S (2024) Impact of droughts on served drinking water disparities in California, 2007–2020. Am J Public Health 114(9):935–945. https://doi.org/10.2105/AJPH.2024.307758

Terry G (2009) No climate justice without gender justice: an overview of the issues. Gend Dev 17(1):5–18. https://doi.org/10.1080/13552070802696839

The Nation's Health: A Publication of the American Public Health Association (November/ December 2024) Hispanic areas risk water contamination. Retrieved from the Nation's Health, www.thenationshealth.org

Tol RS (2001) Equitable cost-benefit analysis of climate change policies. Ecol Econ 36(1):71–85. https://doi.org/10.1016/S0921-8009(00)00204-4

Tompkins EL, Adger WN (2005) Defining response capacity to enhance climate change policy. Environ Sci Pol 8(6):562–571. https://doi.org/10.1016/j.envsci.2005.06.012

United Nations (n.d.) Climate action. https://www.epa.gov/environmentaljustice/learn-about-environmental-justice. Accessed 24 Sept 2024

U.S. Environmental Protection Agency (2024) Climate change indicators: heat-related deaths. https://www.epa.gov/climate-indicators/climate-change-indicators-heat-related-deaths. Accessed 30 Nov 2024

Vairavamoorthy K, Gorantiwar SD, Pathirana A (2008) Managing urban water supplies in developing countries–Climate change and water scarcity scenarios. Phys Chem Earth, Parts A/B/C 33(5):330–339. https://doi.org/10.1016/j.pce.2008.02.008

Van Vliet MT, Franssen WH, Yearsley JR, Ludwig F, Haddeland I, Lettenmaier DP, Kabat P (2013) Global river discharge and water temperature under climate change. Glob Environ Chang 23(2):450–464. https://doi.org/10.1016/j.gloenvcha.2012.11.002

Wang XJ, Zhang JY, Shahid S, Guan EH, Wu YX, Gao J, He RM (2016) Adaptation to climate change impacts on water demand. Mitig Adapt Strateg Glob Chang 21:81–99. https://doi.org/10.1007/s11027-014-9571-6

Watkins K (2006) Human Development Report 2006-Beyond scarcity: power, poverty and the global water crisis. UNDP Human Development Reports 2006

Watts N, Adger WN, Ayeb-Karlsson S, Bai Y, Byass P, Campbell-Lendrum D, Colbourn T, Cox P, Davies M, Depledge M, Depoux A (2017) The Lancet Countdown: tracking progress on health and climate change. Lancet 389(10074):1151–1164. https://doi.org/10.1016/S0140-6736(16)32124-9

Whitehead PG, Wilby RL, Battarbee RW, Kernan M, Wade AJ (2009) A review of the potential impacts of climate change on surface water quality. Hydrol Sci J 54(1):101–123. https://doi.org/10.1623/hysj.54.1.101

Wilson SM, Richard R, Joseph L, Williams E (2010) Climate change, environmental justice, and vulnerability: an exploratory spatial analysis. Environ Justice 3(1):13–19. https://doi-org.scsu.idm.oclc.org/10.1089/env.2009.0035

World Health Organization – WHO (2023) October 12 'Climate change'. https://www.who.int/news-room/fact-sheets/detail/climate-change-and-health. Accessed 26 Nov 2024

Part IV
A Global Agenda with Community-Based Solutions

Chapter 15
Impact of Climate Change and War on Water, the Environment, and Health in Ukraine

Boris Faybishenko, Chad Cox, Oksana Halych, Volodymyr Korniichuk, Yevgen Matseliuk, Susan Murcott, Volodymyr Panteleyev, Mykhailo Romaschenko, Roman Saydak, Anatoliy Shevchenko, Stefanie Sobel, and Mykhailo Yatsuik

1 Learning Objectives

1. Role of Climate Change in Water Scarcity: Using Ukraine as a specific example, this section explores how climate change exacerbates water scarcity, leading to negative consequences for people.
2. Water Stress to Water Security: Learn how Ukrainians are attempting to turn a situation of water stress into water security by securing sufficient safe drinking

B. Faybishenko (✉)
Lawrence Berkeley National Laboratory, Berkeley, CA, USA
e-mail: bafaybishenko@lbl.gov

C. Cox
GZA GeoEnvironmental, Inc., Brookline, MA, USA
e-mail: chad.cox@gza.com

O. Halych
Department of Hydropower, Thermal Power and Hydraulic Machines, Institute of Automation, Energy and Water Engineering, National University of Water and Environmental Engineering, Rivne, Rivne Region, Ukraine
e-mail: o.o.halych@nuwm.edu.ua

V. Korniichuk
Department of Hydrotechnical Engineering and Hydraulics, Institute of Automation, Energy and Water Engineering, National University of Water and Environmental Engineering, Rivne, Rivne Region, Ukraine
e-mail: v.i.korniichuk@nuwm.edu.ua

Y. Matseliuk
Water and Sanitation Laboratory of the Department of Water Resources, Kyiv Agrarian University of the National Academy of Agrarian Sciences of Ukraine, Kyiv, Ukraine

A. R. Molla et al. (eds.), *Climate Change, Drinking Water Security, and Public Health*, https://doi.org/10.1007/978-3-032-12074-8_15

water and an additional supply for irrigated agriculture to enable food production and economic and environmental stability.
3. The Environmental Toll of Warfare: Understand that war inflicts significant adverse environmental impacts in addition to its direct human toll.
4. Vision to Reality—the Planning Process: Realize how planners, designers, policymakers, engineers, and ordinary citizens work with existing conditions to determine the vision, possibilities, and feasibilities of building new solutions under conditions of transformation brought on by climate change and war.
5. Learning Courage to Build a Better Future: Be inspired by how, through learning and engagement, people can take courage and build better futures even in the most challenging conditions.

2 Introduction

This chapter examines the association between climate change, war, water, environmental degradation, and health in the specific context of Ukraine. It outlines a plan to address significant threats caused by climate change, war, and the Russian destruction of the Kakhovka Dam, which abruptly deprived over one million people of access to adequate water resources.

S. Murcott
D-Lab, Massachusetts Institute of Technology, Cambridge, MA, USA
e-mail: murcott@mit.edu

V. Panteleyev
Scientific Department, Institute of Water Problems and Land Reclamation of the National Academy of Agrarian Sciences of Ukraine, Kyiv, Ukraine

M. Romaschenko
Institute of Water Problems and Land Reclamation of the National Academy of Agrarian Sciences of Ukraine, Kyiv, Ukraine

Kyiv Agrarian University of the National Academy of Agrarian Sciences of Ukraine, Kyiv, Ukraine

R. Saydak
Department of Agrarian Resources, Kyiv Agrarian University of the National Academy of Agrarian Sciences of Ukraine, Kyiv, Ukraine

A. Shevchenko
Department of Water Resources, Institute of Water Problems and Land Reclamation of the National Academy of Agrarian Sciences of Ukraine, Kyiv, Ukraine

S. Sobel
Ukraine Water Association, Kyiv, Ukraine

M. Yatsuik
Institute of Water Problems and Land Reclamation of the National Academy of Agrarian Sciences of Ukraine, Kyiv, Ukraine

At the time of publication of this book (2025), Ukraine is still at war, confronting daily life and death challenges. This chapter explores some of the most complex challenges facing the Ukrainian people. We learn how climate change and war exacerbate water scarcity, accessibility, and quality problems and cause enormous negative health consequences for millions. Moreover, irrigation networks, agricultural productivity, livelihoods of farmers and businesses, and food prices on the domestic and international markets depend on a reliable and clean water supply. Not only is the health and well-being of every Ukrainian at risk, but the indirect consequences of the war are experienced on a regional and global scale. Moreover, acts of war are often acts of ecocide that have long-lasting planned or unintended environmental consequences.

Research and analysis of human and environmental impacts are critical for developing approaches to managing, conserving, and sustaining the use of surface and groundwater resources, implementing urban and agricultural land restoration, and providing energy and food security in and beyond Ukraine.

The authors of this chapter, a diverse team of international scientists, engineers, urban planners, development workers, lawyers, diplomats, and practitioners from academic, business, government, and non-profit sectors have come together to envision, research, design, and analyze the feasibility of a post-war, renewable energy-powered, decentralized drinking water supply, treatment, and distribution system in Ukraine. We envision a future country-wide water sufficiency and safety plan for Ukraine that enhances water security, accessibility, and quality. This chapter outlines two specific examples of that vision: a strategy for reconstructing the Kakhovka Dam and Reservoir with new modern features and the inter-basin Danube River water transfer to the Dnipro River. This plan is not the entire story for achieving Ukraine's sustainable, water-secure future, but it offers an essential goal in the broader context of rebuilding a unified, independent country.

3 Impact of Climate Change on Water Resources in Ukraine

Ukraine has experienced dramatic climate changes over the past century. The climate crisis has caused a decline in surface and groundwater availability and a deterioration of water quality. The most dramatic changes began in the 1980s. Continued climate change has led to a rise in temperature, which, in turn, has increased evaporation and transpiration. Yet precipitation in Ukraine has remained virtually unchanged or even decreased. Ukraine is facing an ever-growing shortage of both surface water and groundwater.

Climate change, primarily through rising temperatures, significantly increases evapotranspiration by accelerating the evaporation of water from open water surfaces and soil and the transpiration of water from plants. Evapotranspiration is the sum of plant transpiration and evaporation (Fig. 15.1). This speeds up the water cycle and leads to potential impacts such as increased drought in some regions and/or intense precipitation in others. Because warmer air holds more moisture, water

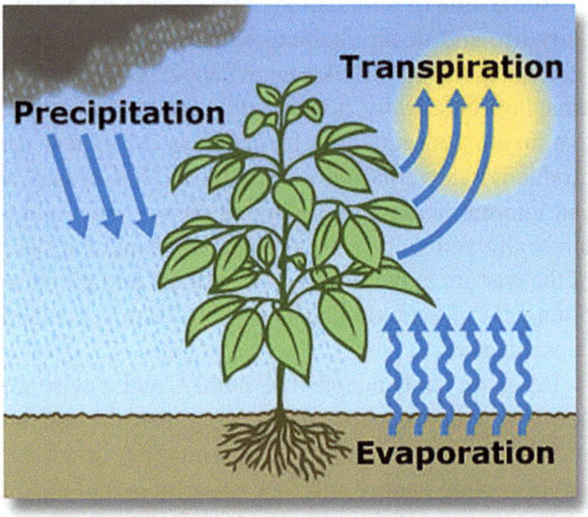

Fig. 15.1 Evapotranspiration is the sum of plant transpiration and evaporation (USGS 2018)

moves from the Earth's surface into the atmosphere. In a warming climate, increased evapotranspiration may shift the fraction of precipitation that runs off as surface water or infiltrates to the subsurface as groundwater recharge. Long-term shifts in recharge patterns can change groundwater levels and subsequently groundwater and surface water interactions with soil moisture and plants.

Using the temperature in 1901 as a reference, Fig. 15.2a shows the overall increase in temperature in Ukraine from 1901 to the present. A drastic temperature rise started in the 1980s (Faybishenko et al., 2024). Based on the calculated *Z*-score of temperature, the hot phase of climate warming started in the 1980s, as shown in red in Fig. 15.2b. During the hot phase, the temperature increases from 8 °C to over 11 °C.

Under these conditions, what happens with the discharge of the Dnipro, Ukraine's most significant river? Between the 1940s and the early 2020s, the average annual outflow of the Dnipro River to the Black Sea decreased by 23%—a considerable decrease. And what caused this? This was probably due to two factors: (1) climate change and (2) the construction of six major water reservoirs along the Dnipro in the 1940s and the 1950s. Temperature increases due to climate change can increase transpiration and thus reduce inflow into the tributaries and main stem of the Dnipro River. The construction of these reservoirs increased evaporation because of the increased open surface water area of these reservoirs. Due to increased evaporation by 40 to 80 millimeters per year (mm/year), the overall water balance in Ukraine decreased by 20 to 40 cubic kilometers per year (km^3/year) compared to 1990. The increase in evapotranspiration and a decrease in the river flux (from 10 to 20% in the north to 30 to 50% in the south) have caused a reduction in infiltration and groundwater recharge, thereby reducing groundwater operational reserves and further exacerbating water insecurity.

Some additional statistics underline the severity of the water crisis in Ukraine:

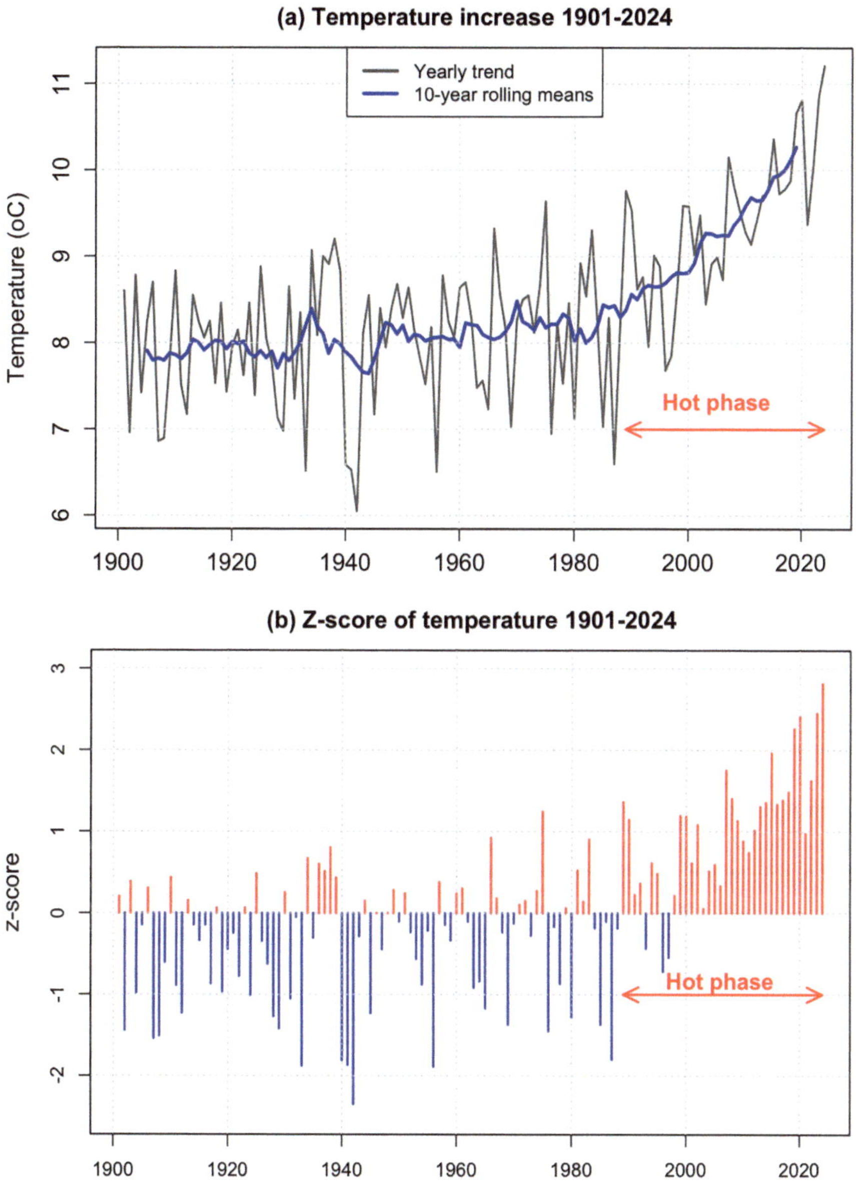

Fig. 15.2 (**a**) Temporal trend of annual temperatures in Ukraine and the 10-year rolling average; (**b**) Z-score of temperature for 1901–2024. The source of the annual temperature record is the Climate Change Knowledge Portal, World Bank Group (n.d.) (https://climateknowledgeportal.worldbank.org/country/ukraine/climate-data-historical, licensed under the terms of the Creative Commons Attribution 4.0 International License (http://creativecommons.org/licenses/by/4.0/)). The 10-year rolling average and the Z-score calculations were made using the R code *zoo* (Zeileis and Grothendieck 2005)

- Almost 75% of the water in Ukraine comes from its rivers. Yet only 25% of those river resources originate in Ukraine—most of Ukraine's river water comes from the upstream countries of Belarus and Russia. This illustrates that transboundary water sources are subject to geopolitical factors. Ukraine is also riparian to the final 120 kilometers (km) of the Danube River before its discharge to the Black Sea; however, the Danube's water, which is 80 km^3/year, has thus far not been practically used in Ukraine.
- Ukraine's per capita renewable freshwater resources, defined as internal river flows and groundwater from rainfall, indicate that it is a water-stressed country. Its renewable freshwater resources are far lower than those of the United States of America (USA), France, or the United Kingdom (UK). In Fig. 15.3, Ukraine has less than 1250 cubic meters (m^3) of water per person per year (GCDL and FAO 2024).
- Ukraine ranks 32nd among the 40 most water-scarce countries in Europe.
- Globally, Ukraine ranks 111th of 152 countries in terms of domestic water resources per capita (GCDL and FAO 2024).

Ukraine is generally a water-poor country. Romashchenko et al. (2020) assessed that most of Ukraine falls into severe water-stressed categories: "Catastrophic," "Very Low," and "Low" renewable freshwater resources. Catastrophic renewable freshwater resources are less than 1.0 km^3 /person/year. Very low is 1.0 to 1.7 km^3 / person/year. Low is 1.7 to 2.5 km^3 /person/year. Southern Ukraine is more water-stressed than the rest of the country, down to 0.5 km^3/person/year.

For those privileged to have grown up in a society where we have always been able to turn on the tap and have plentiful and safe drinking water, the situation in Ukraine or other water-stressed countries is hard to imagine. The human toll of these challenges is hard to grasp just by looking at the numbers. We might glimpse this reality when we read about drought in the American West, or of massive lakes, like Lake Chad or the Aral Sea, drying up. Or we may know the story of victims of huge hurricanes, such as Katrina (Louisiana, 2005), Maria (Puerto Rico, 2017), Harvey (Texas, 2017), or Helene (North Carolina, 2024). In the case of Helene, the

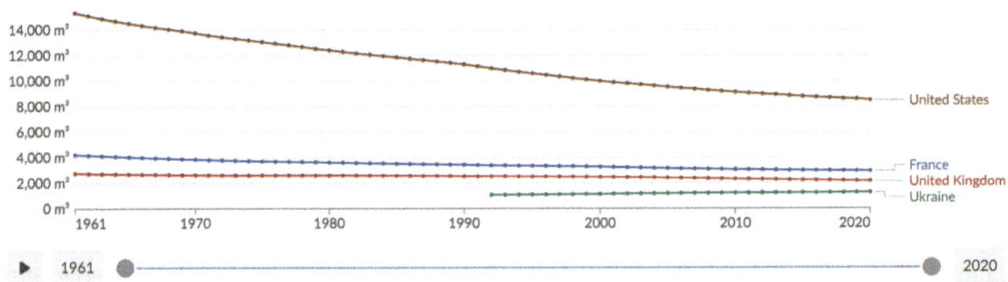

Fig. 15.3 Comparing Renewable Freshwater Resources of USA, France, UK, and Ukraine (GCDL and FAO 2024) (https://ourworldindata.org/grapher/renewable-water-resources-per-capita, licensed under the terms of the Creative Commons Attribution 4.0 International License (http://creativecommons.org/licenses/by/4.0/))

town of Ashville, NC suffered 50 days without piped water coming into homes. Families stood in line for bottled water hand-outs, cleaned themselves via sponge baths, and flushed their toilets with dirty dishwater.

The difference between hurricane victims and Ukrainians experiencing water stress is that the absence of piped water in Ashville, N.C., for example, lasted 50 days. In Ukraine, it has been a year or years, depending on the location. This is a huge physical and emotional burden, hard to grasp only in numbers. Some version of this kind of hardship is the everyday reality for many Ukrainians.

Based on European Union (EU) regulations, water withdrawal from rivers should be less than or equal to 10% of the river flow. Due to chronic water shortages, Ukraine's water withdrawals exceed the limit, putting additional pressure on other life forms, plants, and animals.

According to the Koppen-Geiger climate classification, between 1980 and 2016, most of Ukraine was within a "cold, no dry season, warm summer" climatic zone (in blue), with a small section of the south as an "arid steppe" zone (in yellow). However, by 2100, the "arid steppe" zone is expected to increase dramatically and cover half of the country (Beck et al. 2023; Semenova Vicente-Serrano 2024). Ukraine may become desert-like unless efforts are directed toward mitigation and adaptation to climate change.

Besides temperature, precipitation is another important climatic and meteorological parameter. Figure 15.4 shows a graph of the annual average precipitation in Ukraine. During the hot phase, precipitation initially increased, and starting from the year 2000, precipitation decreased, indicating a considerable imbalance in the meteorological conditions. Evapotranspiration is increasing without an offsetting increase in precipitation. This adds to Ukraine's burden of water insecurity. Likewise, Europe has been experiencing a multi-year drought in the 2020s, indicating an urgent need for proactive climate adaptation and mitigation strategies (Rakovec et al. 2022).

4 Impact of War in Ukraine on Water Resources

The Russian military campaign against Ukraine's internationally recognized territory, which began in 2014, has had severe negative impacts on Ukraine's water resources.

Military actions have caused damage and destruction of water infrastructure, including the Kakhovka Hydroelectric Dam and the Oskil Reservoir Dam. Furthermore, there has been major damage to municipal water supplies, water treatment, and wastewater treatment systems in cities such as Mykolaiv, Kryvyi Rih, Nikopol, Marganets, Melitopol, and Mariupol. Irrigation has ceased across almost 600,000 hectares (6000 square kilometers, km^2) in southern Ukraine, profoundly affecting soils, crops, food production, and livelihoods within and beyond Ukraine.

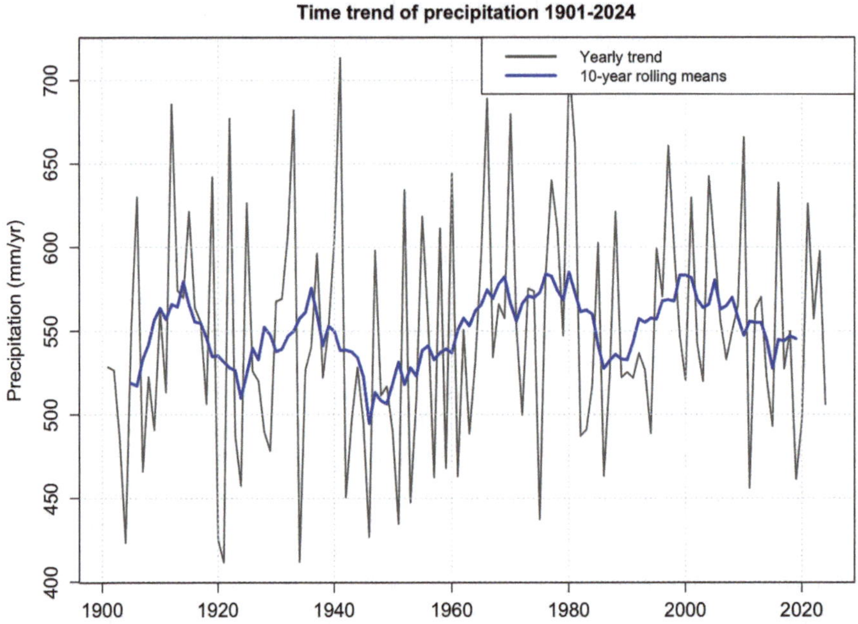

Fig. 15.4 Graphs showing changes over time in the annual precipitation in Ukraine from 1901 to 2024 and the 10-year rolling average of precipitation. Annual data downloaded from the Climate Change Knowledge Portal of the World Bank Group (n.d.) (https://climateknowledgeportal.worldbank.org/country/ukraine/climate-data-historical, licensed under the terms of the Creative Commons Attribution 4.0 International License (http://creativecommons.org/licenses/by/4.0/)) Calculations of the 10-year rolling precipitation were made using the R code *zoo* (Zeileis and Grothendieck 2005)

As of 2025, the war in Ukraine has not ended. With Ukrainians subjected to a fourth year of violence, Russian forces continue to decimate communities, deliberately targeting civilian infrastructure, destroying basic services and livelihoods, and leaving millions of people dependent on humanitarian assistance. According to the Ukraine Humanitarian Needs and Response Plan (OCHA 2024), nearly 15 million Ukrainian civilians out of a total population of 38 million continue to rely on the Humanitarian Needs and Response Plan as of December 2023 for humanitarian support (OCHA 2023a).

Of these, 9.6 million—well over half—need access to essential water, sanitation, and hygiene (WASH) services. The scale of attacks has been devastating. As of December 2023, Ukraine's WASH infrastructure has sustained an estimated United States Dollars (US$) 7 billion of damage, including 1046 kilometers of water supply networks, 327 kilometers of sewage networks, 82 pumping stations, 12 water treatment plants, and eight dams (OCHA 2023b).

Prior to the escalation of the war in February 2022, Ukraine's water sector was already negatively affected by decades of underinvestment and neglect dating to the Soviet era, with approximately 40% of water supply and treatment facilities in critical condition. Especially in the south and east of the country, decades of groundwater and surface water contamination from agriculture and industry have limited the use of treatment facilities.

In addition, attacks on Ukraine's energy infrastructure further compromise water availability to the communities, industry, and agriculture that depend on power stations to fuel Ukraine's network of *Vodakanals* (a public corporation responsible for water supply and sanitation utilities in Ukraine), centralized water treatment and delivery systems.

The destruction of the Kakhovka Dam and Reservoir in June 2023 exemplifies how the deliberate targeting of Ukraine's water and sanitation infrastructure has impacted civilian lives. Within hours, nearly a million people lost access to safe drinking water and became reliant on humanitarian assistance.

At the humanitarian response level, the latest WASH report highlights the involvement of 43 reporting partners and 101 implementing partners, including United Nations (UN) agencies, International Non-Governmental Organizations (NGOs), and National NGOs. It is also unclear how long the myriad entities will have funding and resources to provide emergency WASH assistance. Despite these efforts, a unified understanding of water availability at the national and lower basin/catchment levels and a data-driven rationale for targeted interventions are lacking. The ongoing conflict and widespread destruction of water infrastructure exacerbates this challenge (OCHA 2023b).

5 Effects of War on Human and Environmental Health and Well-Being

Acts of aggression meant to obliterate the Ukrainian people, their culture, and their environment occur daily, with profound effects on human environmental health and well-being. A catalog of those health effects to date can be arranged thematically into two broad categories of statistics, "Human Health" and "Environmental Health," which are inextricably linked.

5.1 Human Health

- One million Ukrainians and Russians are dead or injured.
- 87,000 deaths in Mariupol alone (Bowen 2022).
- Over 70,000 Ukrainians with missing limbs.
- One-third of the country is facing hunger (IHPL 2024).
- At least 20,000 unaccompanied, forcibly displaced children (likely a significant underestimate; IHPL 2024).
- Forced labor in prisoner camps, torture, and executions.

- Like COVID-19 pandemic fatigue, war fatigue in Ukraine has opened the door for Russian misinformation campaigns to flourish (IHPL 2024).

5.1.1 Medical Facilities

- 1336 attacks on healthcare infrastructure to date (Haque et al. 2024).
- There were 720 hospitals in Ukraine before the Feb 2022 Russian invasion. By April 2023, just 450 are still operating (Haque et al. 2024).
- Hospitals report decreased in-laboratory testing, cancer screening, gynecological services, rehabilitation services, and supply chain disruptions while dealing with increased emergency services (Haque et al. 2024).

5.1.2 Displacement

- Ten million people are displaced (IOM UN Migration 2024).
- 6.5 million refugees and 3.7 million internally displaced people (IOM UN Migration 2024).

5.1.3 Mental Health

- ten million Ukrainians at risk of mental health issues (IHPL 2024).
- 3.9 million people experiencing moderate to severe mental health-related symptoms (IHPL 2024).
- 80% of children with post-traumatic stress disorder (Khomenko 2024).

5.1.4 Chemical and Radiological Contamination of Water, Land, and Air

- Forests, agricultural fields, soil, water, and air are contaminated.
- Toxics include nitrogen, phosphorus, salinity, arsenic, lead, manganese, uranium, PFAS (Per- and polyfluoroalkyl substances, commonly referred to as "forever chemicals"), sewage, and agricultural run-off.
- Significant greenhouse gas emissions caused by military actions in Ukraine are estimated to have increased by 77 metric tons of carbon dioxide-equivalent due to bombs, missiles, mines, petroleum products, fires at petroleum depots and buildings, forest fires, and emissions from garbage and waste (Bun et al. 2024).

5.1.5 Kakhovka Dam Destruction Impact on Human Health

- Lost water access for up to one million people (Stone 2024).

- The number of people killed by the destruction of the Kakhovka Dam is estimated to be at least in the hundreds, but the exact number may never be known.

5.1.6 Nuclear Power Plant (NPP): Potential Release of Radiological Materials Due to War Actions

The ongoing military activities in Ukraine are raising concerns related to potential radiological releases from the nuclear power plant facilities if these are hit by missile strikes or otherwise affected by the ongoing military operations. These releases could lead to widespread contamination of the surface water, groundwater, and soil, resulting in short- and long-term impacts on people and the environment (Faybishenko et al. 2022).

5.2 Environmental Health

5.2.1 Kakhovka Dam Destruction: Environmental Impacts

- Flooding of 620 km^2, affecting 80 settlements.
- Flooding of nature reserves and agricultural land.
- Damage to a unique sturgeon breeding facility.
- Death among organisms adapted to brackish estuaries near the Dnipro and Black Sea confluence.
- Billions of dead mussels.
- Deaths of dolphins and porpoises in the Black Sea off Ukraine's southern coast.

5.2.2 Water Sector Impacts (Shumilova et al. 2023)

- Water transfer interruption.
- Surface water pollution due to military action—six cases, with four due to sunken military objects and two due to the release of chemicals from shelling.
- Dam damage—four instances at reservoirs and one along the North Crimean Canal.
- Mines overflooding.
- Bacteriological pollution due to mass poultry death.
- Interrupted operation of Kakhovka Hydroelectric Station and Zaporizhzhia Nuclear Power Station.
- Disrupted centralized water supply.
- Disrupted operation of wastewater treatment facilities.

5.2.3 National and International Impacts Due to the Water/Irrigation/Agriculture/Food Nexus

- Ukrainians' livelihoods are disrupted.
- Global food supply implications.

5.2.4 Land Impacts

- One-third of the territory of Ukraine estimated to be mined.

5.2.5 Potential Extinction of Endangered Species (Stone 2024)

- Sandy blink mole-rat (*Spalax arenarius).*
- Nordmann's birch mouse *(Sicista loriger).*
- A sand gadfly.
- Two species of pearl knapweeds.
- A rare birch tree *(Betula borysthenica).*
- Fish species: Sea roach (*Rutilus Frisii*), Dnipro barbel (*Barbus borysthenicus*), and Sarmatian bleak (*Alburnus sarmaticus*).
- Estuarine perch (*Sander marinus*).

In addition to the above-mentioned impacts from the war, it is likely that Ukraine will experience long-term effects of naturally occurring and anthropogenic chemical contaminants in drinking water. The following are examples of health impacts to be monitored:

- Non-threshold effects—carcinogens.
- Arsenic: skin, lung cancer.
- DDT (organic pesticide): breast cancer.
- Chloroform: kidney cancer.
- Threshold effects.
- Neurotoxins.
- Lead: IQ reduction, behavior problems.
- Neonicotinoids: memory loss.
- Endocrine disrupters.
- Uranium: fertility changes in mice.
- PFAS: thyroid hormone disruption.
- Developmental effects.
- TCE: heart defects.
- Other issues.
- Fluoride: dental fluorosis.
- Organochlorine pesticides: diabetes.

6 On the Need for Restoration of the Kakhovka Reservoir

The Dnipro River Basin includes a watershed area of 531,817 square kilometers (205,336 square miles), with upstream portions of the basin within southeastern Belarus and southwestern Russia and the downgradient regions covering much of eastern Ukraine. The river is reported to account for approximately 80% (percent) of Ukraine's surface freshwater resources (without consideration of the Danube water resources).

Until June 6, 2023, the Kakhovka Dam impounded the largest reservoir in Ukraine and the second largest in Europe. The Dam is located at the southern end of the Dnipro River, which flows north to south into the Black Sea. The dam is the final in a series of Ukrainian hydropower projects called the Dnipro Cascade, which consists of six hydroelectric dams along the Dnipro River.

The Kakhovka Dam was constructed from 1950 to 1956 when Ukraine was a part of the former Soviet Union. The dam had a maximum height of 30 meters (98 feet, ft) and a total length of approximately 3273 meters (10,738 ft.). The dam consists of earth fill embankments (accounting for most of the structure), a concrete gravity spillway with control gates, and a powerhouse. The hydropower plant at the dam included six turbine/generator units with a capacity of 60 megawatts (MW) each, for a total generation capacity of 360 MW and an average annual energy production of 1400 gigawatt (GW).

The dam was constructed to provide essential benefits, including municipal, agricultural, and industrial water supply. Until 2014, the Kakhovka Dam and the Reservoir provided drinking water to between one and two million people in southern Ukraine, including Russian-occupied Crimea. Water was used to irrigate agricultural land covering over 1.2 million hectares. The Kakhovka Reservoir supplied water to consumers through the North Crimean, Kakhovka, and Dnipro-Kryvyi Rih canals. As of 2021, 584,000 hectares were irrigated with water from the Kakhovka Reservoir. Water from the Kakhovka Reservoir was supplied to Russian-occupied Crimea via the 402 km long North-Crimean Canal to irrigate 113,000 hectares of agricultural land and meet that population's domestic and drinking water needs. The Kakhovka Reservoir was also a source of the Zaporizhzhia Nuclear Power Plant (ZNPP) Cooling Pond. The ZNPP is the largest nuclear power plant in Europe. The Kakhovka Dam has also provided downstream flood control. A ship lock constructed concurrently with the dam supplied navigation into and out of the Kakhovka Reservoir. The destruction of the Kakhovka Hydropower Plant is the most significant act of ecocide that Russia has committed during the full-scale invasion of Ukraine.

After the Russian invasion of Ukraine in February 2022, Russian troops seized control of the dam. Ultimately, the Armed Forces of Ukraine regained control of the right (west) bank of the Dnipro River in the Kherson region; however, Russian forces retained control of the dam. On June 6, 2023, the Kakhovka Dam was catastrophically breached, destroying most of the powerhouse and spillway and completely emptying the reservoir. While the Russian government has denied responsibility, subsequent investigations of available evidence by the Government of Ukraine and independent international entities, including the New York Times, indicate that the dam was deliberately destroyed by explosives placed and detonated by Russian forces.

There were numerous and widespread consequences from the deliberate breaching of the dam. Most immediately, it is believed that several hundred people drowned in the flood wave released from the reservoir. Exact numbers are difficult to quantify because of the ongoing military conflict after the dam's destruction. A United Nations Environment Program report estimates that the extent of the downstream flooding was 620 square kilometers (240 square miles), and up to 110,000 people were potentially affected. Thousands of structures were flooded in riverside towns, including Nova Kakhovka, Oleshky, and others downstream to the delta at the Dniprovska Gulf. Inundated areas were reported to have suffered extensive environmental damage. Water withdrawals from the reservoir were no longer possible after the dam breach, which impacted multiple sectors. The irrigation of farmland was curtailed, resulting in an estimated decrease in agricultural exports from Ukraine, valued at US$1.5 billion in 2023 alone. Water supplies to numerous cities were disrupted or eliminated.

To the west of the dam is at Mykolaiv, a city with a pre-war population of 470,000, which relied on the Kakhovka Reservoir as its primary water supply. Following the dam's destruction and additional attacks on the city's aqueducts, the city was forced to turn to groundwater for its municipal water supply. However, because of the salinity of the groundwater in the area, drinking water must first be desalinated via point-of-use reverse osmosis systems at communal distribution points throughout the city. The unwatering of the reservoir also eliminated the primary source of cooling water for the Zaporizhzhia Nuclear Power Plant. Because the power plant had been in shutdown mode due to the war, locally stored cooling water supplies were sufficient to prevent an immediate failure of the reactors; however, multiple groundwater wells had to be rapidly completed to provide adequate water supply for even ongoing standby-mode operations at the facility.

The 1949 Geneva Conventions (Article 56 of the 1977 Additional Protocol I) prohibit the intentional destruction of civil works such as dams "if such an attack may cause the release of dangerous forces and consequent severe losses among the civilian population." This prohibition applies even when these objects are military objectives. In addition to the direct loss of life, a joint report by the Government of Ukraine and the United Nations in 2023 estimated the total value of damage and losses resulting from the destruction of the Kakhovka Dam at US$14 billion.

Reconstruction of the Kakhovka Dam is impossible until the ongoing military conflict ends. Planning for post-war restoration will require cooperation between the Ukrainian Government, *Ukrhydroenergo* (a Ukrainian state-owned enterprise that manages hydro power plants along the Dnieper and Dniester rivers, and is Ukraine's primary hydropower generating company), local and national stakeholders, nongovernmental organizations, and international partners. Typically, the planning phase of such an effort would involve an alternatives analysis, which might consider options including:

1. No action.
2. Formal dam breach and river restoration.
3. Construction of multiple smaller dams.
4. Temporary reconstruction.
5. Reconstruction of the dam in-kind (i.e., using equivalent materials).
6. Reconstruction of the dam with new features.

The "No action" alternative is unlikely to be preferred by any stakeholders, but it will be the status quo until a plan is developed and funding is secured. Advocates against reconstructing the dam and promoting river restoration will point to ongoing ecological restoration in the river channel, natural vegetation reestablishment in the former upstream impoundment, the ability to replace benefits previously provided by the dam through other means, and the vulnerability of a new structure to future acts of aggression. Reconstruction alternatives could range from in-kind replacement to constructing new or multiple new structures. The alternative analysis offers the opportunity to consider means of addressing problems identified at the Kakhovka Dam, such as downstream foundation scour, reservoir water quality, and evaporation losses. One example is the area known as the Great Meadow, which was inundated by the original filling of the reservoir. This area had historical and cultural importance but was flooded by a relatively shallow portion of the reservoir of limited utility. Under a reconstruction plan, there is an option to create a levee that would protect the Great Meadow from flooding. Another feature that could be added to any potential new structure is fish passage in the form of provisions for upstream and downstream transit by native and commercial fish species. Location of the Dnipro River, Kakhovka Dam, Zaporizhzhia Nuclear Power Plant, and the Great Meadow are illustrated in the article by Stone (2024).

All alternatives should be evaluated concerning multiple criteria that include but are not limited to dam safety and engineering feasibility; benefits to public health, ecological impacts, renewable energy production, infrastructure security, direct financial costs, and benefits; overall economic benefits to society, and resiliency considering anticipated climate change. The destruction of the Kakhovka Dam was also a direct affront to Ukrainian sovereignty and national pride. Therefore, all decisions regarding the dam's future will be political. There will be significant interest from the entire Ukrainian population worldwide as the Ukrainian Government chooses its path to post-war reconstruction at the Kakhovka Dam and other priority sites.

Thus, first and foremost, Ukraine requires the restoration of control over all its territories and the end of hostilities. Subsequent conditions include:

1. Development and adoption of plans for the post-war reconstruction of Ukraine that create favorable conditions for the return of the population to the territories currently under occupation.
2. Development of feasibility studies of projects pertaining to the restoration of the Kakhovka Reservoir and Kakhovka Hydroelectric Power Plant.
3. Establish the main parameters of the Kakhovka reservoir, including a reduction of the flood area by 15 to 20% due to previous damming of the Great Meadow area.

4. Take measures to restore the supported normal water level (NWL) of 16 m, which will allow the North Crimean Canal to maintain its gravity mode of operation.
5. Install anti-seepage measures at the embankment structure to reduce water losses from seepage and mitigate flooding of adjacent territories.
6. Provide structural solutions in the reservoir dam to ensure navigation and passage of fish for spawning.
7. Advance policies to facilitate the use of Danube River water in 3 to 5 km^3/year to meet the entire demand of the southern regions.

Develop a feasibility study and a capital project for the supply of Danube River water to the southern regions of Ukraine.

7 Prospects for Using the Danube Water in Southern Ukraine

Freshwater is naturally unevenly spatially distributed in Ukraine. Climate change and war exacerbate this problem in Ukraine, drastically increasing the pressure on water resources and the ecosystem. Water transfer megaprojects (WTMP) may play an essential role in sustaining the water-food-energy nexus, providing water for irrigation, drinking water supply, and industrial development (Sternberg 2016). The inter-basin water transfer is "the transfer of water from one geographically distinct river basin to another (Davies et al. 1992; Gupta and van der Zaag 2008). According to the International Commission on Irrigation and Dams (ICID 2005), inter-basin water transfer accounted for 540 km^3 per year, 14% of global water withdrawals. Shumilova et al. (2018) presented a summary of water transfer megaprojects, which are defined as large-scale engineering interventions to divert water within and between river basins that meet one of the following criteria: construction costs over US$1 billion, transfer distance of more than 190 km, or water volume transfer exceeding 0.23 km^3 per year. These projects are generally associated with large-scale agricultural and energy development schemes. A schematic of the proposed protective dam across the Kakhovka Reservoir within the "Green Meadow" area can be found in Stone (2024) and Romashchenko et al. (2024).

Ukrainian scientists have performed an expert evaluation of the possibility of water transfer from the Danube River to the Dnipro River to rehabilitate the Kakhovka Reservoir, which will help restore irrigation in southern Ukraine (Ponomarenko 1982; Romashchenko et al. 2017, 2024). The Danube River is one of the most significant waterways in Europe. The Danube River provides an essential source of drinking water, hydroelectric power generation, irrigation, and navigation. It originates in Germany and flows into the Black Sea via the Danube Delta across ten European countries. Most of the Danube Delta lies in Romania. A small part is in Odesa Oblast,[1] Ukraine. Ukraine has formally engaged in international cooperation with other European countries. Ukraine has been a Signatory State of the Danube River Protection Convention since 1994.

The proposed project for using water resources of the Danube River in the southern regions of Ukraine includes constructing one or several large-diameter pipelines of different lengths. The project implies that the Danube's water will first flow through the existing Danube-Sasyk Lake canal, built during the Soviet Union era. Then, it will be transferred through a series of pipelines to provide drinking and domestic water supply facilities and irrigation systems in Odesa (500,000 hectares), Mykolaiv (300,000 hectares), and the western and southern parts (50,000 hectares) of the Kherson region. The endpoint of the Danube water supply should be the restored Kakhovka Reservoir. Implementation of this project can begin before the end of the war. The number and parameters of the pipelines necessary for project implementation will be determined during the feasibility study. This option also involves the construction of a pipeline to supply water from the Kakhovka canal to the city of Mariupol. It is estimated to cost between US$10 billion and US$15 billion, which meets the megaproject criteria. Megaprojects are considered high-risk projects because they require significant financial investments, demand long time frames from planning to completion, and may have substantial socio-economic and environmental ramifications (Flyvbjerg 2007; Sternberg 2016). Shumilova et al. (2018) underscored the extent to which humans must re-engineer the global hydrological network and flow through water transfer megaprojects, creating a network of "artificial rivers." They emphasized the need to ensure the inclusion of these projects in international and basin hydrological models, and to develop internationally agreed-upon criteria to assess the ecological, social, and economic impacts of WTMP.

8 Concluding Remarks

In this chapter, we learned about the impacts of climate change and war on Ukraine. We have learned that in the face of an overwhelming challenge, an initial strategy to proceed is to take steps each day toward the best possible vision.

Let's conclude by reviewing the learning objectives we have considered in this chapter:

1. Role of Climate Change in Water Scarcity: We have learned that climate change affected Ukraine before the current war, and that the country has experienced a "hot phase" that began in the 1980s and continues into the present. Moreover, we have learned that this "hot phase" is expected to expand the arid parts of the Ukrainian territory to cover about 1/2 of its territory by 2050.

[1] Oblast can be translated into English as "province" or "region." Ukraine currently has 24 oblasts.

2. Water Stress to Water Security: This chapter envisions transforming water stress into water security, ensuring sufficient and safe drinking water, along with enough for irrigated agriculture to enable food production and economic and environmental stability. This is not today's reality in Ukraine, but it is a possibility we work toward.
3. The Environmental Toll of Warfare: Imagine that you are one of the million people living downstream of the destroyed Kakhovka Dam and Reservoir who are now without water for everyday activities of drinking, cooking, washing, irrigating your fields, and supporting the life of animals, plants, and trees. This chapter presents several examples to help the reader understand how war is hell for humans and the environment that supports all life. The Kakhovka Dam and Reservoir cannot be rebuilt until the war ends. Still, planning processes can begin to explore the projected costs and benefits of alternatives in the context of competing national priorities and limited resources.
4. Vision to Reality—the Planning Process: This chapter is the realization of the work of Ukrainians and friends of Ukrainians who are planners, designers, policymakers, engineers, and ordinary citizens working with existing conditions to articulate the vision of building new solutions under the transformation brought on by climate change and war. Climate change will affect every living thing in the coming century. Similarly, disruptions due to climate change and war, such as human displacements, are our present and future realities. We propose a path to avoid the worst impacts of climate change and war. Then, we work toward those possibilities through a collective planning process that is, at its heart, an inclusive and democratic one. It is also essential to use the opportunity of any large-scale capital investments to build back better with a modern green infrastructure and position Ukraine for economic development. International partners should also share their experiences and best practices to help Ukrainians make informed decisions about massive reconstruction projects.
5. Learning Courage to Build a Better Future: We hope this learning journey will inspire the reader. Through collaboration and engagement, our team has tried to show how to find the courage to dream of a bright future even in Ukraine's most challenging present conditions of climate change and war.

9 Discussion Questions

1. Where is the water-stressed area of Ukraine located, and what does it mean to be water-stressed?
2. How do people manage without sufficient water?
3. Will Ukraine be able to rebuild the Kakhovka Dam and Reservoir?
4. Is it possible to reverse negative climatic trends before they overwhelm society? How?
5. What would you do if you lived in Ukraine today?

6. Discuss the causes of surface water and groundwater deficiency and deterioration of water quality in Ukraine caused by climate change.
7. What are the reasons for the deterioration of surface water and groundwater quality caused by military actions?
8. Provide several scenarios of how Ukraine can rebuild and reverse negative trends in climate change and the war.
9. In your opinion, will inter-basin water transfer (Danube to Dnipro) help resolve water scarcity issues in Ukraine? Based on a review of articles related to inter-basin water transfer, discuss several potentially suitable scenarios for Ukraine.
10. Propose at least one climate adaptation scenario for rehabilitating agriculture in Ukraine.

Acknowledgments We thank the MIT-Ukraine Water Project Group for the opportunity to discuss the research and practical issues related to the environmental consequences of climate change and war in Ukraine. BF's work was partially supported by the Laboratory Directed Research and Development Program of LBNL under U.S. DOE's Contract No. DE-AC02-05CH11231. The co-authors are sincerely thankful to Timothy Buckley for his careful review and comments on the manuscript, and to the book editors, Seth Frisbie and Azizur Molla, for their invitation and final editing of the manuscript. The opinions expressed in the text are those of the authors and do not necessarily reflect the positions of their institutions.

References

Beck HE, McVicar TR, Vergopolan N et al (2023) High-resolution (1 km) Köppen-Geiger maps for 1901–2099 based on constrained CMIP6 projections. Sci Data 10:724. https://doi.org/10.1038/s41597-023-02549-6

Bowen AS (Congressional Research Service) (2022) Russia's war in Ukraine: military and intelligence aspects. https://crsreports.congress.gov/product/pdf/R/R47068. Accessed 5 December 2024

Bun R, Marland G, Oda T, See L, Puliafito E, Nahorski Z, Jonas M, Kovalyshyn V, Ialongo I, Yashchun O, Romanchuk Z (2024) Tracking unaccounted greenhouse gas emissions due to the war in Ukraine since 2022. Sci Total Environ 914:169879. https://doi.org/10.1016/j.scitotenv.2024.169879

Davies BR, Thoms M, Meador M (1992) An assessment of the ecological impacts of inter-basin water transfers, and their threats to river basin integrity and conservation. Aqua Conserv Mar Freshw 2(4):325–349. https://doi.org/10.1002/aqc.3270020404

Faybishenko B, Dwivedi D, Ryan D, Nico P, Birkholzer J, Zavarin M, Kersting A, Slessarev E, Cherlinka V, Dmytruk Y, Romashchenko M, Nezgoduk A (2022) GIS- and modeling-based assessment of soil and groundwater radioecological vulnerability in Ukraine: proof-of-concept and directions of future research. BNL-2001498 draft report Lawrence Berkeley National Laboratory. 30 November 2022. https://profiles.lbl.gov/19709-boris-faybishenko/publications?favouritesFirst=true&perPage=25&sort=dateDesc&startFrom=0. Assessed 21 Nov 2024

Faybishenko B, Romashchenko MI, Shevchenko AM (2024) State and prospects of groundwater use for irrigation in Ukraine. Presentation to the Massachusetts Institute of Technology (MIT) Ukraine Water Project, Cambridge, MA. 19 July 2024

Flyvbjerg B (2007) Policy and planning for large-infrastructure projects: problems, causes, cures. Environ Plann B Plann Des 34:4. https://doi.org/10.1068/b32111

Global Change Data Lab and Food and Agriculture Organization of the United Nations (GCDL and FAO; via World Bank) (2024) Renewable freshwater resources per capita. https://ourworldindata.org/grapher/renewable-water-resources-per-capita?country=USA~UKR~FRA~GBR. Accessed 20 November 2024

Gupta J, van der Zaag P (2008) Interbasin water transfers and integrated water resources management: where engineering, science and politics interlock. Phys Chem Earth Pt A, B, C 33(1–2):28–40. https://doi.org/10.1016/j.pce.2007.04.003

Haque U, Bukhari MH, Fiedler N, Wang S, Korzh O, Espinoza J, Ahmad M, Holovanova I, Chumachenko T, Marchak O, Chumachenko D, Ulvi O, Sikder I, Hubenko H, Barrett ES (2024) A comparison of Ukrainian hospital services and functions before and during the Russia-Ukraine War. JAMA Health Forum 5(5):e240901. https://doi.org/10.1001/jamahealthforum.2024.0901

International Commission on Irrigation & Drainage (ICID) (2005) Experience in inter-basin water transfers for irrigation, drainage or flood management (3rd draft 15 August 2005). http://www.icid.org/ibwt_draft_aug05.pdf. Accessed 1 November 2024

International Organization for Migration (IOM UN Migration) (2024) Ukraine. ION U.N. Migration. https://ukraine.iom.int/ Accessed 21 November 2024

Khomenko I (2024) 80% of Ukrainian children suffer post-traumatic stress due to Russian invasion. https://united24media.com/latest-news/80-of-ukrainian-children-suffer-from-ptsd-due-to-russian-invasion-2665. Accessed 5 December 2024

Loma Linda University Health, Institute for Health Policy and Leadership (IHPL) (2024) The Impact of war in Ukraine on the environment and health. https://ihpl.llu.edu/blog/impact-war-ukraine-environment-and-health. Accessed 21 November 2024

Ponomarenko VD (1982) Selection of the route of inter-basin transfer of runoff (on the example of the Danube-Dnipro canal). Hydraulic Eng Reclam 8:34–38. (in Russian)

Rakovec O, Samaniego L, Hari V, Markonis Y, Moravec V, Thober S, Hanel M, Kumar R (2022) The 2018–2020 multi-year drought sets a new benchmark in Europe. Earth's Fut 10(3):e2021EF002394. https://doi.org/10.1029/2021EF002394

Romashchenko MI, Shevchenko AM, Lyutnytskyi SM (2017) State and ways of enhancing the water supply of the southern region of Ukraine by using the water resources of the Danube River. Reclam Water Manage 105(3):12–21

Romashchenko MI, Gusev YV, Shatkovskiy AP, Saydak RV, Yatsyuk MV, Shevchenko AM, Matyash TV (2020) The impact of current climate change on water resources and agricultural production. Land Reclam Water Manag 1:2020

Romashchenko MI, Panteleyev V, Saidak R, Muzyka O, Usatyi S, Usata L, Shevenko I, Knysh V (2024) Is it necessary to restore the Kakhovka Reservoir? Presentation to the Massachusetts Institute of Technology (MIT) Ukraine Water Project, Cambridge, MA. 20 August 2024

Semenova I, Vicente-Serrano S (2024) Long-term variability and trends of meteorological droughts in Ukraine. Int J Climatol 44:1–18. https://doi.org/10.1002/joc.8416

Shumilova O, Tockner K, Thieme M, Koska A, Zarfl C (2018) Global water transfer megaprojects: a potential solution for the water-food-energy nexus? Front Environ Sci 6. https://doi.org/10.3389/fenvs.2018.00150

Shumilova O, Tockner K, Sukhodolov A, Khilchevskyi V, De Meester L, Stepanenko S, Trokhymenko G, Antonio J, Gleick P (2023) Impact of the Russia–Ukraine armed conflict on water resources and water infrastructure. Nat Sustain 6:578–586. https://doi.org/10.1038/s41893-023-01068-x

Sternberg T (2016) Water megaprojects in deserts and drylands. Int J Water Resource Dev 32(2):301–320. https://doi.org/10.1080/07900627.2015.1012660

Stone R 2024 Laid to waste. Science 383(6678). https://www.science.org/content/article/ukrainian-scientists-tally-grave-environmental-consequences-kakhovka-dam-disaster. Assessed 21 November 2024

United Nations Office for the Coordination of Humanitarian Affairs (OCHA) (2023a) Ukraine humanitarian relief and response. https://reliefweb.int/report/ukraine/ukraine-humanitarian-needs-and-response-plan-2024-december-2023-enuk. Accessed 21 November 2024

United Nations Office for the Coordination of Humanitarian Affairs (OCHA) (2023b) Ukraine water, sanitation, hygiene. https://response.reliefweb.int/ukraine/water-sanitation-and-hygiene. Accessed 21 November 2024

United Nations Office for the Coordination of Humanitarian Affairs (OCHA) (2024) Ukraine humanitarian needs and response plan 2024. https://reliefweb.int/report/ukraine/ukraine-humanitarian-needs-and-response-plan-2024-december-2023-enuk. Accessed 5 December 2024

USGS (2018) Water Science School by the United States Geological Survey, Evaporation and the water cycle. https://www.usgs.gov/media/images/evapotranspiration-sum-plant-transpiration-and-evaporation, Accessed June 11, 2025

World Bank Group (n.d.) Climate change knowledge portal. https://climateknowledgeportal.worldbank.org/country/ukraine/climate-data-historical (data downloaded on September 7, 2025)

Zeileis A, Grothendieck G (2005) Zoo: S3 Infrastructure for regular and irregular time series. J Stat Softw 14(6). https://doi.org/10.18637/jss.v014.i06

Chapter 16
Waterless and Low-Water Sanitation Technologies that Improve Quality of Life and Conserve Water Resources

Libby Hsu

1 Learning Objectives

1. Understand the trajectory of large-scale sanitation since the late nineteenth century, why infrastructure projects have favored sanitation technologies that use large amounts of water, and why this has disrupted the nitrogen cycle.
2. Explain the concept of the compost toilet, a solution that requires no water to process waste.
3. Explain the concept of the vermifiltration septic system, a solution that requires water to process waste, but treats and returns that water to the local ecosystem at a household level.
4. Propose community participation methods that build buy-in for novel sanitation technologies in the Global South and help countries achieve SDG6 targets for equitable access to water resources and water-use efficiency.

2 Large-Scale Sanitation Infrastructure and Its Impacts

Sanitation is essential to human health and quality of life, and the United Nation's Sustainable Development Goals (UN's SDG) Target 6.2 states that by 2030, the world should "achieve access to adequate and equitable sanitation and hygiene for all and end open defecation, paying special attention to the needs of women and

L. Hsu (✉)
MIT D-Lab, Massachusetts Institute of Technology, Cambridge, MA, USA
e-mail: slhsu@mit.edu

A. R. Molla et al. (eds.), *Climate Change, Drinking Water Security, and Public Health*, https://doi.org/10.1007/978-3-032-12074-8_16

girls and those in vulnerable situations." The World Health Organization/United Nations (WHO/UNICEF) Joint Monitoring Programme for Water Supply, Sanitation, and Hygiene (JMP) assesses progress toward this goal, with its most recent report published in 2021. The JMP has developed a sanitation ladder, shown in Fig. 16.1, that clarifies what level of service qualifies as "safely managed" sanitation (UN 2024; WHO and UNICEF 2021).

The UN measures progress toward Target 6.2 through Indicator 6.2.1 by measuring the percentage of individuals who are using a safely managed sanitation system—that is, an improved facility (such as a septic tank, an improved pit latrine, or connection to a centralized or decentralized wastewater treatment system) that is not shared with other households, ensuring that human waste is stored, transported, and treated according to widely accepted standards. Figure 16.2 shows progress toward this goal in 2020. It is important to realize that data is not available for many countries, making it more difficult to assess worldwide progress toward the Sustainable Development Goals, but some important conclusions can still be made. For example, the JMP estimates that the population using safely managed sanitation increased from 47% to 54% between 2015 and 2020, and the number of people practicing open defecation decreased by one-third, with the majority of the decrease occurring in rural areas, where this practice is more prevalent. However, access to sanitation remains a major challenge around the world: 3.6 billion people still lack safely managed sanitation services according to the JMP's definition. Only eight countries are projected to achieve universal access to safely managed sanitation by 2030 (WHO and UNICEF 2021).

SERVICE LEVEL	DEFINITION
SAFELY MANAGED	Use of improved facilities that are not shared with other households and where excreta are safely disposed of in situ or removed and treated offsite
BASIC	Use of improved facilities that are not shared with other households
LIMITED	Use of improved facilities that are shared with other households
UNIMPROVED	Use of pit latrines without a slab or platform, hanging latrines or bucket latrines
OPEN DEFECATION	Disposal of human faeces in fields, forests, bushes, open bodies of water, beaches or other open places, or with solid waste

Fig. 16.1 The Joint Monitoring Programme for Water Supply, Sanitation, and Hygiene (JMP) sanitation ladder (WHO and UNICEF 2021) (Reproduced from WHO and UNICEF 2021, Figure 49, p. 48. https://data.unicef.org/wp-content/uploads/2022/01/jmp-2021-wash-households_3.pdf, licensed under the terms of the Creative Commons Attribution-NonCommercial-ShareAlike 3.0 IGO (CC BY-NC-SA 3.0 IGO) licence (https://creativecommons.org/licenses/by-nc-sa/3.0/igo). Used with permission)

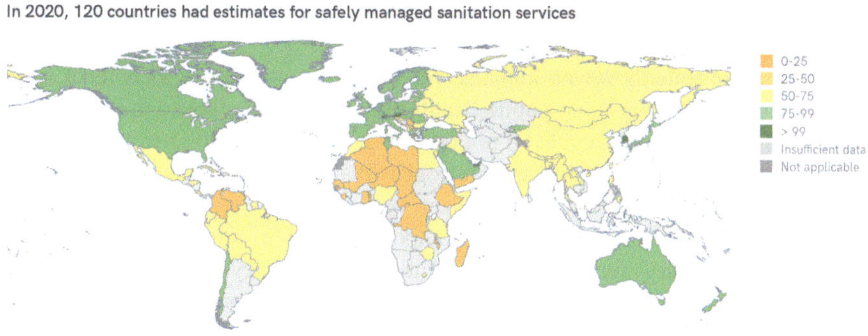

Fig. 16.2 Percentage of population using safely managed sanitation services in 120 countries in 2020, as defined by Sustainable Development Goal (SDG) Indicator 6.2.1 (WHO and UNICEF 2021) (Reproduced from WHO and UNICEF 2021, Figure 7, p. 9. https://data.unicef.org/wp-content/uploads/2022/01/jmp-2021-wash-households_3.pdf, licensed under the terms of the Creative Commons Attribution-NonCommercial-ShareAlike 3.0 IGO (CC BY-NC-SA 3.0 IGO) licence (https://creativecommons.org/licenses/by-nc-sa/3.0/igo). Used with permission)

2.1 The Impact of Human Waste on the Nitrogen Cycle

The nitrogen cycle refers to the processes through which nitrogen is circulated through ecosystems on Earth. Nitrogen from the atmosphere enters ecosystems through biological nitrogen fixation (BNF) in organisms. Organisms then deposit nitrogen into ecosystems through their outputs and decomposition, and the process of denitrification returns nitrogen to the atmosphere from soil, water, and sediments. BNF produces organic compounds that contain reactive nitrogen (N_r), which have implications for the ecosystems in which they are found. For example, one of these compounds is ammonia (NH_3), which occurs naturally in the environment and is produced by bacteria and decomposing organisms, but is toxic to aquatic life in excessive quantities (Fowler et al. 2013).

Animals are a component of the nitrogen cycle. An animal consumes material that contains nitrogen and phosphorus obtained from the environment, and then deposits it back in the environment through its waste products and its eventual decomposition. Human agriculture has made use of this concept for thousands of years by applying feces and urine from humans and animals to crops as a fertilizer, adding nitrogen to the process and effectively harnessing the BNF of these organisms for agricultural production (Delwiche 1970; Widdison and Burt 2008).

The use of untreated feces in agriculture presents health challenges. Not only does it put agricultural workers in direct contact with the pathogens in feces, it also renders many crops unsafe for consumption, particularly those that are eaten raw. At the start of the twentieth century, the invention of the Haber-Bosch process reduced the need for animal waste as fertilizer, which also resolved the problem of pathogen exposure from manure use in agriculture. This process converted atmospheric nitrogen to ammonia through a reaction with hydrogen, and allowed for drastic increases in agricultural production and, thus, growth in human population (Bosch 1908; Delwiche 1970; Widdison and Burt 2008).

At the same time, however, humans were starting to sequester their feces from the environment for the same public health reasons, contributing to the increasing disruption of the nitrogen cycle by not returning this material to terrestrial ecosystems. Other forms of anthropogenic nitrogen fixation, such as the combustion of fossil fuels and the application of fertilizer to increasingly large areas, contributed to the increase in the production of N_r compounds, such as nitrous oxide (N_2O), in the twentieth century, and the quantity of nitrogen in the cycle has doubled since pre-industrial times. The negative impacts of all these additional nitrogen inputs on terrestrial and marine ecosystems did not begin to be realized until the 1970s, through the identification of the so-called "nitrogen cascade" and the effects of fertilizer runoff in notable locations such as Long Island, New York (Fowler et al. 2013; Galloway et al. 2003; Stoner 1977).

Anthropogenic nitrogen fixation is a biogeochemical variable affecting the so-called planetary boundaries, which propose limits to how much Earth can tolerate the anthropogenic modification of its environment. According to the creators of the concept, "transgressing one or more planetary boundaries may be deleterious or even catastrophic due to the risk of crossing thresholds that will trigger non-linear, abrupt environmental change within continental-scale to planetary-scale systems" (Rockström et al. 2009). The boundary value set for anthropogenic nitrogen fixation has already been crossed, indicating that the nitrogen cycle is unbalanced and will continue to have cascading consequences for the environment and humans' ability to survive in it (Earth's boundaries? 2009).

2.2 *Modern Sanitation Infrastructure Uses Treated Water and Disrupts the Nitrogen Cycle*

Before the Industrial Revolution, most humans disposed of their urine and feces at a location in or near their own homes, with minimal to no containment and no treatment processes to neutralize pathogens or odors. In more densely populated areas, services sometimes existed to collect feces from homes and deposit it in a more centralized location, but it was unlikely to have been any better contained from exposure to the populace and the environment. Waste was often disposed in open cesspools, which allowed liquids to slowly drain into the soil. In some cases, the remaining solids were collected for sale to agricultural workers, perpetuating the nitrogen cycle through agriculture. In both rural and urban areas, urine and feces was directly or indirectly deposited into soil and waterways, returning nitrogen and phosphorus to the ground, while also spreading disease (Ashley 1909; Stoner 1977; George 2008).

This state of affairs became untenable with the major expansion of urban areas in the nineteenth century. London's population grew from 959,000 in 1801 to 2.3

million by 1851, and New York City's population grew more than tenfold from 1800 to 1870. This rapid expansion necessitated new solutions to disposing of waste. Concurrently with this growth, the first flush toilet was patented in 1775, making the possibility of flushing waste farther away from the user a reality for a greater proportion of humans in some parts of the world. However, this use of water also increased the volume of waste to be processed. Toilets produced before 1992 used between 3.5–7 gallons per flush, while toilets built to newer standards use no more than 1.6 gallons per flush. Now, water was needed to remove waste where little to none had been used before (Rosenwaike 1972; Stoner 1977; Halliday 2007; George 2008).

To address the need, cities in the United States and Europe began developing sewer systems to direct waste from toilets flushed in buildings to a centralized location, and overdesigned them to account for future population growth—though not by nearly enough. However, waste was still being dumped in the local environment after being piped away from homes and businesses. London sewers sent sewage into the River Thames, from which water was concurrently taken for human consumption in the nineteenth century; and sewage sludge was dumped into Boston Harbor until 1991, well past the enactment of the 1972 Clean Water Act in the United States. These practices damage marine ecosystems and needlessly disrupt the contributions of human waste to the nitrogen cycle (Levy and Connor 1992; George 2008). As Victor Hugo memorably wrote in *Les Misérables*:

> What is done with [humans'] golden dung? It is swept into the gulf. We send at a great expense fleets of ships to collect at the Southern Pole the guano of petrels and penguins, and cast into the sea the incalculable element of wealth which we have under our hand. All the human and animal manure which the world loses, if returned to the land instead of being thrown in the sea, would suffice to nourish the world. (Hugo 1862)

Almost two centuries later, manure alone may no longer meet the demands of human agriculture, but Hugo's point still stands in parts of the world that have been unable to afford centralized wastewater treatment infrastructure: humans waste their waste.

2.3 Modern Centralized Sanitation Infrastructure Cannot Meet Demand

Urban wastewater infrastructure has evolved such that a building has one input and one output for water. That is, one source of water piped into a building may be used for human consumption, to wash one's hands, to take a shower, or to flush a toilet; and the used water from all these sources is combined upon exit. If water entering a building has been treated to safe drinking water standards, it follows that this treated water is being used to flush toilets. Thus, the entire volume of wastewater exiting a building must be treated as "blackwater" containing fecal matter. This situation needlessly contaminates vast quantities of water that must be contained and treated at significant cost and energy use (Anand and Apul 2014; Stoner 1977).

The problem is compounded in some locations by the design of storm drains. In some municipalities, sewer pipes and storm drains are separate. For reasons of cost and scale, however, other municipalities have created combined drains for both sewage and stormwater. During heavy rain events, the system overflows into outfall pipes that may drain into nearby bodies of water, contaminating them with raw sewage. These systems will experience more and more overflow events as climate change affects precipitation patterns (Tarr 1979).

Regions of the world facing water scarcity and population growth simply cannot afford to contaminate and treat water at these scales. At the same time, they urgently need sanitation solutions that improve public health in order to meet the goals of SDG6 and achieve universal safely managed sanitation. In the absence of large-scale centralized sanitation infrastructure, decentralized, household- or neighborhood-level improved sanitation systems can reduce the complexity of blackwater treatment on site, or even eliminate the need for water use (Stoner 1977; UN 2024; WHO and UNICEF 2021). These systems have the added benefit of safely returning nitrogen to local ecosystems. This chapter will describe two such systems, waterless compost toilets and vermifiltration.

3 Compost Toilets as a Waterless Sanitation Solution

Composted waste has been used by various cultures for millennia. Civilizations in Mesopotamia, China, Europe, and Egypt show evidence of composting feces for use as an agricultural product, based on the understanding that composted human waste products would increase agricultural productivity. Human civilizations throughout history have contributed to the nitrogen cycle by transferring human and animal waste back to agricultural fields (though not always safely, when waste is not composted before application). When modern plumbing infrastructure became widely available in North America and Europe in the late nineteenth and twentieth centuries, this infrastructure supplanted other forms of sanitation in these areas. Yet innovations were still being made in the area of so-called "dry" toilets that did not require water to flush, such as a prototype developed by Henry Moule in 1860 (Rockefeller 1998; Anand and Apul 2014; Stoner 1977).

Compost toilet systems differ from other off-grid sanitation systems, such as pit latrines, in two notable ways. First, they require the separation of liquids and solids (i.e., urine must be separated from feces) in order to facilitate the composting process. Second, they use no water to flush waste products, although water should still be used to clean the facility. Innovations in compost toilet design in the last 50 years have promoted design components that appeal to modern users' cultural expectations and personal hygiene needs (Stoner 1977).

These systems reduce the potential for odor and pest proliferation in latrines. They can be an effective solution in geographical areas that are prone to flood risk, because the waste is effectively contained and protected from overflow during heavy rains and flooding. Compost toilet systems can be installed at low cost at a decentralized household level, making them attractive to stakeholders who are eager to move away from the use of pit latrines. However, they take effort to maintain, and they require significant behavior changes from all users to be successful (Stoner 1977; Mihelcic et al. 2009).

3.1 Separating and Processing Liquids and Solids in Compost Systems

A compost toilet system may be installed inside an existing building or in a dedicated structure. Regardless of where it is situated, it must include a toilet seat that separates urine and feces effectively, two or more aerobic chambers to hold feces in various stages of composting, and a receptacle with which to divert and/or store urine (Stoner 1977; Mihelcic et al. 2009; Berger 2011).

A compost system destroys pathogens using four key design criteria:

1. Temperature: the chamber must be kept at a minimum of 42 degrees Celsius to promote pathogen destruction;
2. pH: the pH of the compost must be at least 9;
3. Moisture: the moisture content of the compost must be as low as possible, and ideally below 25%; and.
4. Time: the compost must be allowed to sit in the chamber for at least six months and up to two years, under the conditions listed above, to destroy pathogens; lower ambient temperatures at the site will require longer composting times (WHO 2006).

Separating urine and feces is therefore critical to the functioning of the system, because urine would disrupt the moisture and pH requirements for the feces to compost successfully. Additionally, urine's high nitrogen contact would interfere with the carbon to nitrogen (C:N) ratio needed for optimal composting of any material, which is 30:1, since urine's C:N ratio is 0.8:1. The steps needed to meet these criteria are described below (Mihelcic et al. 2009).

Urine-diversion toilet seats resemble existing toilet seats, with one major difference: a barrier is constructed separating the front and back halves of the seat opening. The front of the seat is usually designed to resemble a funnel, so urine is directed into a pipe or tube. The back half of the seat is usually open, with a composting chamber positioned directly underneath the installed toilet seat. There are products available on the market that can be inserted into an existing toilet seat/bowl to convert it for urine diversion, or the toilet can be constructed from scratch (Stoner 1977; Mihelcic et al. 2009).

When someone uses the toilet, they must make sure they are positioned correctly on the seat to ensure that urine and feces are diverted effectively. This can require practice and reinforcement of how the system works to ensure success. Menstrual blood can exit on either side of the urine-diversion barrier without harm to the system, although menstrual products such as pads and tampons cannot be disposed in the compost chamber, so a separate form of waste disposal must exist for these products. However, toilet paper can be added to the compost chamber without issue (Berger 2011).

After depositing feces in the compost chamber through the urine-diversion toilet, the user must add 1–2 cups of dry material such as cooking ashes, sawdust, lime, or grass clippings into the compost chamber. This step controls odor and pests and advances the composting process by desiccating the chamber's contents. The choice of desiccant may also aid in increasing pH (e.g., ash) or add carbon to the compost (e.g., grass). It is useful for users of the system to occasionally view the contents, with the help of a light source, to identify wet areas that need to have more desiccating material added to them. Some sources recommend stirring or turning over the compost weekly (Mihelcic et al. 2009).

The compost chamber must be well-sealed to prevent the incursion of moisture, insects, and ambient air. It is usually constructed above ground level to avoid flooding. In a standalone building, this requires the toilet chamber to be constructed with stair or ramp access above the compost chambers. It is recommended that two compost chambers be constructed side by side in the structure. Thus, while one chamber is being actively filled, the other is sealed to accelerate the compost process until it needs to be used again. The latrine might contain two toilets positioned over each chamber, or it could contain movable elements to transfer one toilet seat back and forth as each chamber is filled and then sealed in turn. The compost chamber must have an access door so that compost can be removed when the process is complete, making that chamber ready to be filled again (Mihelcic et al. 2009; Berger 2011).

Urine can be directed by pipe or tube to a receptacle or drainage system, depending on whether the user wants to collect the urine for use or not. The easiest way to dispose of unwanted urine is to dig a pit behind the latrine, fill it with gravel, and let the urine drain into the pit so it does not pool on ground surfaces. Because the urine has not mixed with feces, it is safe to drain it into soil in this way. Alternately, the urine can be stored in a tank or container for future use (Mihelcic et al. 2009; Berger 2011). Figure 16.3 shows a compost chamber under construction in El Salvador, with two openings left in the brick wall for access doors, and the urine pit visible in the foreground prior to being filled with gravel.

3.2 *Uses for Composted Feces and Urine*

Compost toilets turn feces into a soil amendment, applied by mixing it with soil or spreading it around plants. However, compost immediately removed from the compost chamber may benefit from further treatment. One option is to perform solar drying by spreading the compost out in a sunny area for one week before applying it to fields. The climate in which the compost is used can also affect its safety, as

Fig. 16.3 Construction in progress of a compost chamber for a toilet in El Salvador (Photo courtesy of the author)

cool, rainy climates may not destroy pathogens as quickly as dry, sunny climates. Timing the compost completion can also help ensure pathogens have been destroyed, by removing it from the chamber during a drier part of the year. Regardless of the climate or use, there should be a one-month withholding period between the time of compost application and the harvesting of crops, which ensures that additional pathogens have been destroyed through contact with the soil (Stoner 1977; Mihelcic et al. 2009).

Some cultures are accustomed to using urine as a fertilizer, since it contains high amounts of nitrogen and phosphorus and can be useful for non-nitrogen-fixing plants. Urine can contain some pathogens, but this does not affect its suitability as a household fertilizer as long as there is a one-month time lapse between the application of urine to the crop and the harvesting of foods that are eaten raw. Stored urine is usually diluted before application in this way (Drangert 1998; WHO 2006).

There may be other local uses for urine, such as leather tanning, that can earn a user income for the urine stored from a compost toilet. Many users, however, would prefer to simply let the urine drain into the ground safely, due to cultural barriers regarding its use.

3.3 Cost and Materials

Compost systems can be constructed with basic, cost-effective materials available all over the world. The materials need to be structurally sound, appropriate to the local environment and climate, should be easy for users to keep clean, and uphold the conditions needed for composting described above.

The compost chambers must be constructed of waterproof materials to ensure moisture control inside the chambers. Appropriate materials include fired clay bricks, concrete blocks, or poured concrete, with clay bricks probably the most cost-effective. The chambers should be topped with a structurally appropriate concrete or plastic slab, and stairs or a ramp must be constructed out of similar materials to provide access to the toilets (Mihelcic et al. 2009).

The walls, door, and roof of the toilet chamber can be made out of a variety of materials that promote ventilation, easy-to-clean surfaces, and comfort inside the space. Compost toilets constructed by the author with students and community members in El Salvador have used adobe bricks, bamboo, and sheet metal, which were chosen for their local availability and low cost.

The toilets themselves can use prefabricated or self-constructed parts. Plastic toilets and urine-diversion toilet seat inserts are available for purchase in many regions, and are easy and straightforward to keep clean, but are often prohibitively expensive in the Global South. Toilets can be constructed out of other materials such as wood, plaster, adobe, or concrete instead. Figure 16.4 shows inexpensive poured

Fig. 16.4 Concrete urine-diversion toilet seats prepared from a mold in El Salvador (Photo courtesy of the author)

concrete urine-diversion toilets created using a mold, for installation in a compost toilet project. These were painted for durability and ease of cleaning, and paired with a plastic toilet seat for comfort.

3.4 *Strategies for Uptake and Acceptance of Compost Systems*

Users who are unfamiliar with compost systems and urine-diversion toilets will require training and support to use these systems effectively. Many users have positive experiences with compost toilets and understand their value to personal and community health, conservation of local water resources, and agricultural benefits. However, the barriers to acceptance of these systems can come from cultural or personal opposition to coming into contact with feces, misunderstanding of how the system works, lack of time to maintain the system, or personal difficulty using the system.

Cultures around the world have long-standing taboos against handling feces that stems from an understanding of its dangers to health. The decision to introduce the idea of compost toilets must be grounded in an understanding of local taboos and whether the strategies for promoting compost toilets will address and mitigate the specific concerns in a community or culture. For example, a community that already composts other waste materials may have an understanding of how composting works that makes it possible to extend their knowledge to the topic of composting feces. Additionally, the sources of information about compost toilets should be trusted, and depending on who is trusted, could come from a local public health organization, a governmental health ministry, or community members who have had a positive experience with the technology in another location (Anand and Apul 2014; Ahmed and Ahmed 2017; Brewer and Wutich 2019).

The author worked on a compost toilet design in rural El Salvador using several trusted sources to run a pilot project that built awareness and buy-in. Working with students from the author's class at Massachusetts Institute of Technology (MIT) D-Lab, the team partnered with a Salvadoran non-governmental health organization, Asociación Salvadoreña Pro Salud Rural (ASAPROSAR), which had already established a presence in rural communities with inadequate sanitation access, and had built trust through the creation of local women's groups. The team obtained blueprints for a compost toilet design published by the Salvadoran Ministry of Health and visited existing toilets built according to these blueprints. These toilets had been built using expensive materials, and the team consulted local stakeholders in two communities where ASAPROSAR has a strong presence, to engage them in the design process for a cheaper version of the toilet. One family in each community volunteered to have a toilet built at their home, to serve as a community model and provide feedback over time. Through this process, the concept of a compost toilet was introduced to the community through trusted pilot users, providing an opportunity for others to learn about how compost toilets might address their sanitation needs.

Two barriers to the success of compost toilets discovered by the team are: (a) a lack of understanding of how the system must be maintained and (b) a shortage of time to perform that maintenance. When constructing compost toilets, it is important to identify who will be cleaning the structure, who will be checking the compost for moisture, and who will be removing the compost from the chamber when it is ready for use in agriculture. These tasks may be performed by the same person or by different people, and each person must understand why they are performing these tasks in order to underscore their importance to the system's functioning. The users must also have the time to perform these tasks regularly. If they do not have enough time, it may be worth exploring whether multiple people can share the responsibility for the task.

Finally, some users struggle to adapt to using a urine-diversion toilet. It is important to engage both users who stand and those who sit to urinate, as well as children (who are still learning consistent hygiene practices) and users with mobility restrictions, in the toilet design, and incorporate their needs. During the pilot project in El Salvador, users who stand to urinate provided feedback that they often did not use the compost toilet if they only needed to urinate, because the opening to collect urine was too small. Instead, they would urinate in the open environment surrounding the toilet, which was a behavior that proponents had wanted to reduce by building the compost toilets. Upon receiving this feedback, the team decided to add plaster-molded urinals to the pilot toilet structures, which were diverted into the same gravel pit as the urine from the toilets themselves. This solution satisfied the users who had complained.

4 Vermifiltration Systems as a Water-Recycling Sanitation Solution

Vermifiltration is a waste processing system that incorporates the use of *Eisenia fetida* earthworms, or closely related species of composting worms, to process human feces. These worms have many common names, including manure worms, tiger worms, and red worms. Native to Europe, these worms have been introduced to every continent except Antarctica, and have been used across continents in the development of vermifiltration solutions at household and community scales (Furlong et al. 2014; Jiang et al. 2016).

When added to a series of proven wastewater treatment steps, vermifiltration is effective at processing wastewater at a household or neighborhood level so the water can be returned to the local ecosystem. Moreover, the compost produced by the worms is low enough in pathogens to be used in agricultural applications. These systems can be introduced to systems with existing plumbing and flush toilets, or constructed from scratch where no prior septic system exists. This flexibility has allowed for pilots in diverse climates for a variety of user needs. They require little to no behavior change at the user level, but require more significant initial construction work than compost toilets, and need buy-in to maintain the steps in the system. When successful, toilets using vermifiltration need to be emptied and refurbished every three to five years, compared to pit latrines that must be emptied every two to three years; and while their initial construction costs are similar to those of pit latrines or other common solutions, their maintenance costs over time are lower (Eastman et al. 2001; Furlong et al. 2014; Snoad 2019).

4.1 Components of an Effective Vermifiltration System

Vermifiltration systems include multiple stages of blackwater treatment, and can be combined with other forms of household water treatment. *Eisenia fetida* must be provided with a suitable bedding material to receive and consume feces flushed from a toilet. The tiger worm chamber may be positioned directly underneath or next to a toilet, or it may be offset from the latrine or building in a different location. Urine and water must be allowed to filter through the medium in which the worms live, so it can be processed by other methods down the line. The most common second step in the treatment is a slow-sand filtration system or other arrangement of drainage layers below the worm bedding layer. Figures 16.5 and 16.6 show two possible designs for vermifiltration systems depending on the site needs (Furlong et al. 2014; Snoad 2019). These are only two examples of many possible permutations and material choices that should be explored based on the site specifications, space available, and local material variations.

A concern identified in some vermifiltration chambers is the effective spreading of waste across the entire area of the bedding layer in which the worms live. The author worked with MIT students and two non-governmental organizations (NGOs) in Colombia, Diversa and Ecoconcientizate, to design an insert that would help spread waste more effectively across the area of a pilot vermifiltration system in Cundinamarca, Colombia. The pyramid-shaped distributor is shown in Fig. 16.7 and addressed the issue of waste building up unevenly in the middle of the plastic drum. Depending on the size and shape of the collection chamber, creative solutions to distribute the waste can be added.

After filtration through sand, gravel, and other media, the wastewater may require additional retention time in a closed chamber or secondary treatment. Soak pits can be used to collect filtered water and facilitate its slow absorption into the soil. Alternately, a small horizontal flow constructed wetland can provide the treatment needed to neutralize additional pathogens before the water is released back into the ecosystem. A lined basin filled with gravel and sand is planted with native vegetation that will filter additional pathogens and particles, promote the colonization of anaerobic bacteria that perform additional filtration, and remove nitrate, phosphorus, and ammonia from the water. Figure 16.8 shows a diagram of such a constructed wetland. An example of a constructed wetland in the same systems in Colombia described above is shown in Fig. 16.9. The wetlands contain papyrus, lily,

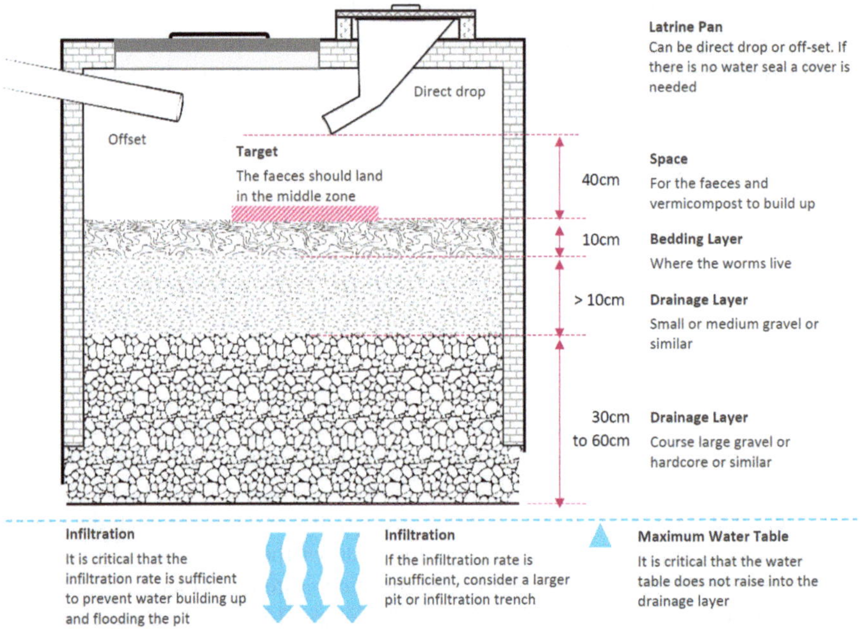

Fig. 16.5 Example of a vermifiltration system design positioned under a latrine, which includes two drainage layers below the worm bedding layer (Snoad 2019) (The material "*Tiger worm toilets design manual – C Snoad – 2019: 2 figures on pages 4 and 19*" is reproduced with the permission of Oxfam, Oxfam House, John Smith Drive, Cowley, Oxford OX4 2JY, UK www.oxfam.org.uk. Oxfam does not necessarily endorse any text or activities that accompany the materials)

and banana species, among others, which effectively filter water for households of 2–6 users (Mihelcic et al. 2009; Milonova 2020).

4.2 *Effectiveness of Water Treatment in Vermifiltration Systems*

Field experiments have tested the ability of vermifiltration, also referred to in the literature as vermicomposting, in removing pathogens with the goal of meeting or exceeding United States Environmental Protection Agency standards for Class A stabilization in domestic wastewater biosolids. The proper use of *E. fetida* reduces pathogens including fecal coliforms, salmonella, enteric viruses, and helminth eggs to acceptable levels for the vermicompost to be used in agricultural settings. Vermifiltration matches conventional septic systems and activated sludge processes in its effectiveness at wastewater treatment (Eastman et al. 2001; Li et al. 2011; Furlong et al. 2014).

Vermifiltration will not be effective unless a stable environment can be maintained for the worms. This means that temperature and humidity must be

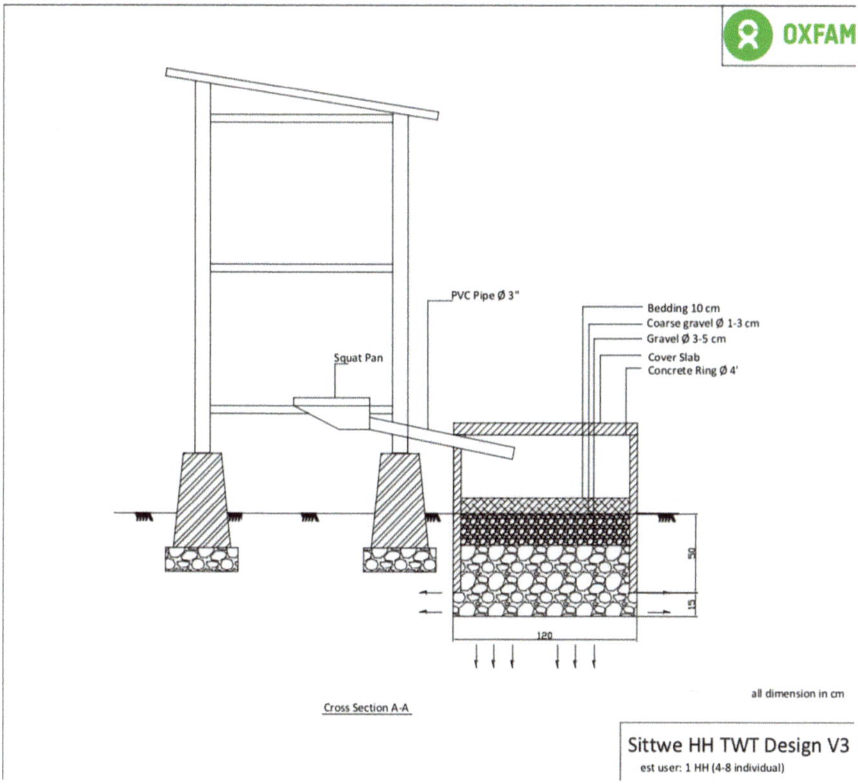

Fig. 16.6 Example of an offset vermifiltration system design employed in Myanmar (Snoad 2019) (The material "*Tiger worm toilets design manual – C Snoad – 2019: 2 figures on pages 4 and 19*" is reproduced with the permission of Oxfam, Oxfam House, John Smith Drive, Cowley, Oxford OX4 2JY, UK www.oxfam.org.uk. Oxfam does not necessarily endorse any text or activities that accompany the materials)

maintained in the system, and products such as bleach cannot be used in toilets that connect to the system. Field experiments have shown the necessity of being able to monitor temperature consistently and ensure the worms' survival. Flushing a toilet can affect the temperature of the worms' environment, so the temperature may fluctuate at different times of day depending on how often the toilet is being used. Temperature and humidity will affect how quickly the worms process the waste and produce compost. This also means that vermifiltration will not work in environments where the temperature fluctuates too greatly and climate control cannot be ensured in the system (Furlong et al. 2016; Jiang et al. 2016).

The worms may not be able to process fluctuations in use. If the toilet is not used for an extended period of time (e.g., the toilet is not used for more than a day or two), the worms will need to be kept alive by other means. Conversely, the worms may tolerate occasionally higher use than the system was designed for, but not enough data exists on how much overuse the system can sustain. Ideally, after the

Fig. 16.7 Pyramid-shaped waste distributor added to a vermifiltration system in Colombia in 2023 (Photo courtesy of the author)

worms acclimatize to the system, they will convert all newly introduced human feces into vermicompost over a 24-hour period, which ensures odor control and processing of pathogens at an appropriate pace (Furlong et al. 2014, 2016).

4.3 Strategies for Uptake and Acceptance of Vermifiltration Systems

The idea of using earthworms to process human feces may be a surprising and unusual idea to some. Proponents of vermifiltration have assessed community willingness to engage with vermifiltration using small trials in a variety of locations to obtain user feedback and buy-in before expansion.

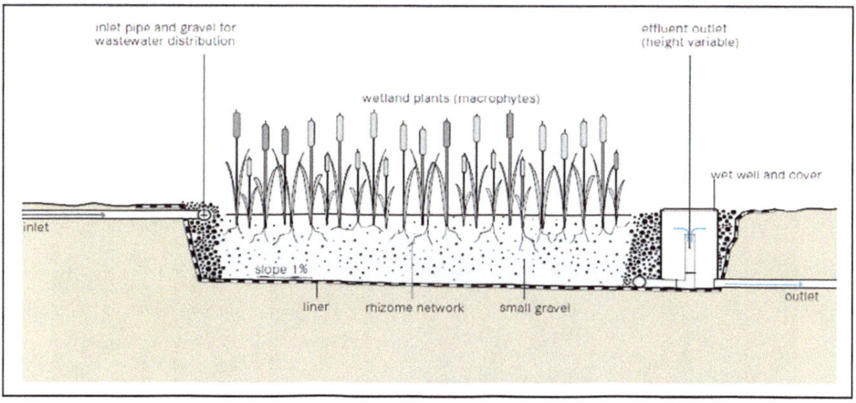

Fig. 16.8 Horizontal flow constructed wetland (Milonova 2020) (The material "*WASH Handbook for Protracted Emergencies: The OXSI Experience in Myanmar – Sonya Milonova – 2020: Figure 4.3.2, p. 16 (p. 117 in the PDF)*" is reproduced with the permission of Oxfam, Oxfam House, John Smith Drive, Cowley, Oxford OX4 2JY, UK www.oxfam.org.uk. Oxfam does not necessarily endorse any text or activities that accompany the materials)

Oxfam has tested vermifiltration as a solution in refugee camps in Sittwe, Myanmar. The designers established that 200 grams of worms were required per user of the toilet, and that each kilogram of worms required 0.5 m^2 of surface area in the compost toilet. Due to dense populations in the refugee camps, which intensified due to the ongoing arrival of displaced friends and family members, Oxfam distributed materials to ensure the maximum capacity of 20 users per latrine would not be exceeded. Monitoring ensured that the constructed latrines were being used and maintained properly, through the use of community engagement teams (Snoad 2019; Milonova 2020).

A trial in Maharashtra, India, tested the use of vermifiltration in a pilot of 10 households, each with a maximum number of 10 toilet users. Each vermicomposting toilet was constructed with brick and provided with *Eisenia andrei* worms. Because the users of these toilets had previously practiced open defecation, they were generally satisfied with the opportunity to have access to more private, effective toilets. After 12 months, they were satisfied with the vermifiltration toilets due to the convenience, lack of odor and mosquitoes, and the technology itself; however, they were dissatisfied with the entrance of water into the system during the rainy season, which could be adjusted in future designs. The participants in the study unanimously preferred the use of vermifiltration toilets over their previous open defecation practices, making the expansion of vermifiltration attractive as an affordable and culturally acceptable option (Furlong et al. 2016).

Fig. 16.9 Constructed wetland in a household-scale vermifiltration system in Aguabonita, Colombia (Photo courtesy of the author)

5 Using Participatory Design and Development to Promote Community Buy-In

The human sense of disgust and cultural taboos are major barriers to the uptake of novel human waste management strategies that conserve water. The two systems described above cannot be successful without addressing users' concerns about handling material that they knew contained pathogens before it was processed, and it is essential to build trust in the processes by engaging users at multiple stages of planning (Furlong et al. 2016; Brewer and Wutich 2019). Participatory design provides a framework for engaging stakeholders in the development of sanitation strategies that meet sociocultural needs as well as environmental, health, and economic goals.

Since its formation in 2001, MIT D-Lab has developed and influenced the use of participatory design methods in international development contexts, in order to facilitate the development of low-cost technologies that improve quality of life and livelihoods for people living in poverty. D-Lab's work in participatory design teaches professionals, practitioners, and students how to discern when (or if) to use a participatory design process to solve a challenge. Through participatory design, D-Lab facilitators learn how to connect more meaningfully with people of diverse perspectives and to make decisions with them collectively and quickly. This can lead to more desirable and sustainable solutions. In cases where participatory design is considered an appropriate methodology, D-Lab advocates taking the time to include the people who are facing the challenge that the innovation will address (Smith and Adams 2022; MIT D-Lab 2024).

D-La's mission is to advance the effective use of three types of participatory design in international development:

- Use inclusive practices when designing **FOR** people living in poverty (also known as user-centered design);
- Engage in effective co-creation when designing **WITH** people living in poverty (also known as co-design); and.
- Build confidence and capacity to promote design **BY** people living in poverty (also known as user-generated design) (Leith et al. 2022).

Throughout any participatory design process, the use of a design cycle promotes a process that iterates on the concepts of learning, imagining, creating, and testing ideas. Figure 16.10 shows the design cycle D-Lab uses to engage students, partners, and community members in participatory design projects related to water and sanitation, renewable energy, healthcare, education, and many other topics. Using such a process, practitioners and students at D-Lab are encouraged to listen deeply to users' needs and engage them in effective design work that solves a problem they or their community are experiencing (Smith and Adams 2022; MIT D-Lab 2024).

When facilitating user-generated design, D-Lab uses an additional methodology called creative capacity building, which provides users with the confidence and a selection of skills to be active creators of technology, rather than passive recipients. A randomized controlled trial performed in 2013 examined the quantitative economic impacts on participants in the Global South when they were engaged in a creative capacity building workshop, run by MIT D-Lab, that focused on technologies that would improve their quality of life. The technologies they developed during the workshop reduced their personal labor expenditures by more than 50%. A follow-up study in 2017 assessed the qualitative personal impacts on users from the same workshop, finding meaningful outcomes in three categories: *design for income*, by empowering users to create new businesses related to the technologies they developed; *design for resilience*, by enabling additional strategies for participants to absorb economic shocks; and *design for development*, by providing participants with strategies to progress on community goals (Childs 2017).

The author helped run a participatory design summit in southern Ethiopia around the topic of sanitation, in partnership with Population Services International Ethiopia, in 2018. Charged with improving access to sanitation in a water-scarce region, PSI and MIT D-Lab brought together residents of water-stressed communities, local suppliers and fabricators, educators, government officials, students, and healthcare workers to brainstorm and prototype solutions that would meet cultural and environmental sanitation needs at low cost (Vogel 2021).

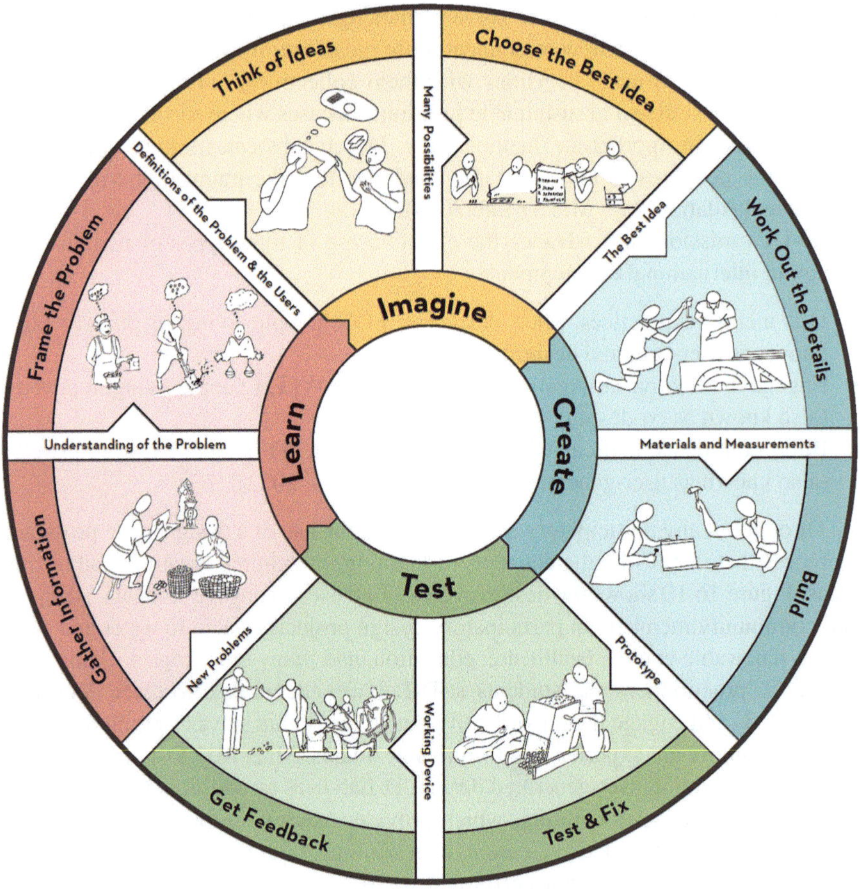

Fig. 16.10 MIT D-Lab design cycle (MIT D-Lab 2024). Used with permission

The participants were divided into six groups to address broad needs that had been pre-identified by a smaller group of stakeholders, including handwashing, latrine superstructure materials, child-friendly toilet designs, and making women's participation in latrine construction easier and more culturally acceptable. At the end of the 10-day summit, the group had not only brainstormed and tested early prototypes related to these broad categories, but they had also changed mindsets around many of these topics, including understanding the importance of handwashing and correcting misconceptions about the potential dangers of toilets to small children. PSI Ethiopia ended up with a set of ideas and many new collaborators for its work to provide affordable toilet options to users in this region (Jones et al. 2018; Vogel 2021).

Participatory design processes need time to implement, and skilled facilitators increase its chances of success. Additionally, participation is only as effective as the diversity of the stakeholders who have been invited, so effort must be made to include as many types of stakeholders in the process as possible. Services such as translation and visual education tools (for participants who are illiterate or did not complete formal schooling) add to the cost and types of expertise needed to facilitate an effective process. However, the benefits of participatory design can expand user access to innovation and technologies that will increase their quality of life, including in the face of changing conditions due to climate change (Vogel 2021; Leith et al. 2022).

Given the challenges faced by sanitation professionals in cultural acceptance and behavior change, participatory design offers a path forward for achieving buy-in and sustained use of waterless or low-water toilets in regions that need it most.

6 Conclusions

SDG6 aims to provide universal access to safely managed sanitation around the world by 2030, but there is significant work to be done to achieve this goal. Worldwide, 3.6 billion people still lack access to sanitation strategies that meet SDG goals. Existing large-scale infrastructure in the Global North is ineffective for meeting this need because it contributes to anthropogenic disruption of the nitrogen cycle and needlessly uses treated water resources.

SDG6 can become more attainable by considering sanitation technologies that recycle water or use no water at all. Two technologies, compost toilets and vermifiltration systems, have been effective in the Global South and could be successfully implemented at larger scales at reasonable costs—if effective maintenance strategies are promoted and sociocultural barriers are addressed in the process of implementation. Participatory design principles can be used to engage users throughout the process of designing, constructing, and maintaining these sanitation systems, and lessons from MIT D-Lab's development and use of a participatory design cycle, together with communities that need access to sustainable sanitation solutions, can inform future implementation strategies for these technologies.

This chapter is dedicated to the memory of Elaine Kung, MIT Class of 2015, who helped the author advance compost toilet projects during her studies at MIT D-Lab.

7 Discussion Questions

1. Explain why the development of modern sanitation systems led to the unnecessary overuse of treated drinking water for sanitation.
2. Discuss the benefits and drawbacks of a compost toilet system as a decentralized, water-saving sanitation system.
3. Discuss the benefits and drawbacks of a vermifiltration system as a decentralized, water-recycling sanitation system.

4. Explain how participatory methods can help overcome barriers to implementing water-saving sanitation systems in the Global South.

References

Ahmed SK, Ahmed SS (2017) Socio-cultural acceptability of urine diverted composting toilets: a review of literature for possible adoption in peri-urban areas as a sustainable sanitation solution. AIP Conf Proc 1919(1). https://doi.org/10.1063/1.5018561

Anand CK, Apul DS (2014) Composting toilets as a sustainable alternative to urban sanitation – a review. Waste Manag 34(2):329–343

Ashley BJ (1909) The cesspool and its dangers. In: Anonymous (ed) Municipal and county engineering, vol 37. Gillette Publishing Company, Urbana-Champaign, pp 240–245

Berger W (2011) Technology review of composting toilets – Basic overview of composting toilets (with or without urine diversion). Deutsche Gesellschaft für Internationale Zusammenarbeit (GIZ) GmbH, Eschborn

Bosch C (1908) Process of producing ammonia. US Patent 990:091A

Brewer A, Wutich A (2019) Lazy, crazy, and disgusting: stigma and the undoing of global health. Johns Hopkins University Press, Baltimore, pp 23–73

Childs T (2017) Creative capacity building in Uganda: qualitative case research into the impact of CCB on individuals and communities. Available via International Development Innovation Network. https://www.idin.org/resources/research/creative-capacity-building-uganda-qualitative-case-research-impact-ccb. Accessed 17 Oct 2024

Delwiche CC (1970) The nitrogen cycle. Sci Am 223(3):136–147

Drangert J (1998) Fighting the urine blindness to provide more sanitation options. Water SA 24(2):157–164

Earth's boundaries? (2009) Nature 461:447–448. https://doi.org/10.1038/461447b

Eastman BR, Kane PN, Edwards CA et al (2001) The effectiveness of vermiculture in human pathogen reduction for USEPA biosolids stabilization. Compost Sci Util 9(1):38–49

Fowler D, Coyle M, Skiba U et al (2013) The global nitrogen cycle in the twenty-first century. Philos Trans R Soc Lond Ser B Biol Sci 368(1621):20130164. https://doi.org/10.1098/rstb.2013.0164

Furlong C, Templeton MR, Gibson WT (2014) Processing of human faeces by wet vermifiltration for improved on-site sanitation. J Water Sanit Hyg Dev 4(2):231–239. https://doi.org/10.2166/washdev.2014.107

Furlong C, Gibson WT, Oak A et al (2016) Technical and user evaluation of a novel worm-based, on-site sanitation system in rural India. Waterlines 35(2):148–162

Galloway J, Aber JD, Erisman JW et al (2003) The nitrogen cascade. Bioscience 53(4):341–356. https://doi.org/10.1641/0006-3568(2003)053[0341:TNC]2.0.CO;2

George R (2008) The big necessity. Metropolitan Books, New York, pp 15–37

Halliday S (2007) The great stink of London: Sir Joseph Bazalgette and the cleansing of the Victorian capital. Sutton Publishing, Stroud, p 45

Hugo V (1862) Les misérables. J Hetzel

Jiang L, Liu Y, Hu X et al (2016) The use of microbial-earthworm ecofilters for wastewater treatment with special attention to influencing factors in performance: a review. Bioresour Technol 200:999–1007. https://doi.org/10.1016/j.biortech.2015.11.011

Jones A, Fentie M, Tegenfeldt K (2018) Ethiopian summit focuses on participatory co-design to develop low-cost WASH solutions. GlobalWaters.org, USAID. https://www.globalwaters.org/resources/blogs/ethiopian-summit-focuses-participatory-co-design-develop-low-cost-wash-products. Accessed 6 Oct 2024

Leith K, Vogel S, Mehta K (2022) The participatory design spectrum: design for, with, and by. In: Régnier P, Frey D, Pierre S et al (eds) Handbook of innovation & appropriate technologies for international development. Edward Elgar Publishing, Northampton, MA, pp 125–139
Levy PF, Connor MS (1992) The Boston Harbor cleanup. New Engl J Publ Policy 8(2):91–104
Li YS, Xiao YQ, Qiu JP et al (2011) Continuous village sewage treatment by vermifiltration and activated sludge process. Water Sci Technol 60(11):3001–3010
Mihelcic JR, Fry LM, Myre EA, Phillips LD, Barkdoll BD (2009) Field guide to environmental engineering for development workers. ASCE Press, Reston, pp 14–27, 381–408, 427–253
Milonova S (2020) WASH handbook for protracted emergencies: the OXSI experience in Myanmar. HARP-F, Oxfam International, Oxford
MIT D-Lab (2024) MIT D-Lab Approach. https://d-lab.mit.edu/approach. Accessed 17 Oct 2024
Rockefeller AA (1998) Civilization and sludge: notes on the history of the management of human excreta. Capital Nat Social 9(3):3–18
Rockström J, Steffen W, Noone K et al (2009) A safe operating space for humanity. Nature 461:472–475. https://doi.org/10.1038/461472a
Rosenwaike I (1972) Population history of New York City. University Press, Syracuse, NY
Smith A, Adams N (2022) MIT D-Lab: designing for a more equitable world. In: Régnier P, Frey D, Pierre S et al (eds) Handbook of innovation & appropriate technologies for international development. Edward Elgar, Northampton, MA, pp 7–21
Snoad C (2019) Tiger worm toilets design manual. Oxfam GB. Available via Elrha. https://www.elrha.org/researchdatabase/tiger-worm-toilet-manual-globally-relevant-learnings-from-myanmar/. Accessed 29 Sept 2024
Stoner CH (ed) (1977) Goodbye to the flush toilet. Rodale Press, Emmaus, PA
Tarr JA (1979) The separate vs. combined sewer problem: a case study in urban technology design choice. J Urban Hist 5(3):308–339
United Nations Department of Economic and Social Affairs (2024) Goal 6. Available via United Nations. https://sdgs.un.org/goals/goal6#targets_and_indicators. Accessed 17 Oct 2024
Vogel S (2021) Co-creating a more equitable world: the transformative benefits of participatory design. NextBillion. https://nextbillion.net/transformative-benefits-participatory-design/. Accessed 6 Oct 2024
Widdison PE, Burt TP (2008) Nitrogen cycle. In: Jorgensen SE (ed) Global ecology. Elsevier Academic Press, San Diego, CA, pp 190–197
World Health Organization (WHO) (2006) Safe use of wastewater, excreta and greywater, Volume IV: excreta and greywater use in agriculture. WHO Press, Geneva
World Health Organization, United Nations Children's Fund (WHO, UNICEF) (2021) Progress on household drinking water, sanitation and hygiene 2000–2020: five years into the SDGs. WHO and UNICEF, Geneva

Chapter 17
Climate Change and Its Effect on Bangladesh's Drinking Water: Status, Projections, and Community-Based Integrated Solutions

Azizur R. Molla and Afnan Rahman

1 Learning Objectives

1. Discuss the status and projections of how climate change impacts drinking water in Bangladesh.
2. Describe climate-related outcomes on Bangladesh's freshwater reserves.
3. Explain community-based solutions to these challenges, such as reducing emissions, sequestering carbon, building more resilient cities and communities, and fighting disinformation.

2 Introduction

Bangladesh, located in the delta of the Ganges-Brahmaputra-Meghna (GBM) river system, is one of the most climate-vulnerable countries in the world. The country's extensive river networks and reliance on groundwater for drinking water make it particularly susceptible to climate-induced changes (Fig. 17.1). This paper explores the current status of drinking water resources in Bangladesh, evaluates projections under future climate scenarios, and examines community-based integrated solutions to mitigate these impacts. Bangladesh is in the Bengal Delta Plain, one of the

A. R. Molla (✉)
College of Health Professions, School of Interdisciplinary Health, Public Health Program, Grand Valley State University, Grand Rapids, MI, USA
e-mail: mollaziz@gvsu.edu

A. Rahman
Department of Computational Mathematics, Science and Engineering, College of Natural Science, Michigan State University, East Lansing, MI, USA
e-mail: rahman44@msu.edu

A. R. Molla et al. (eds.), *Climate Change, Drinking Water Security, and Public Health*, https://doi.org/10.1007/978-3-032-12074-8_17

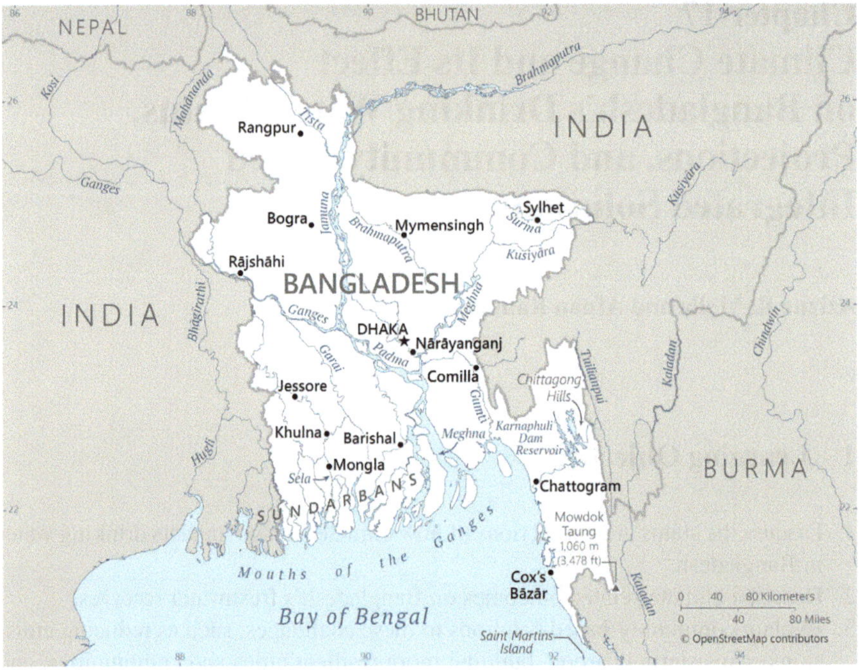

Fig. 17.1 Bangladesh map showing major rivers (Central Intelligence Agency 2024) (Bangladesh map showing major cities as well as parts of surrounding countries and the Bay of Bengal. In: *The World Factbook 2024*. Washington, DC: Central Intelligence Agency. https://www.cia.gov/)

world's largest river deltas (Monan 1995). Global climate change will drastically reduce the amount of freshwater that could be readily used for drinking water and agricultural production in Bangladesh. The low-lying coastal topography, annual monsoons, and frequent cyclones are expected to be severely impacted by flooding as sea levels continue to rise in Bangladesh (Dewan et al. 2003; Pethick and Orford 2013).

It is projected that with climate change increasing sea levels globally, Bangladesh could lose up to 17% of its land area due to sea level rise, and about 30% of land will be lost by 2050. The rising sea level will displace about 20 of the expected 192 million people by this time (IMF 2019). The progression of climate change will increase sea level rise, ultimately expanding the area and duration of yearly floods (Frisbie et al. 2024).

It has long been identified for regional water management; the plan is to specify areas where water exploration can be monitored and controlled, have a sustainable method for water use, and establish regulations beyond Bangladesh's political boundaries to preserve its ecosystem and water resource management (de Sherbinin 1997).

Climate change significantly impacts safe water availability, ultimately contributing to health problems. Variations in climate parameters, such as salinity, droughts during the dry season, and floods during the wet season, dramatically reduce access to clean drinking water, contributing to increased waterborne diseases. Notably, existing healthcare facilities and safe drinking water practices are inadequate to address the growing health issues (Ashrafuzzaman et al. 2023).

For Bangladesh, access to safe drinking water and sustaining water resources in rural Bangladesh require local planning, a program, and community-engaged implementation. Undoubtedly, education programs effectively ensure safe drinking water use and conserve resources. It is important to note that such a program must include local stakeholders who are aware of socioeconomic and cultural resources and challenges. Waterborne diseases vary across localities and villages. Molla (2008) found a strong village effect on disease incidence caused by water contamination. The villages are characterized mainly by socioeconomic and educational status and cultural practices.

Many interventions have been shared to prevent disease. Batmanghelidj (2003), a medical doctor, identified four vital steps for better health: balancing the content of water and salt in our bodies. Safe drinking water can prevent waterborne diseases; Ahmed (2000) recommended low-cost drinking water supply technological options, including shallow shrouded tubewells, ponds and filters, household filters, solar desalination, and rainwater harvesting.

3 Status of Drinking Water in Bangladesh

3.1 Water Resources Overview

Bangladesh's drinking water resources primarily come from surface water (rivers and lakes) and groundwater (tubewells). The Ganges-Brahmaputra-Meghna (GBM) river basin is crucial for replenishing both sources. However, the quality and availability of these water resources are compromised by various factors, including pollution, overexploitation, and natural contamination. The GBM delta poses a significant challenge to climate change adaptation for about 200 million inhabitants living in Bangladesh (Becker et al. 2020).

It is argued that globally, most disasters are water-related; climate change worsens natural calamities like floods, landslides, storms, heat waves, wildfires, extreme colds, droughts, and waterborne diseases. All these are becoming more frequent and primarily intense due to climate change (UN 2023). Climate scientists suggest we develop a resilient climate change healthcare system to ensure improved accessibility for vulnerable people. And to reduce health risks, access to safe drinking water must be secured for everyone (Chowdhury et al. 2022).

3.2 Key Challenges

3.2.1 Arsenic Contamination

Arsenic contamination is a severe issue, affecting approximately 20 million people in Bangladesh (Rahman et al. 2022). This contamination arises from natural geological sources and is exacerbated by the use of deep tubewells installed in the 1970s and 1980s. The rising sea level would increase salt concentration in the underlying aquifer, increasing the release of arsenic from sediment into drinking water wells (Frisbie et al. 2024). Arsenic exposure poses serious health risks, including cancer and cardiovascular diseases (Smith et al. 2020).

High concentrations of arsenic are found in the shallow aquifers (30–150 m depth), which coincides with the optimal well depth. A 2019 survey indicated that 18.6% had above 10 micrograms per liter of arsenic while 11.8% of households had arsenic levels above 50 micrograms per liter, which equates to approximately 31 million people at varying levels of risk. The wealthiest quintile of the population had the lowest exposure to arsenic (Roman et al. 2021).

3.2.2 Salinity Intrusion

Coastal regions are experiencing increased salinity due to rising sea levels and decreased upstream freshwater flow (Hossain et al. 2023). Salinity intrusion impacts both surface and groundwater, making them unsuitable for drinking and agriculture, which exacerbates food and water insecurity. Farmers have reported that rising salinity has led to decreased soil fertility and crop failures (Saddaf et al. 2024). Groundwater salinity is a significant issue in coastal areas where shallow aquifers are contaminated with varying salinity levels, forcing communities to drill deeper wells to reach the primary aquifer (Roman et al. 2021). The expansion of shrimp farms along the coast contributes to the salinization of water and soil (Shammi et al. 2019).

3.2.3 Flooding and Waterlogging

Seasonal flooding and waterlogging, intensified by heavy monsoon rains and cyclones, lead to contamination of water sources with pathogens and pollutants. While floods can recharge groundwater, they often disrupt access to safe drinking water and sanitation facilities.

4 Projections Under Climate Change

4.1 Impact on Water Availability

4.1.1 Increased Frequency of Extreme Weather

Climate change is projected to increase the frequency and severity of extreme weather events, such as heavy rains and prolonged droughts. These events will exacerbate variability in water availability, causing shortages in some areas while leading to excess water in others. Climate change projections for rainfall in Bangladesh remain uncertain, but all scenarios point to increased irrigation water demand, causing declines in groundwater levels. This issue is especially critical in the Northwest region of Bangladesh, where it could exacerbate existing drinking and irrigation water shortages during the dry season. Additionally, upstream water diversions from transboundary rivers, such as the Ganges and Teesta, pose an even greater threat to water availability than climate change, as they have already significantly reduced dry season flows (Kirby et al. 2016).

4.1.2 Glacial Melt

The accelerated melting of Himalayan glaciers, which feed major rivers in Bangladesh, is expected to increase river flows initially but could eventually result in reduced water availability as glaciers recede and seasonal flow patterns become more erratic (Immerzeel et al. 2010).

4.2 Impact on Water Quality

4.2.1 Salinity Increase

Rising sea levels and altered river flow patterns are expected to worsen salinity intrusion in coastal areas (Frisbie et al. 2024). Increased salinity levels will compromise water quality and reduce the viability of groundwater and surface water sources for consumption.

4.2.2 Contamination Risks

Higher temperatures and increased precipitation may enhance the spread of contaminants from agricultural runoff and industrial discharges, which poses significant risks to water quality and increases the likelihood of waterborne diseases. Studies have shown elevated sodium, chloride, sulfate, and other ions in water samples exceeding safe drinking water standards (Shammi et al. 2019).

4.3 Socioeconomic Implications

4.3.1 Health Risks

Deteriorating water quality and availability are likely to increase the incidence of waterborne diseases, such as cholera and dysentery, putting additional strain on public health systems and diminishing overall quality of life.

4.3.2 Economic Impact

Issues of water scarcity and quality will negatively impact agricultural productivity, fisheries, and livelihoods, worsening poverty and economic instability (World Bank 2023). These changes will mainly affect the agricultural sector, which heavily depends on water. The time and effort required to collect water from distant sources (1–2 kilometers, km) also impose an economic burden, especially for women who primarily bear this responsibility (Talukder et al. 2020). Decreased agricultural productivity leads to financial instability and affects livelihoods (Saddaf et al. 2024).

4.3.3 Innovation at the Grassroots Level

The agricultural sector is implementing climate-friendly interventions to continue producing farm products. Different vegetables are cultivated on a gloating seedbed (see picture below). Pond Sand Filters (PSFs) are promoted as a solution for safe water provision. They are effective in reducing microbiological contamination but less effective in removing salinity. The functionality and maintenance of PSFs are often inadequate, raising concerns about their sustainability (Talukder et al. 2020).

Agroforestry practices, specifically planting flood-tolerant tree species, help stabilize soil, protect crops during floods, and provide additional income sources (Ali et al. 2024). Modern floating gardens of hyacinths have been used as new farming land (Saddaf et al. 2024).

In Bangladesh, climate change poses both environmental and developmental challenges. Patterns of behavior, human practices, and international actions influence the country's adaptations. Coordination is crucial for addressing climate change initiatives, integrating planning, and managing information (Rashid and Islam 2007). This coordination will accelerate the processes of climate mitigation and adaptation. Various factors, including seasonal characteristics and climatic variables such as temperature, rainfall, humidity, and day length, shape the overall climate situation in Bangladesh. Responding to climatic and related uncertainties is imperative for the nation (Asaduzzaman et al. 2016).

4.4 *Impact on Public Health*

Climate change and water-related challenges disproportionately impact vulnerable populations, worsening poverty, food insecurity, and displacement, all of which contribute to the mental health burden in Bangladesh (Wahid et al. 2024). Climate change has increasingly impacted human health. Many believe that it affects the health of the physical, mental, and environmental domains. Higher education is more correlated with climate change effects (George et al. 2024).

Mora et al. (2022) found that about 58% of infectious diseases worldwide have been aggravated by climatic hazards at some point. The empirical evidence suggests that the wide range of diseases and transmission pathways worsened by climatic hazards shows how much climate change threatens human health and emphasizes the need for strong efforts to reduce greenhouse gas emissions (Mora et al. 2022).

Long-term exposure to saline water can contribute to chronic health issues like hypertension, further straining healthcare systems (Talukder et al. 2020). Studies have shown a direct correlation between elevated levels of drinking water sodium (DWS) and increased risks of gestational hypertension, preeclampsia, and postpartum infant morbidity and mortality. High DWS intake during pregnancy is linked to higher infant mortality rates in coastal Bangladesh (Shammi et al. 2019).

One study found 83.05% of people living in the Noakhali district in Bangladesh believe that diseases have increased in recent years due to climate change. However, 13% of the study population did not note any change in the disease pattern due to the impacts of climate change, while the remaining 4% could not respond to the question (Alam et al. 2024). There is a relationship between climatic parameters like heatwaves, precipitation, tree cover, and Cholera in Bangladesh. Malaria and dengue fever are widespread in Bangladesh. Under climate change, it is projected that pathogens-caused diseases are expected to rise (Wu et al. 2018).

5 Community-Based Integrated Solutions

5.1 *Enhanced Water Management Practices*

5.1.1 Rainwater Harvesting

Establishing community-based rainwater harvesting systems can offer a dependable alternative to groundwater and surface water sources, especially in contaminated areas. Rainwater harvesting decreases reliance on traditional sources and enhances water security. Water from different rooftops can be collected through a piped network and stored for later use. Figure 17.2 shows the schematic view of a rainwater harvesting system (Rahman et al. 2014).

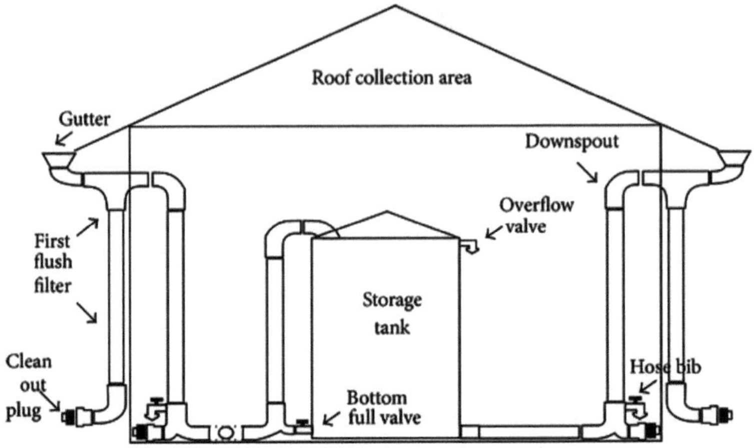

Fig. 17.2 Schematic of a rainwater harvesting system. (Reproduced from Rahman et al. 2014) (Reproduced from Rahman et al. 2014, Figure 8. https://doi.org/10.1155/2014/721357, licensed under the terms of the Creative Commons Attribution 3.0 Unported (CC BY 3.0) licence (https://creativecommons.org/licenses/by/3.0/))

5.2 *Alternative Water Sources*

5.2.1 Desalination

In areas facing salinity intrusion, desalination technology offers a viable solution for providing fresh water. Community-scale desalination projects can address local water needs, though they require significant investment and operational management. Desalination plants are gaining popularity in coastal Bangladesh as a means of treating brackish groundwater (Roman et al. 2021).

5.2.2 Deep Tube Wells and Piped Water Systems

Investment in deep tube wells that access arsenic-free aquifers and expansion of piped water supply systems can improve water safety. Ensuring these systems are well-maintained and accessible is essential for effective water management.

5.3 Community Engagement and Education

5.3.1 Awareness Programs

Educating communities about water conservation, the dangers of contamination, and the benefits of alternative water sources is crucial for effective water management. Community education initiatives can empower individuals to adopt better water practices.

5.3.2 Community Radio

Community radio in Bangladesh has been significantly involved in educating people on climate change, disaster mitigation, and women's empowerment (Rahman 2023). It is argued that awareness of respecting local cultural practices would enhance the effectiveness of educational interventions (Datta and Kibria 2025).

5.3.3 Local Water Management Committees

Establishing local water management committees can enable communities to take ownership of their water resources. These committees can oversee local water systems, address contamination issues, and manage water distribution.

Community-based and community-owned interventions are more effective in addressing climate change threats. There is an instance of a local girl reading Bangladesh's Flood Forecasting and Warning Centre in the Sylhet district (NPR 2024). Considering the climate and rising water situation, tubewells located above the flood level can serve as a safe water source. Such intervention is an adaptation for climate change, focusing on water, sanitation, and hygiene (WASH), as per the WHO (2024). Figure 17.3 shows building adaptions for climate change focusing on water, sanitation and hygiene (WASH; WHO 2024).

One assessment shows that countries with an existing high burden of climate-sensitive diseases, including WASH-related diseases, and the extent through which climate variability and change are expected to adversely impact health and WASH. An overview of the expected outputs from this project is presented in Fig. 17.3 (WHO 2024).

Fig. 17.3 Key outputs from the DFID (Department for International Development)-funded project "Building adaptation to climate change in health in LDCs (Learning and Capacity Development) through resilient WASH (water, sanitation and hygiene)." (Reproduced from WHO 2024) (Reproduced from WHO 2024, Figure 2. https://www.who.int/publications/m/item/bangladesh-climate-change-wash, licensed under the terms of the Creative Commons Attribution-NonCommercial-ShareAlike 3.0 IGO (CC BY-NC-SA 3.0 IGO) licence (https://creativecommons.org/licenses/by-nc-sa/3.0/igo). Used with permission)

5.4 Government and Non-Governmental Organization (NGO) Collaboration

5.4.1 Integrated Water Resource Management

Implementing Integrated Water Resource Management (IWRM) practices involves coordinating efforts across different sectors and stakeholders to manage water resources sustainably. Effective collaboration between government agencies, non-governmental organizations (NGOs), and local communities is essential for addressing complex water challenges. It is a time when the government should not only focus on relief and rehabilitation but rather emphasize building community resilience, which will be more effective in addressing disaster risks caused by climate change (Abir et al. 2024).

NGOs promote safe water technologies like PSFs, support pond excavation and maintenance, and facilitate community engagement in water management. To address water resource management challenges, the government has formed long-term plans like the Bangladesh Delta Plan 2100 (Talukder et al. 2020).

Water governance in Bangladesh faces significant challenges, with gaps in implementing policies such as the Safe Drinking Water Supply and Sanitation Policy of 1998, various constitutional provisions, and the Bangladesh Climate Change Strategy and Action Plan (BCCSAP) of 2009. Despite having these frameworks in place, their objectives remain unmet due to weak enforcement and limited community participation in water management decisions. Additionally, effective transboundary water cooperation with neighboring countries, especially India, is vital for equitable and sustainable use of the 57 shared rivers, requiring joint monitoring, data sharing, and adherence to international water laws (Alamgir et al. 2024). Surface water availability contributes to safe drinking water sources; due to the Farakka barrage built by India in 1975, there has been a significant reduction of water flow from the primary water source, the Ganges. The barrage caused water scarcity and security in most parts; specifically, the southwestern region of Bangladesh has been suffering from environmental degradation. The hydrological flow of the Ganges River during the pre-Farakka (1935–1975) and post-Farakka (1976–2015) periods is noticeably low (Rahman and Rahaman 2018).

5.4.2 Nature-Based Solution Strategies

Strategies such as floating gardens, crop diversification, and salt-tolerant crops provide sustainable approaches to address the impacts of climate change (Saddaf et al. 2024). It is well documented how ecosystem and socioeconomic vulnerabilities contribute to social-ecological vulnerability and connect with nature-based solutions. Agroforestry initiatives, including the use of flood-tolerant trees, the integration of fruit trees, bamboo plantations, community-based agroforestry efforts, and community-resilient tree species, would be practical and straightforward to implement (Ali et al. 2024).

It is evident that nature-based solutions, such as planting *Sonneratia apetala* and *Nypa fruticans* trees, will benefit coastal communities by reducing salinity, mitigating climate change, and protecting against natural hazards like cyclones and lower temperatures. Additionally, the fruits from these trees can provide alternative livelihoods by producing fruit-based food items such as jam, jelly, pickles, sauce, tea, and more (Uddin et al. 2024).

5.4.3 Funding and Research

Increased funding for research into climate-resilient water technologies and infrastructure improvements is necessary. Government incentives and support for community-based projects can drive innovation and enhance water management efforts (World Bank 2023).

Government support and collaboration with NGOs are crucial for addressing farming communities' water-related challenges. However, inadequate support and corruption are significant obstacles that hinder the implementation of practical solutions (Saddaf et al. 2024).

Asaaga et al. (2024) recommended a cross-sectoral collaboration to inform effective climate-health interventions better. They argued that research needs to shift paradigms from a single pathogen or disease to multiple disease stressors, along with their burdens and impacts, to support the development and targeting of early warning systems.

6 Conclusions

Climate change significantly threatens Bangladesh's drinking water resources, affecting availability and quality. The challenges of arsenic contamination, salinity intrusion, and flooding are exacerbated by climate-induced changes, posing risks to public health and economic stability. Addressing these challenges requires a multifaceted approach that includes improved water management practices, exploration of alternative water sources, and robust community engagement. By integrating scientific research with community-based solutions, Bangladesh can enhance its resilience to climate change and ensure sustainable drinking water supplies. Collaborative efforts and proactive measures will be crucial in navigating the complexities of climate change and securing water resources for future generations.

7 Discussion Questions

1. How does climate change worsen the drinking water crisis in Bangladesh, and what are the main factors contributing to this problem?
2. What community-based solutions can effectively tackle the challenges of accessing safe drinking water in Bangladesh, especially in areas most affected by climate-related pollution?
3. How can lessons from Bangladesh's strategies for addressing climate-related water issues inspire and empower other countries facing similar public health and environmental challenges?

References

Abir TM, Das T, Ferdous F (2024) Building community resilience and adaptation to climate change: integrating disaster risk reduction in the coastal regions of Bangladesh. J Coast Res 40(4):792–802. https://doi.org/10.2112/JCOASTRES-D-23-00043.1

Ahmed MF (2000) Water supply & sanitation: Rural and low income urban communities. ITN-Bangladesh, Centre for Water Supply and Waste Management. Bangladesh University of Engineering and Technology (BUET), Dhaka, Bangladesh

Alam E, Uddin MI, Johnstone PD, Molla MH, Islam MM, Miah S, Nasim TM, Hoque MA, Afrin N (2024) Perceptions of climate change and its impacts on local communities along the Noakhali Coast in Bangladesh. In: Singh N, Babu SA (eds) Climate crisis and sustainable solutions. Springer, Singapore, pp 235–253. https://doi.org/10.1007/978-981-97-7110-3_16

Alamgir M, Hashim M, Uddin MD, Mahmud M, Islam A, Shahid S (2024) Enhancing water governance and water resources management in Bangladesh. Int J River Basin Manag. https://doi.org/10.1080/15715124.2024.2356219

Ali MM, Baro MS, Hanif MA, Rahman MS, Ahmed B, Rahman MS (2024) Charland resilience: agroforestry innovations as nature-based solutions for climate-vulnerable communities in Bangladesh. NbS 6:100165. https://doi.org/10.1016/j.nbsj.2024.100165

Asaaga FA, Tomude ES, Rickards NJ, Hassall R, Sarkar S, Purse BV (2024) Informing climate-health adaptation options through mapping the needs and potential for integrated climate driven early warning forecasting systems in South Asia—a scoping review. PLoS One 19(10):e0309757. https://doi.org/10.1371/journal.pone.0309757

Asaduzzaman M, Qamar Munir M, Ghulam Hussain SK, Khan MFA, Abdur Rashid M, Rahaman AZ, Akand MK (2016) An action plan for adaptation in Bangladesh agriculture under climate change. Center for Environmental and Geographic Information Services (CEGIS) Report. Center for Environmental and Geographic Information Services, Dhaka, Bangladesh. https://hdl.handle.net/10568/82900. Accessed 29 Dec 2024

Ashrafuzzaman M, Gomes C, Guerra J (2023) The changing climate is changing safe drinking water, impacting health: a case in the Southwestern Coastal Region of Bangladesh (SWCRB). Climate 11(7):146. https://doi.org/10.3390/cli11070146

Batmanghelidj F (2003) You're not sick, you're thirsty! – water for health, for healing, for life. Warner Books, New York

Becker M, Papa F, Karpytchev M, Delebecque C, Krien Y, Khan JU, Ballu V, Durand F, Le Cozannet G, Islam AKMS, Calmant S, Shum CK (2020) Water level changes, subsidence, and sea level rise in the Ganges-Brahmaputra-Meghna delta. Proc Natl Acad Sci 117(4):1867–1876. https://doi.org/10.1073/pnas.1912921117

Central Intelligence Agency (2024) Bangladesh map showing major cities as well as parts of surrounding countries and the Bay of Bengal. In: The World Factbook 2024. Central Intelligence Agency, Washington, DC. https://www.cia.gov/

Chowdhury MA, Hasan MK, Islam SLU (2022) Climate change adaptation in Bangladesh: current practices, challenges and the way forward. J Clim Chang Health 6:100108. https://doi.org/10.1016/j.joclim.2021.100108

Datta R, Kibria A (2025) Indigenous women-led climate crisis solutions: a decolonial perspective from the Garo Indigenous community in Bangladesh. Polit Geogr 117:103258. https://doi.org/10.1016/j.polgeo.2024.103258

Dewan AM, Nishigaki M, Komatsu M (2003) Floods in Bangladesh: a comparative hydrological investigation on two catastrophic events. J Faculty Environ Sci Technol, Okayama University 8(1):53–62. https://core.ac.uk/reader/12549250

Frisbie SH, Mitchell EJ, Molla AR (2024) Sea level rise from climate change is expected to increase the release of arsenic into Bangladesh's drinking well water by reduction and by the salt effect. PLoS One 19(1):e0295172. https://doi.org/10.1371/journal.pone.0295172

George N, Mahendran P, Rajkumar VL, Vivekanandan VS, Thangavel V, Elangovan V (2024) Climate change influence on health domains, a public perception from a district in South India – a cross-sectional study. Indian J Community Med 49:752–758. https://doi.org/10.4103/ijcm.ijcm_655_23

Hossain MA, Khan AR, Rahman M (2023) Assessing the impact of climate change on coastal salinity intrusion and water resources in Bangladesh. Clim Risk Manag 40:101504. https://doi.org/10.1016/j.crm.2023.101504

Immerzeel WW, Beek LPHV, Bierkens MFP (2010) Climate change will affect the Asian water towers. Science 328(5984):1382–1385. https://doi.org/10.1126/science.1183188

International Monetary Fund (IMF) (2019) Bangladesh: selected issues. Country report No. 2019/300. https://www.imf.org/en/Publications/CR/Issues/2019/09/17/Bangladesh-Selected-Issues-48683#:~:text=The%20outlook%20for%20Bangladesh%20is,of%20food%20production%20by%202050

Kirby JM, Mainuddin M, Mpelasoka F, Ahmad MD, Palash W, Quadir ME, Shah-Newaz SM, Hossain MM (2016) The impact of climate change on regional water balances in Bangladesh. Clim Chang 135:481–491. https://doi.org/10.1007/s10584-016-1597-1

Molla AR (2008) Water, sewage, and disease in Bangladesh – a medical anthropology. The Edwin Mellen Press, New York

Monan J (1995) Bangladesh the strength to succeed. Oxfam, Oxford. p 4,5,63

Mora C, McKenzie T, Gaw IM (2022) Over half of known human pathogenic diseases can be aggravated by climate change. Nat Clim Chang 12:869–875. https://doi.org/10.1038/s41558-022-01426-1

National Public Radio (NPR) (2024) Facing floods: what the world can learn from Bangladesh's climate solutions. https://www.npr.org/sections/goatsandsoda/2023/03/26/1165779335/facing-floods-what-the-world-can-learn-from-bangladeshs-climate-solutions. Accessed 28 Dec 2024

Pethick J, Orford JD (2013) Rapid rise in effective sea-level in southwest Bangladesh: its causes and contemporary rates. Glob Planet Change 111:237–245. https://doi.org/10.1016/j.gloplacha.2013.09.019

Rahman M (2023) Bangladesh's community radios: what they broadcast and who listen. J Community Posit Pract 2(2023). https://doi.org/10.35782/JCPP.2023.2.06

Rahman MM, Rahaman MM (2018) Impacts of Farakka Barrage on hydrological flow of Ganges River and environment in Bangladesh. Sustain Water Resour Manag 4:767–780. https://doi.org/10.1007/s40899-017-0163-y

Rahman S, Khan MT, Akib S, Din NB, Biswas SK, Shirazi SM (2014) Sustainability of rain-water harvesting system in terms of water quality. Sci World J 2014(1):721357. https://doi.org/10.1155/2014/721357

Rashid MH, Islam MS (2007) Adaptation to climate change for sustainable development of Bangladesh agriculture. https://un-csam.org/sites/default/files/2021-01/01bd.pdf. Accessed 29 Dec 2024

Roman O, Hoque SF, Ford L, Salehin M, Alam MM, Hope R, Hall JW (2021) Optimizing rural drinking water supply infrastructure to account for spatial variations in groundwater quality and household welfare in coastal Bangladesh. Water Resour Res 57:e2021WR029621. https://doi.org/10.1029/2021WR029621

Saddaf N, Sultana R, Anjum B (2024) Vulnerability and effectiveness of Nature-based Solutions (NbS) in the farming communities of coastal Bangladesh. Environ Challenges 14(1):100863. https://doi.org/10.1016/j.envc.2024.100863

Shammi M, Rahman MM, Bondad SE (2019) Impacts of salinity intrusion in community health: a review of experiences on drinking water sodium from coastal areas of Bangladesh. Healthcare 7(1):50. https://doi.org/10.3390/healthcare7010050

de Sherbinin A (1997) Water and population dynamics: local approaches to a global challenge. Report on a collaborative initiative and workshop at IUCN's World Conservation Congress, Montreal, Canada, October 1996. https://portals.iucn.org/library/node/7294. Accessed 28 Dec 2024

Smith L, Glaser AW, Greenwood DC, Feltbower RG (2020) Cumulative burden of subsequent neoplasms, cardiovascular and respiratory morbidity in young people surviving cancer. Cancer Epidemiology 66, 101711, ISSN 1877–7821. https://doi.org/10.1016/j.canep.2020.101711

Talukder M, Rutherford S, Islam K, Chu C (2020) Water scarcity and challenges for access to safe water: a case of Bangladesh's coastal area vulnerable to climate change. Asian J Water Environ Pollut 17(4):1–9. https://doi.org/10.3233/AJW200045

Uddin MR, Khandaker MU, Ahmed S, Abedin MJ, Hossain SMM, Al Mansur MA (2024) Assessment of coastal river water quality in Bangladesh: implications for drinking and irrigation purposes. PLoS One 19(4):e0300878. https://doi.org/10.1371/journal.pone.0300878

United Nations (UN) (2023) Water and disasters. https://www.unwater.org/water-facts/water-and-disasters. Accessed 29 Dec 2024

Wahid SS, Islam MN, Raza WA (2024) Climate change and mental health in Bangladesh: vulnerability, inequality, and the crucial need for intervention on the frontlines of the climate crisis. In: Arafat SMY (ed) Mental health in Bangladesh. Springer, Singapore. https://doi.org/10.1007/978-981-97-0610-5_15

World Bank (2023) Global water security and sanitation partnership: annual report 2023. https://www.worldbank.org/en/topic/water/publication/global-water-security-and-sanitation-partnership-annual-report-2023. Accessed 29 Dec 2024

World Health Organization (WHO) (2024) Bangladesh climate, climate change & WASH. https://cdn.who.int/media/docs/default-source/climate-change/bangladesh-climate-change-wash.pdf?sfvrsn=dc20a09e_3&download=true. Accessed 29 Dec 2024

Wu J, Yunus M, Ali M, Escamilla V, Emch M (2018) Influences of heatwave, rainfall, and tree cover on cholera in Bangladesh. Environ Int 120:304–311. https://doi.org/10.1016/j.envint.2018.08.012

Chapter 18
Climate Change Mitigation Strategies for Drinking Water

Lakshmi Raghu Nagendra Prasad Rentachintala

1 Learning Objectives

1. Review available literature on climate change mitigation strategies for drinking water.
2. Present a case study to assess the impacts of climatic variation on various water quality parameters.
3. Describe various future research needs for climatic change mitigation strategies for drinking water.

2 Introduction

Climatic variation is instigating an irregular pattern of precipitation and intense famine situations and thus necessitates a vital focus to outline the protective methods to mitigate in the water-scarce region as well as zones with more groundwater diminution or area contributory to the recharge. The maximum number of the Indian rivers is contaminated, and thus the amputation of contaminants is an extra activity to make a catchment robust. There is a need to assess urban water to better define and advance the Water Quality Information Platform (WQIP). The present requirement for a method for sustainable water supply must be produced with thoughtful projected climatic variations (Kumar et al. 2019).

Climatic variation can result in vital modifications in the variables and processes that impact water quality and freshwater biological diversity. These comprise

L. R. N. P. Rentachintala (✉)
Department of Civil Engineering, Bapatla Engineering College,
Bapatla, Andhra Pradesh, India
e-mail: rlr.nagendraprasad@becbapatla.ac.in

A. R. Molla et al. (eds.), *Climate Change, Drinking Water Security, and Public Health*, https://doi.org/10.1007/978-3-032-12074-8_18

physical variations, such as more water heating; less river and lake ice cover; higher steady vertical stratification and low mixing of water of deep-water lakes; and variations in water flow impacting water level and retention time; chemical variations, such as more nutrient concentrations and water colour (DOC); and less oxygen content and biologic variations, consisting of northwards movement of species and amendment of habitats, affecting the structure and working of freshwater ecology. Variations in these variables govern the impact on all the socio-monetary and conservational goods and amenities that rely on these schemes directly or indirectly (Solheim et al. 2010).

Climatic variation situations propose more recurrent famines in summer, as well as flash-flooding, drive to uninhibited discharges from urban regions to receiving waters and estuaries. Lesser flows, decreased velocities and, thus, more water residence times in rivers and lakes will augment the possible toxic algal blooms and decrease dissolved oxygen scales (Whitehead et al. 2009). Significant actions to decrease climate risks comprise the integration of methods of climate resilience into water safety strategies, as well as upgraded accountancy and management of water sources. Policy preparations for methods to serve delivery and variations in management models provide probable lessons of risks, specific to low-income sets. Water and sanitation services lead to greenhouse gas releases. Selection of wastewater treatment methods, enhanced pumping efficiency, usage of renewable sources of energy, and within-system development of energy provide capacity to reduce releases (Howard et al. 2016).

There is a necessity to sustain and advance timely scrutiny and swift epidemiological responses to outbreaks and advent of new waterborne pathogens in all countries. Though the main problem of waterborne ailments is in developing countries, there needs to be both methodical and fiscal methods to confirm suitable quantities of best quality water, sewage removal, and hygiene to all. This will be indispensable to prevent surplus sickness and mortality in extents that will affect from significant variations in climate (Nichols et al. 2018).

Thus, the present study is aimed to review available literature on mitigation strategies for climatic change with focus on drinking water. Also, a case study of Vijayawada city of India is considered. As Vijayawada city is in the vicinity of the Krishna River, the impacts of urbanisation and climatic variation can exacerbate the water quality of the Krishna River. Further, the extent of impacts of climate change and urbanisation on water quality of Krishna River is assessed as trends of various water quality parameters, such as pH, BOD, DO, etc.

3 Literature Review

Mitigation is defined as an anthropogenic intercession which is a response to adverse consequences or an effort of making the consequences less severe. Adaptation accounts for modification of a scheme in reply to actual or projected stimuli or their outcomes, with modest impairment which exploits beneficial chances.

3.1 Impacts and Mitigations of Climate Change on Water Quality and Supply

Probable technical resilient specific small drinking water systems are evaluated (Rajib et al. 2012). Groundwater will be key to lessen some of the intense drought conditions. Flooding, contaminated water supply, and more severe weather occasions are probable to rise enhancing the risk of infectious sickness (Mishra 2023). Problems with freshwater conservation, subjects with drinking water quality and obtainability, and concerns with climatic variation adapting and mitigation are reviewed (Kunwar 2019). Rise in food productivity, variation in diet, and reduction in food loss and waste can decrease the need for land conversion, so free-up land and making chances for enhanced execution of other approaches can resolve land related aspects (Smith et al. 2020).

Fabian et al. (2023) aimed to (1) recognise present systematic methods of water quality modelling and projection for flood and drought evaluation, (2) debate the challenges and barriers, and (3) suggest plausible results to advance the interpretation of impacts of climate extremes on water quality and mitigate their adverse influences. The associations amid climate indices and water quality indicators were established for better interpretation of the link among climate extremes and water quality for a chosen catchment (Fabian et al. 2023).

The impacts on water quality of sources (rivers and lakes) that can modify parameters (physico-chemical, micropollutant, and biological parameters) are assessed. Also, the likely impacts on drinking water quantity and quality are discussed. A degrading trend of drinking water quality attributed to climate change drives the rise of at-risk conditions associated with probable health impact (Delpla et al. 2009).

The impacts of climate change on water quality mostly comprise the aggravation of eutrophication, variations in the flow, hydrological and thermal situations, and the obliteration of ecology and biodiversity. Further investigation of the influencing mechanism of climatic variation on cyanobacteria is an evolving research area. Also, the water quality situations of shallow lakes and drinking water are capable future research aspects. For climatic variation, the general regulations of management of water quality and the technical preparation of land use are of prodigious implication and need to be studied further (Li et al. 2020).

Experts interpret risk factors impacting both drinking water quality and obtainability as compound and systemically interconnected (Boholm and Prutzer 2017). A lack of political salience on drinking water as a public amenity is acknowledged as a barrier to the growth of strong adapting strategies (Boholm and Prutzer 2017). Another area of alarm relates to the topographical, structural, and organised limits amid the plethora of public actors with partly lapping and occasionally indistinct tasks to provide safe drinking water (Boholm and Prutzer 2017). Climatic variation adapting for drinking water will need a novel integration of the familiarity of systematic risk relations in association with more efficient work alliance on a distinct delineation of accountability amongst actors (Boholm and Prutzer 2017).

Climatic variation may drive enhanced requirements for water, over and above that rise which is projected for non-climatic causes, attributed to enhanced use of watering for garden (Arnell 1998). These augmented pressures on water sources will affect not only the dependability of water supply, but also the various other systems, such as navigation, and will have consequences for water quality management (Arnell 1998). It is essential for the government to launch an effective water policy, with distinct, sustainable strategies to mitigate the impacts of climatic variation on sustainable freshwater (Faye et al. 2021).

The details of influences of climatic variation impacts on water quality consist of:

- Rise in water heating
- Enhanced NOM concentrations
- Rise in turbidity, taste and odour compounds, and problems from related algae
- Increased microbial quantities in water source
- Quicker and more intense raw water quality variations
- Enhanced concentrations of organic and inorganic microcontaminants:
- The cause for depreciated quality
- The type of quality loss
- The susceptibility of the water source (Raspati et al. 2013).

Impacts of climate change on water quality for raw water supply perception include relevant risks, present operational practices, detecting quick variations in risks, remedial response, and curative methods (Mesquita et al. 2013). Impacts of climatic variation on inland waters are mostly through modifications in temperature and associated physico-chemical amendments. Indirect effects are developed through processes, such as extreme weather occurrences in the watershed, and comprise enhanced nutrient loading amid others. The combination of all influences will drive intense and considerable variations in the physical and biological configuration of the waters in climate situations (Dokulil and Teubner 2011).

Strengthening systems and capability for long-term climatic risk management is needed. Many communities will need to endure to participate in water management to go further for water security and generate maximum monetary, social, and environmental benefits that can be obtained from cleverer water use (Sadoff and Muller 2009). The impacts of climatic variation on urban water supply, management, and planning of minor towns in the coastal region were reviewed (Islam et al. 2017). Adapting methods and associated influences are to be suggested to foster sustainable growth strategies for water management in the context of eutrophication and climatic variation (Xia et al. 2016). Contaminant amounts were more inclined by variations in socio-fiscal factors than by climatic variation (Islam et al. 2018).

There is a risk that water may surpass crucial total organic carbon (TOC) levels, driving to probable deviations of disinfection by-product directives under probable forthcoming scenarios. The upcoming risk, however, can be accomplished with operational modifications at the water treatment plant, such as enhancing the rate of granular activated carbon reactivation (Li et al. 2014). Organising sensitivity analysis to discover the inferences of multiple climate scenarios, and the significance of accounting for policy portfolios instead of separate policy choices is vital (Miller and Belton 2014).

Steady monitoring and execution of mitigation strategies can lessen the probable risks related to the practices. It is crucial to recognise the likely perils resulting from the alternate water supply and delivery practices to advance tentative and enduring observation and mitigation proposals and decrease the microbial and chemical pollution of potable water distributed to the users (Salehi 2022).

3.2 Impacts and Mitigations of Climate Change on Temperature and Environment

Temperature rise is initially accountable for a general rise of the water column constancy and for a decrease of the mass transfer amid deep and surface waters with straight insinuations on the oxygen and nutrient levels (Copetti et al. 2013). The value of the temperature rise suffices to affect the advance of phytoplankton quantity. Nearly 70% of the nitrogen load is from diffuse source(s) while 65% of the phosphorus load instigates from point source(s). Nutrient loads estimated at 2100 (under various IPCC scenarios) quantify that both nitrogen and phosphorus loads are firmly reliant on the resident populace, which is accountable for a 61% and 41% rise, respectively, for nitrogen and phosphorus (Copetti et al. 2013). Expected nutrient load changes are insignificant when maintaining a fixed resident quantity. The phosphorus load is obviously impacted by the efficacy of the wastewater treatment plants (WWTPs) (Copetti et al. 2013).

More heating itself can evaporate more water, augment the requirement for water supply, cause added heating stress, and exacerbate prevailing high levels of air contamination (El-Din 2013). Upstream can face more dissolved organic carbon and colour scales, needing action at water treatment plants to obstruct toxic by-products inflowing into water supplies. Storms that end famine will flush nutrients from urban and rural regions or develop acid pulses in acidified upstream watersheds. Policy responses to climatic variation, such as the growing of bio-fuels or release controls, will highly influence the quality of freshwater (Whitehead et al. 2009).

Temperature rise from 16 °C to 24 °C can change the biofilm community structure and enhance the probability of removing colour. With only flush or flush with hyperchlorine is depicted as a decrease in discolor capacity and decreases the abundance of microorganisms which can compromise the quality and safety of water (Calero Preciado et al. 2022).

As higher evaporation rate and extreme precipitation are attributed to climatic variation, it further leads to urban flooding. In the case of urbanisation, impervious surface zones cause runoff and reduce the number of wetlands, lakes, and ponds. Improper urban planning describes inadequate drainage scheme designs, which form the basis for stormwater runoff from the ground and make the water more contaminated. Climatic variation and urbanisation are vital factors impacting the water quality in urban watersheds (Salerno et al. 2018; Sandeep et al. 2020).

There remain knowledge gaps about the direct impacts of climatic variation on water bodies and their indirect impacts on areas that employ huge amounts of water. Ecohydrological and alternate sewage treatment methods can signify an authoritative tool to mitigate climate change impacts (Garnier et al. 2015).

3.3 Impacts and Mitigations of Climate Change on Groundwater

The strategic significance of groundwater to global water and food security will deepen under climatic variation, as more recurrent and severe climatic extremes (droughts, floods) raise variance in soil moisture and surface water (Taylor et al. 2013). Sterner nutrient management will probably be the most viable and practical method for long-term CyanoHAB regulator in a heater, as well as rainier and more peak climate (Paerl and Paul 2012). Hands-on management efforts will lessen the risks to ecology and society and may be less costly than reactive exercises taken only once problems have arisen (Palmer et al. 2008).

3.4 Impacts and Mitigations of Climate Change on Livestock

The livestock sector produces 14.5% of worldwide greenhouse gas (GHG) releases, leading to further climatic variation. Further, the livestock sector will be a vital driver in the mitigation of GHG releases and to improve worldwide food security. Thus, in the evolution to sustainable livestock raising, there is a necessity for: (a) evaluations related to the usage of adapting and mitigation methods apt for the position and livestock raising system in use, and (b) methods that aid and enable the execution of climate change adaptation and mitigation approaches (Rojas-Downing et al. 2017).

4 Case Study

4.1 Study Area

Vijayawada city is in the vicinity of the capital of the state of Andhra Pradesh in India. Also, the city is in the vicinity of the Krishna River. The area of Vijayawada city is 61.88 km^2 and is positioned at 16.5062° N latitude and 80.6480° E longitude.

The range of various water quality parameters of the Krishna River at the study area, i.e., capital area in the vicinity of Vijayawada city, were investigated to identify the impacts of urbanisation (Rentachintala et al. 2022). Urbanisation was initiated in the capital area from 2014. The parameters, pH, dissolved oxygen (DO), and total coliform (TCol) exceeded the maximum tolerance limits per BIS standards from 2014 (Rentachintala et al. 2022). The range of pH values was 7.2 to 8.57 before urbanisation and 6.5 to 8.7 during initial urbanisation (2014–2019). However, as per BIS standards, tolerance limits of pH are 6.5 to 8.5 (Rentachintala et al. 2022). The range of total dissolved solids was 168 to 650 mg/l before urbanisation and 223 to 504 mg/l during initial urbanisation. However, tolerance limits of total dissolved solids are 500 to 1500 mg/l (Rentachintala et al. 2022). DO had a range of 5.66 to 8.27 mg/l before urbanisation and 4 to 7.9 mg/l during initial urbanisation. However, the permissible range for DO is 4 to 6 mg/l (Rentachintala et al. 2022). TCol ranged from 2 to 2400 MPN/100 ml before urbanisation and 330 to 16,000 MPN/100 ml during initial urbanisation. However, tolerance limits of TCol range from 50 to 5000 MPN/100 ml (Rentachintala et al. 2022). Faecal coliform was in the range of 2 to 800 MPN/100 ml before urbanisation and 80 to 16,000 MPN/100 ml during initial urbanisation; however, the tolerance limit of faecal coliform range is 40% of TCol (Rentachintala et al. 2022). Thus, the present case study found that imperviousness, population density attributed to urbanisation, variations in precipitation patterns and subsequent changes in flow characteristics, and variations in runoff, especially surface runoff quantity and quality, were the impacting drivers of drinking water quality in urban environments. Various anthropogenic activities, which were causing more contamination in urban regions, resulted in adverse impacts on water quality. They need to be managed efficiently to meet the requirements and standards of safe drinking water.

5 Future Research Needs

There are a significant number of adapting methods that can be considered to resolve climatic variation impacts on water resources, and the adapting capacity of the water sector has considerably enhanced. The overall susceptibility remains high, as water demand is still not contented at certain areas and/or for certain uses during drought. Further, there would be a move to a more balanced and integrated method for managing water (Papadaskalopoulou et al. 2015). Spatially variable patterns of snow or summer precipitation that are related to local climatic variation will have vital impacts on watershed bio-geo-chemical methods and surface water quality in interfaces with local geography, land use amendment, or acid depositing (Park et al. 2010).

Climatic variance has vital influences on BMP efficiency with effects on catchment hydrology, soil erosion, and nutrient losses. The significance of optimal outline of site-specific BMPs at sub-basin level is key to adapting to climatic variation (Qiu et al. 2019). Drinking water managers should permit their participation in the adapting decision-making process, to confirm their receiving and collaboration, and to avert conflicts (Garnier and Holman 2019). Strategies to cope with uncertainty when adapting to climatic variation should replicate the kind of the noncertainty causes and how they intermingle with risk scale and decision creation: (i) epistemic noncertainties can be lessened by gaining more acquaintance; (ii) noncertainties related to indistinctness can be decreased by discussion and information sharing amid the various shareholders; and (iii) aleatory noncertainty is, by its form, non-reducible. Thus, even large noncertainties may have small values for decision-making, because multiple bases of evidence provide adequate acquaintance to defend action in climatic variation adaptation (Refsgaard et al. 2013).

More action to resolve global inequities is needed, along with more inclusive techniques to climatic variation mitigation (Delany-Crowe et al. 2019). The prioritising of the findings with regard to their significance, probability, and assurance is key to dealing with climatic variation in the context of drinking water (Kundzewicz et al. 2008). Yet, there are water systems that are intrinsically resilient to few degrees of hydrologic variability and climatic variability, are not substantially impacted, or the consequences of influence can be readily alleviated (Alavian et al. 2009).

6 Conclusions

The following conclusions may be drawn from the present study:

- A review of available literature on climatic change mitigation strategies for drinking water is carried out.
- A case study of Vijayawada city in India is considered to determine the impacts of various drivers on water quality, including climatic variation.
- A study on water quality indicates that pH, DO, and total coliform are beyond maximum allowable limits as per standards for the study area.
- Various factors that may adversely impact water quality, such as climatic variation, urbanisation, and contamination, need to be addressed to mitigate their effects and cater to safe and quality drinking water.

7 Discussion Questions

1. What are the key drivers for climatic change mitigation strategies for drinking water?
2. Is mitigation or adaptation governing needed for drinking water to cope with climatic variation?
3. Which water quality parameters are most critical to assess drinking water in view of climatic variation?

References

Alavian V, Qaddumi HM, Dickson E, Diez SM, Danilenko AV, Hirji RF, Puz G, Pizarro C, Jacobsen M, Blankespoor B (2009) Water and climate change: understanding the risks and making climate-smart investment decisions. World Bank, Washington, DC. 52911 https://citeseerx.ist.psu.edu/document?repid=rep1&type=pdf&doi=70d0eeb2b0f0e235804507d8b163549cf8ffb070

Arnell NW (1998) Climate change and water resources in Britain. Clim Chang 39(1):83–110. https://doi.org/10.1023/A:1005339412565

Boholm Å, Prutzer M (2017) Experts' understandings of drinking water risk management in a climate change scenario. Clim Risk Manag 16:133–144. https://doi.org/10.1016/j.crm.2017.01.003

Calero Preciado C, Soria-Carrasco V, Boxall J, Douterelo I (2022) Climate change and management of biofilms within drinking water distribution systems. Front Environ Sci 10:962514. https://doi.org/10.3389/fenvs.2022.962514

Copetti D, Carniato L, Crise A, Guyennon N, Palmeri L, Pisacane G, Struglia MV, Tartari G (2013) Impacts of climate change on water quality. In: Regional assessment of climate change in the mediterranean: Volume 1: Air, sea and precipitation and water. Springer, pp 307–332. https://doi.org/10.1007/978-94-007-5781-3_10

Delany-Crowe T, Marinova D, Fisher M, McGreevy M, Baum F (2019) Australian policies on water management and climate change: are they supporting the sustainable development goals and improved health and well-being? Glob Health 15:1–15. https://doi.org/10.1186/s12992-019-0509-3

Delpla I, Jung AV, Baures E, Clement M, Thomas O (2009) Impacts of climate change on surface water quality in relation to drinking water production. Environ Int 35(8):1225–1233. https://doi.org/10.1016/j.envint.2009.07.001

Dokulil MT, Teubner K (2011) Eutrophication and climate change: present situation and future scenarios. In: Eutrophication: causes, consequences and control. Springer, pp 1–16. https://doi.org/10.1007/978-90-481-9625-8_1

El-Din MMN (2013) Proposed climate change adaptation strategy for the ministry of water resources & irrigation in Egypt. Joint programme for climate change risk management in Egypt. https://www.eeaa.gov.eg/Uploads/Project/Files/20221123101650781.pdf

Fabian PS, Kwon HH, Vithanage M, Lee JH (2023) Modeling, challenges, and strategies for understanding impacts of climate extremes (droughts and floods) on water quality in Asia: a review. Environ Res 225:115617. https://doi.org/10.1016/j.envres.2023.115617

Faye C, Sow AA, Dieye S (2021) Water management policy for freshwater security in the context of climate change in Senegal. In: Climate change and water resources in Africa: Perspectives and solutions towards an imminent water crisis. Springer, pp 255–276. https://doi.org/10.1007/978-3-030-61225-2_12

Garnier M, Holman I (2019) Critical review of adaptation measures to reduce the vulnerability of European drinking water resources to the pressures of climate change. Environ Manag 64(2):138–153. https://doi.org/10.1007/s00267-019-01184-5

Garnier M, Harper DM, Blaskovicova L, Hancz G, Janauer GA, Jolánkai Z, Lanz E, Porto AL, Mándoki M, Pataki B, Rahuel JL (2015) Climate change and European water bodies, a review of existing gaps and future research needs: Findings of the Climate Water project. Environ Manag 56:271–285. https://doi.org/10.1007/s00267-015-0544-7

Howard G, Calow R, Macdonald A, Bartram J (2016) Climate change and water and sanitation: likely impacts and emerging trends for action. Annu Rev Environ Resour 41(1):253–276. https://doi.org/10.1146/annurev-environ-110615-085856

Islam SN, Reinstädtler S, Ferdaush J (2017) Challenges of climate change impacts on urban water quality management and planning in coastal towns of Bangladesh. Int J Environ Sustain Dev 16(3):228–256. https://doi.org/10.1504/IJESD.2017.085058

Islam MM, Iqbal MS, Leemans R, Hofstra N (2018) Modelling the impact of future socio-economic and climate change scenarios on river microbial water quality. Int J Hyg Environ Health 221(2):283–292. https://doi.org/10.1016/j.ijheh.2017.11.006

Kumar M, Deka JP, Kumari O (2019) Development of Water Resilience Strategies in the context of climate change, and rapid urbanization: a discussion on vulnerability mitigation. Groundw Sustain Dev 10:100308. https://doi.org/10.1016/j.gsd.2019.100308

Kundzewicz ZW, Mata LJ, Arnell NW, Döll P, Jimenez B, Miller K, Oki T, Şen Z, Shiklomanov I (2008) The implications of projected climate change for freshwater resources and their management. Hydrol Sci J 53(1):3–10. https://doi.org/10.1623/hysj.53.1.3

Kunwar SB (2019) Freshwater conservation, drinking water quality & climate change adaptations: a case study on Nepal (Doctoral dissertation, The University of New Mexico). https://digitalrepository.unm.edu/econ_etds/103

Li Z, Clark RM, Buchberger SG, Jeffrey Yang Y (2014) Evaluation of climate change impact on drinking water treatment plant operation. J Environ Eng 140(9):A4014005. https://doi.org/10.1061/(ASCE)EE.1943-7870.0000824

Li X, Li Y, Li G (2020) A scientometric review of the research on the impacts of climate change on water quality during 1998–2018. Environ Sci Pollut Res 27(13):14322–14341. https://doi.org/10.1007/s11356-020-08176-7

Mesquita E, Rosa MJ, Menaia J, Kardinaal E, Eikebrokk B (2013) Adapted operation of drinking water systems to cope with climate change. http://repositorio.lnec.pt:8080/jspui/handle/123456789/1006401

Miller KA, Belton V (2014) Water resource management and climate change adaptation: a holistic and multiple criteria perspective. Mitig Adapt Strateg Glob Chang 19:289–308. https://doi.org/10.1007/s11027-013-9537-0

Mishra RK (2023) Fresh water availability and its global challenge. Br J Multidisc Adv Stud 4(3):1–78. https://doi.org/10.37745/bjmas.2022.0207

Nichols G, Lake I, Heaviside C (2018) Climate change and water-related infectious diseases. Atmos 9(10):385. https://doi.org/10.3390/atmos9100385

Paerl HW, Paul VJ (2012) Climate change: links to global expansion of harmful cyanobacteria. Water Res 46(5):1349–1363. https://doi.org/10.1016/j.watres.2011.08.002

Palmer MA, Reidy Liermann CA, Nilsson C, Flörke M, Alcamo J, Lake PS, Bond N (2008) Climate change and the world's river basins: anticipating management options. Front Ecol Environ 6(2):81–89. https://doi.org/10.1890/060148

Papadaskalopoulou C, Katsou E, Valta K, Moustakas K, Malamis D, Dodou M (2015) Review and assessment of the adaptive capacity of the water sector in Cyprus against climate change impacts on water availability. Resour Conserv Recycl 105:95–112. https://doi.org/10.1016/j.resconrec.2015.10.017

Park JH, Duan L, Kim B, Mitchell MJ, Shibata H (2010) Potential effects of climate change and variability on watershed biogeochemical processes and water quality in Northeast Asia. Environ Int 36(2):212–225. https://doi.org/10.1016/j.envint.2009.10.008

Qiu J, Shen Z, Leng G, Xie H, Hou X, Wei G (2019) Impacts of climate change on watershed systems and potential adaptation through BMPs in a drinking water source area. J Hydrol 573:123–135. https://doi.org/10.1016/j.jhydrol.2019.03.074

Rajib MA, Rahman MM, McBean EA (2012) Evaluating technological resilience of small drinking water systems under the projected changes of climate. J Water Clim Change 3(2):110–124. https://doi.org/10.2166/wcc.2012.019

Raspati G, Menaia J, Raat J, Rosa MJ, González B, Sivertsen E (2013) Assessment of current treatment works to handle climate change related pollutants and options to make current multi-barrier systems climate change proof. http://repositorio.lnec.pt:8080/jspui/handle/123456789/1007785

Refsgaard JCKA-N, Drews M, Halsnæs K, Jeppesen E, Madsen H, Markandya A, Olesen JE, Porter JR, Christensen JH (2013) The role of uncertainty in climate change adaptation strat-

egies—A Danish water management example. Mitig Adapt Strateg Glob Chang 18:337–359. https://doi.org/10.1007/s11027-012-9366-6
Rentachintala LRNP, Muni Reddy MG, Kumar MP (2022) Trends of surface water quality of the Krishna River, India during the urbanization process. Environ Qual Manag 32(3):111–121. https://doi.org/10.1002/tqem.21860
Rojas-Downing MM, Nejadhashemi AP, Harrigan T, Woznicki SA (2017) Climate change and livestock: impacts, adaptation, and mitigation. Clim Risk Manag 16:145–163. https://doi.org/10.1016/j.crm.2017.02.001
Sadoff C, Muller M (2009) Water management, water security and climate change adaptation: early impacts and essential responses. Global Water Partnership, Stockholm. https://hdl.handle.net/10535/4999
Salehi M (2022) Global water shortage and potable water safety; Today's concern and tomorrow's crisis. Environ Int 158:106936. https://doi.org/10.1016/j.envint.2021.106936
Salerno F, Gaetano V, Gianni T (2018)Urbanization and climate change impacts on surface water quality: Enhancing the resilience by reducing impervious surfaces. Water Res 144:491–502
Sandeep V, Khandekar A, Kumar M (2020) Water supply, urbanization and climate change. In: Resilience, Response, and Risk in Water Systems: Shifting Management and Natural Forcings Paradigms. Springer, pp 195–215. https://doi.org/10.1007/978-981-15-4668-6_11
Smith P, Calvin K, Nkem J, Campbell D, Cherubini F, Grassi G, Korotkov V, Le Hoang A, Lwasa S, McElwee P, Nkonya E (2020) Which practices co-deliver food security, climate change mitigation and adaptation, and combat land degradation and desertification? Glob Chang Biol 26(3):1532–1575. https://doi.org/10.1111/gcb.14878
Solheim AL, Austnes K, Eriksen TE, Seifert I, Holen S (2010) Climate change impacts on water quality and biodiversity. Background report for EEA European environment state and outlook report, 1–68. https://citeseerx.ist.psu.edu/document?repid=rep1&type=pdf&doi=124ffa0d61a4661f33475bda55e1e82b435c365d
Taylor RG, Scanlon B, Döll P, Rodell M, Van Beek R, Wada Y, Longuevergne L, Leblanc M, Famiglietti JS, Edmunds M, Konikow L (2013) Ground water and climate change. Nat Clim Chang 3(4):322–329. https://doi.org/10.1038/nclimate1744
Whitehead PG, Wilby RL, Battarbee RW, Kernan M, Wade AJ (2009) A review of the potential impacts of climate change on surface water quality. Hydrol Sci J 54(1):101–123. https://doi.org/10.1623/hysj.54.1.101
Xia R, Zhang Y, Critto A, Wu J, Fan J, Zheng Z, Zhang Y (2016) The potential impacts of climate change factors on freshwater eutrophication: implications for research and countermeasures of water management in China. Sustainability 8(3):229. https://doi.org/10.3390/su8030229

Chapter 19
Case Studies of Waterborne Disease Outbreaks Related to Climate Change

Erika J. Mitchell

1 Learning Objectives

1. Classify agents of waterborne disease according to the type of organism or chemical causing the disease.
2. Discuss how facets of climate change interact with various agents of waterborne disease, resulting in increased outbreaks of disease.
3. Enumerate and describe examples of outbreaks of human waterborne disease that have been linked to facets of climate change.

2 Introduction

Provision of safe drinking water requires ensuring that the water not be contaminated by microbiological pathogens or harmful chemicals. This may necessitate protection of water sources, treatment of water to remove contaminants, and provision of water distribution systems capable of preventing contamination. In addition, water quality monitoring and hazard communication and response can provide further protection from disease outbreaks. All of these components together form a system sometimes termed the multi-barrier approach to ensuring safe drinking water (CCME 2002). Unfortunately, each of these components may be strained or defeated by climate change.

Climate change has caused and will likely continue to cause increases in extreme weather events, including flooding, drought, temperature increases and sea level rise (IPCC 2022). These meteorological impacts may either defeat the barriers

E. J. Mitchell (✉)
Better Life Laboratories, Inc., East Calais, VT, USA
e-mail: em63@cornell.edu

A. R. Molla et al. (eds.), *Climate Change, Drinking Water Security, and Public Health*, https://doi.org/10.1007/978-3-032-12074-8_19

constructed to protect drinking water or facilitate physicochemical changes in the water itself. These changes may promote the growth of harmful microorganisms or augment the content of harmful chemicals in the water.

Recent reviews of the health impacts of climate change have identified water-borne diseases as an area of particular concern, with evidence of links between climate change-related weather events and outbreaks of water-borne disease (Alcayna et al. 2022; IPCC 2022). Extreme precipitation episodes may be responsible for 20–40% of the outbreaks of waterborne diseases in the United States (Rose et al. 2000). Children under age five have been identified as especially vulnerable, as well as low- and middle-income populations, especially in Sub-Saharan Africa and South Asia (Cissé 2019), while most research has been carried out in temperate, high-income countries (Van de Vuurst and Escobar 2023).

This chapter provides case studies of disease outbreaks due to direct ingestion of drinking water that have been or may be ascribed to weather-related causes associated with climate change.

3 Types of Agents of Waterborne Outbreaks

Water-borne illnesses can be caused by several different types of pathogens, including bacteria, protozoa, viruses, and cyanobacteria. Various types of chemicals, both anthropogenic and naturally occurring, can also cause outbreaks of illnesses. Table 19.1 presents examples of specific outbreaks organized by the type of organism or chemical factor responsible, and the type of climatic or meteorological event linked to the outbreak; the case studies that follow are selected from this table.

3.1 Bacterial Pathogens

Bacterial pathogens that have frequently been implicated as the agents causing waterborne outbreaks include *Vibrios* (*V. cholerae* causes cholera), *Salmonellas* (*S. enterica serotype typhi* causes typhoid fever), *Escherichias*, *Shigellas*, *Yersnias*, and *Campylobacters* (Woodall 2009). Examples of outbreaks with bacterial agents linked to climate change have been reported throughout the world, including Canada, Swaziland, Zambia, and Bangladesh (see Table 19.1).

Case Study: Drought and the 1992 *Escherichia Coli* O157 Outbreak in Lubombo District, Swaziland (Effler et al. 2001)

As described by Effler et al. (1990), in the early 1990s, a series of villages centered around a sugar plantation in Lubombo District, Swaziland were afflicted with several years of drought, which became markedly worse from July to September, 1992. In a desperate search for water, local cattle grazed closer to surface water sources and wandered through the remaining trickling streambeds, where they defecated. By August, cattle deaths began to increase, with cattle deaths reported seven times

Table 19.1 Examples of water-borne disease outbreaks linked to manifestations of climate change

	Flooding or extreme precipitation events	Drought	Temperature rise	Sea level rise or sea water intrusion
Bacteria	• Walkerton, Canada, 2000. • 2300 people sickened with *E. coli* and campylobacter after flood (Hrudy et al. 2003; Auld et al. 2010)	• Swaziland, 1992. • >40,000 people sickened with *E. coli* after drought (Effler et al. 2001)[1]	• Lusaka, Zambia, 2003–2006. • 3 outbreaks of cholera with high temperatures (Luque Fernádez et al. 2008)	• Bangladesh, 1993–2007. • Outbreaks of cholera with sea level rise (Hashizume et al. 2011).
Protozoa	• Milwaukee, WI, USA, 1994. • >400,000 sickened with cryptosporidium after heavy rains (Fox and Lytle 1996)	• Isle of Thanet, UK, 1991. • > 47 sickened with cryptosporidium after drought (Joseph et al. 1991)	• Queensland, Australia, 2001. • Increased relative risk of cryptosporidiosis with increasing temperatures (Hu et al. 2010)[2].	
Viruses	• Solomon Islands, 2014. • 6000 sickened with rotavirus after flood (Jones et al. 2016)[3]	• Guinea-Bissau, 2010–2012. • Increased rotavirus prevalence in dry season (Mero et al. 2023).	• Dhaka, Bangladesh, 1996–2001. • Increased risk of rotavirus with temperature (Hashizume et al. 2008).	
Harmful algal blooms	• Queensland, Australia, 2013–2014. • Cyanobacterial blooms affecting water supply after flood (Rose et al. 2018).	• Brazil, 1998. • Increased cyanobacterial blooms in drinking water reservoirs with drought (Bouvy et al. 2000).	• Toledo, OH, USA, 2014. • 110 sickened by cyanobacterial toxins (Benedict et al. 2017); algal blooms increased with temperature (Del Giudice et al. 2021)[4]	
Chemicals	• Andhra Pradesh, India, 2020. • >500 sickened by organochlorine after flood (Sharma et al. 2023)	• Ritopek, Serbia, 2012. • Increased dental fluorosis with increased water fluoride; increased water fluoride with drought (Antonijevic et al. 2015).	• United States, 2000–2002. • Increased disinfection byproducts with increased temperature (Weinberg and Krasner 2002).	• Bangladesh, 2016–2017. • Increased hospital visits with increased salinity (Chakraborty et al. 2019)[5].

[1]Discussed in Sect. 3.1 on Bacterial Pathogens.
[2]Discussed in Sect. 3.2 on Protozoan Pathogens.
[3]Discussed in Sect. 3.3 on Viral Pathogens.
[4]Discussed in Sect. 3.4 on Harmful Algal Blooms.
[5]Discussed in Sect. 3.5 on Chemical Agents.

higher than normal for September and October. Many of the cattle died within the streambeds or watersheds.

Unfortunately, cattle are the major reservoir of the deadly pathogen *Escherichia coli* O157, which can cause severe bloody diarrhea, abdominal cramps in humans, and in some cases, Hemolytic Uremic Syndrome, which may lead to death. When cattle are stressed, they tend to shed more *E. coli* O157 in their feces. Dead and decaying cattle also may release *E. coli* O157 into the environment.

The rains finally arrived in the region on October 29 and 30, bringing 34–36 mm of precipitation to the region over 1–2 days. The first cases of bloody diarrhea among the human inhabitants of the plantation region began approximately 3 days later, fitting the human median incubation period for *E. coli* infection of 3–4 days. Later analyses found *E. coli* in borehole water, standpipe water, and water stored in residents' homes. People who drew drinking water from surface water sources such as streams or rivers had the highest relative risk of developing bloody diarrhea, although other significant risk factors included consuming beef and female gender. All ages of the population were affected, with approximately 42% of the population reporting bouts of bloody diarrhea during November and December. The number of deaths due to this outbreak could not be ascertained with certainty due to limited hospital record keeping. This outbreak of *E. coli* O157 was the first to be documented in the continent of Africa, and the first reported in the developing world (Effler et al. 2001).

In this case, the climatic factor of drought changed the behavior of cattle and decreased their health status, enabling the *E. coli* O157 bacteria to spread from the cattle to water sources. As regional droughts become more common, cattle and wild animals will be forced to adapt to limited water sources, which may lead to additional outbreaks of human disease due to zoonotic spread of bacterial pathogens.

3.2 Protozoan Pathogens

Protozoan pathogens that are common waterborne agents of human disease include *Cryptosporidium parvum*, *Entamoeba histolytica* and *Giardia duodenalis* (Woodall 2009), all of which cause watery diarrhea, nausea, and weight loss. Protozoan pathogens have often been implicated as agents of waterborne outbreaks connected with climate change, especially *Cryptosporidium*, with outbreaks, for example, in Milwaukee, the United Kingdom, and Australia (see Table 19.1).

Case Study: Increased Temperatures Linked to Increased Relative Risk of Cryptosporidiosis in 2001 in Queensland, Australia (Hu et al. 2010)
Queensland is the second largest state in Australia. It comprises the northeastern quarter of the continent and has a sub-tropical climate with average temperatures of 25 Celsius (°C) in summer and 15 °C in winter.

Cryptosporidiosis is a usually self-limiting diarrheal disease caused by the protozoan *Cryptosporidium parvum*. The disease is rife in Queensland, with 20.7 cases per 100,000 population, according to data collected from 2001–2012 (Lal et al. 2015). Sporadic outbreaks occur throughout the year, but most outbreaks occur in the hot summer months (Hu et al. 2010; Lal et al. 2015). Risk factors for developing cryptosporidiosis include drinking water from poorly treated water supplies, swimming in contaminated pools, attending daycare facilities, or caring for farm animals, but the largest outbreaks are waterborne (Hu et al. 2010). Cryptosporidiosis has an incubation period of 2–10 days after exposure to *C. parvum*. In Queensland, children under 4 years old, especially male children, make up the largest group of reported cases, with adult females of child-bearing age a second highly susceptible group (Lal et al. 2015). People with lower socio-economic status tend to be at higher risk of developing the disease (Hu et al. 2010; Lal et al. 2015).

There is no clear geographic distribution of outbreaks of cryptosporidiosis in the region (Greenwood and Reid 2020). Instead, researchers have consistently noted that most outbreaks occur in the hottest months (Hu et al. 2010; Greenwood and Reid 2020). In a detailed statistical analysis designed to uncover which factors are significantly associated with outbreaks, Hu et al. (2010) found that high temperatures are closely linked to the occurrences of outbreaks. In particular, when temperatures exceed 32 °C, the rate of cryptosporidiosis tends to rise, even amongst people from higher socio-economic strata, who usually have a lower incidence rate (Hu et al. 2010).

In this case, high temperatures consistently lead to increasing numbers of cryptosporidiosis cases, with the warmest part of the country being most highly affected, especially in the warmest part of the year. The specific mechanism for the link was not explored, but the researchers note that *C. parvum* thrives in warm waters with moderate salinity (Hu et al. 2010). As climate change brings about higher temperatures, we may expect cases of cryptosporidiosis to increase.

3.3 *Viral Pathogens*

Viral pathogens that commonly cause waterborne disease outbreaks include noroviruses, rotaviruses, and hepatitis A and E (Woodall 2008). While noroviruses and rotaviruses typically cause gastrointestinal symptoms including vomiting and diarrhea that usually resolve within several days to a week, hepatitis viruses involve the liver and usually take several weeks to several months to resolve. While otherwise healthy adults usually recover from gastrointestinal viruses within a few days, such viruses may prove fatal to infants, young children, and elderly individuals if the vomiting and diarrhea leave them severely dehydrated. Viral outbreaks attributable to climate change have been reported throughout the world, especially in developing regions such as Pacific Island nations, Africa, and South Asia (see Table 19.1).

Case Study: 2014 Rotavirus Outbreak after Torrential Floods in the Solomon Islands (Jones et al. 2016)

In 2014, a tropical depression brought sustained torrential rains to Honiara, the capital city of the Solomon Islands. The rains began on March 31 and continued to fall through April 6, bringing a total of 663 millimeters (mm), which was 10 times the mean weekly total rainfall for the previous 3 years. The rains caused extensive flooding in the city of Honiara and throughout the surrounding province of Guadalcanal. More than 10,000 people were displaced from their homes by the floods and forced into temporary evacuation shelters.

An outbreak of diarrhea began in Honiara on April 20th, with fecal samples testing positive for rotavirus. With much of the population crowded into evacuation shelters, the outbreak spread quickly throughout Guadalcanal province, and then on to five other provinces that had not been originally affected by the floods. Eventually, more than 6000 cases of diarrhea were reported, as well as 27 deaths. Children under 5 years old had an attack rate that was 14 times higher than children over 5 years old or adults. Of the 27 deaths, 21 of them (91%) occurred in children under 5 years old (Jones et al. 2016).

This case demonstrates the heightened danger of rapid disease spread after natural disasters linked to climate change. Although the local government put into place Early Warning Alert and Response Surveillance Network (EWARN) protocols, they were not able to prevent the disease from spreading within Honiara to areas that had not been originally affected by the floods. This outbreak also demonstrates how young children are often the most vulnerable population in the face of climate change hazards. As climate change brings more frequent and severe incidents of flooding, emergency systems will be challenged to maintain access to clean drinking water and adequate hygiene, throughout the world, but especially in low-lying coastal or island nations.

3.4 Harmful Algal Blooms

Harmful algal blooms (HABs) include a biologically diverse group of phenomena in which various types of phytoplankton become active or dominant in a water body. The phytoplankton may include dianoflagellates, diatoms, or cyanobacteria. Cyanobacteria are a phylum of autotrophic photosynthesizing bacteria, often referred to as blue-green algae, although they are not true algae. When HABs occur in a drinking water source, they may introduce highly toxic substances into the water, making the water dangerous to drink or touch. Cyanobacteria, in particular, produce microcystins, which are very potent liver toxins. Consuming water containing microcystin toxins can cause acute gastrointestinal symptoms such as vomiting, bloody diarrhea, and anorexia; it can also cause enlargement of the liver or death due to liver failure (Lad et al. 2022).

Case Study: 2014 Harmful Algal Bloom in Toledo, Ohio, USA

In the early evening of Friday, August 1, 2014, workers at the Collins Park water treatment plant for the city of Toledo, Ohio, detected microcystins above safe levels (City of Toledo 2014). A large cyanobacterial bloom had been growing in nearby Lake Erie for some time, which strong winds may have moved to the location of the intake pipe for Toledo's municipal water (Wines 2014; Steffen et al. 2017). Upon the advice of the Ohio Environmental Protection Agency, the city ordered a water ban shortly after midnight on August 2 (City of Toledo 2014). The water ban instructed city residents not to drink the water, use it for cooking, or feed it to pets (City of Toledo 2014). This water ban affected approximately 500,000 people for several days (Fitzsimmons 2014).

The city spread the word about the water ban through print, broadcast, and social media (City of Toledo 2014). Despite these efforts, some residents did not receive word immediately, especially in poorer neighborhoods with residents living without electricity (Gersony 2022). Volunteers and community activists went door to door to warn neighbors not to drink the water (Gersony 2022). By August 4, water tests showed that the microcystin level in the municipal water had dropped to acceptable levels, and the water ban was lifted (City of Toledo 2014). Ultimately, 110 cases of acute microcystin poisoning during this outbreak were reported to the CDC (Benedict et al. 2017); the financial costs of the 2-day outbreak have been estimated at $65 million (Gersony 2022).

This 2014 outbreak motivated much subsequent research on HABs in Lake Erie and the Great Lakes region. It was noted that the HAB was fed by phosphorus and nitrogen runoff from farms, feed lots, and leaking septic systems carried by the Maumee River, which empties into Lake Erie near Toledo (Wines 2014; Bullerjahn et al. 2016; Del Giudice et al. 2021). Modelling data suggest that the frequent blooms on Lake Erie may be closely related to air and water temperature, with higher temperatures leading to earlier starts for blooms, longer growth periods, and potentially greater production of microcystin toxins (Del Giudice et al. 2021). Lake Erie is the smallest and warmest of the Great Lakes, and the area near the Maumee River and Toledo is the warmest part of the lake (Del Giudice et al. 2021).

This case demonstrates the interaction between the ecological factors controlling phytoplankton growth and climatic factors, especially precipitation and air and water temperatures. Anthropogenic releases of agricultural fertilizers and waste provide essential nutrients for HAB growth, while rainfall helps wash these nutrients into lakes where they are made available to harmful phytoplankton (Bullerjahn et al. 2016; Del Giudice et al. 2021). High temperatures, which are expected with climate change, promote cyanobacterial growth and select for more toxic species (Bullerjahn et al. 2016; Del Giudice et al. 2021). In this outbreak, the poorest segment of the population was exceptionally vulnerable since communication efforts by the city about the water ban did not always reach them (Gersony 2022). Following the outbreak, the financial burden of upgrading the infrastructure to protect against future microcystin contamination has been shouldered disproportionately by the poor, with the bottom 20% of lower income residents spending up to 27% of their total income for water (Peterson 2022).

3.5 Chemical Agents

Many types of chemicals can cause outbreaks of waterborne disease, including both naturally occurring and anthropogenic substances. Chemicals such as arsenic or fluoride that are naturally present in the earth's crust can cause serious health problems when excessive amounts are dissolved in drinking water; some aspects of climate change can increase concentrations of these naturally occurring chemicals in water (Frisbie et al. 2024; Antonijevic et al. 2015; Chakraborty et al. 2019). Chemicals with anthropogenic sources, such as pesticides or disinfection byproducts, can also be highly toxic and their presence or concentration in drinking water can be influenced by manifestations of climate change (Sharma et al. 2023; Weinberg and Krasner 2022). Outbreaks of waterborne diseases caused by chemical agents and linked to extreme meteorological events or climate change have been reported throughout the world, including India, Bangladesh, Europe, and the United States (see Table 19.1).

Case Study: Hypertension, Preeclampsia, and Other Health Effects Associated with Increased Salinity and Sea Level Rise in Coastal Bangladesh
Much of coastal Bangladesh is less than 1.5 meters (m) above sea level, making this region exceptionally vulnerable to sea level rise (New Humanitarian 2007); a 1 m rise in global sea level is predicted to inundate 20% of the country's landmass. More than 20 million people living in coastal Bangladesh have already been affected by rising sea levels (New Humanitarian 2007). The rising sea level has increased the salinity of soil, groundwater, and surface water alike (New Humanitarian 2007).

The World Health Organization (WHO) has published a sodium (Na) intake guideline of 2 grams of sodium per day (g Na/day) for adults based on the effects of sodium on blood pressure (WHO 2012). Although this intake guideline could be used to calculate a health-based drinking water guideline of 200 milligrams of sodium per liter (mg Na/L), the WHO has not established a formal drinking water guideline for sodium in water (Mitchell and Frisbie 2023; WHO 2003, WHO 2022). Bangladesh has a national drinking water standard of 200 mg Na/L (DPHE 2019). Several studies have investigated health effects of drinking water with high levels of saline in the coastal region of Bangladesh (Khan et al. 2014; Al Nahian et al. 2018). In these studies, groundwater drinking water sources, especially deep tubewells, which are generally preferred due to lower levels of microbiological contamination, contained mean sodium concentrations of 700 mg/L to 915 mg/L, with some subjects drinking water that contained more 1500 mg/L (Khan et al. 2014; Al Nahian et al. 2018).

Hypertension associated with water salinity in a dose-dependent manner has been widely reported in this area, even among adults under 25 years old (Khan et al. 2011; Talukder et al. 2016; Scheelbeek et al. 2016; Scheelbeek et al. 2017; Al Nahian et al. 2018; Chakraborty et al. 2019). Pregnant women who drink water with high concentrations of sodium have shown not only high rates of hypertension, but also increased rates of preeclampsia and adverse pregnancy outcomes (Khan et al. 2011; Scheelbeek et al. 2016). Rates of hypertension in pregnancy and preeclampsia are higher in these regions than in inland regions, and also seasonally higher during dry seasons when rainwater, which has much lower sodium content than ground or surface water, is less available (Khan et al. 2011). Besides hypertension, subjects consuming high saline water have also reported more frequent hospital visits for cardiovascular disease, diarrhea, and abdominal pain (Chakraborty et al. 2019).

Significantly, there is little awareness of the dangers of high sodium consumption in the region, particularly amongst those whose water contains the most sodium (Rasheed et al. 2016; Chakraborty et al. 2019; Al Nahian et al. 2018). Many local residents believe salt to be a health-supporting substance (Rasheed et al. 2016). Residents who habitually drink water with exceptionally high levels of sodium consider their water "sweet" and are also likely to add more salt to their food (Al Nahian et al. 2018).

This case study demonstrates the extreme vulnerability of low-lying coastal populations to adverse health outcomes from drinking water associated with climate change. It also shows how climate-related changes to drinking water quality are not yet addressed by national and international regulations or guidelines. The regulatory community has not yet recognized that people are drinking water with levels of chemical substances high enough to cause adverse effects, or that their assumptions about water taste being sufficient to protect consumers from unsafe water are unsupported by community practices.

4 Conclusions

Climate change is expected to bring an increase in floods, extreme precipitation events, drought, higher temperatures, and sea level rise. As described in this chapter, each of these meteorological and climatic effects will increase the likelihood of different types of waterborne disease. Outbreaks of waterborne diseases linked to climate change have already occurred throughout the world, from small island nations to very large nations. Both resource-limited and high-income nations have been affected. However, low-lying nations with limited resources will be exceptionally vulnerable to waterborne outbreaks due to climate change. The poorest segments of populations, the very young, pregnant women, and young mothers are likely to be disproportionately affected, as exemplified in these case studies. Protecting public health from waterborne outbreaks in the face of climate change will require extraordinary efforts and cooperation by governments, non-governmental organizations, and citizens alike.

5 Discussion Questions

1. Drawing from the examples presented here or from other cases, discuss ways in which ecological or environmental factors interact with climate change to promote outbreaks of disease.
2. List the populations who are most vulnerable to climate change-related outbreaks of waterborne disease. For each population, describe the mechanisms of their special vulnerability.
3. Explain why a boil water advisory would not have protected the residents of Toledo, Ohio, from microcystins during the cyanobacterial bloom in 2014.
4. Describe the challenges for governments and regulators when addressing drinking water safety with respect to climate change. In light of the case studies presented here, what actions could governments take to prevent or lessen the extent of such outbreaks of waterborne disease?

Competing Interests This study was supported by donations to Better Life Laboratories, Inc. by private individuals. The donors had no role in study design, data collection and analysis, decision to publish, or preparation of the manuscript.

References

Al Nahian M, Ahmed A, Lázár AN, Hutton CW, Salehin M, Streatfield PK (2018) Drinking water salinity associated with health crisis in coastal Bangladesh. Elem Sci Anth 6:2. https://doi.org/10.1525/elementa.143

Alcayna T, Fletcher I, Gibb R, Tremblay L, Funk S, Rao B, Lowe R (2022) Climate-sensitive disease outbreaks in the aftermath of extreme climatic events: a scoping review. OneEarth 5. https://doi.org/10.1016/j.oneear.2022.03.011

Antonijevic E, Mandinic Z, Curcic M, Djukic-Cosic D, Milicevic N, Ivanovic M, Carevic M, Antonijevic B (2015) "Borderline" fluorotic region in Servia: correlations among fluoride in drinking water, biomarkers of exposure and dental fluorosis in schoolchildren. Environ Geochem Health 38:885–896. https://doi.org/10.1007/s10653-015-9769-x

Auld H, MacIver D, Klaassen J (2010) Heavy rainfall and waterborne disease outbreaks: the Walkerton example. J Toxicol Environ Health A 67(20–22):1879–1887. https://doi.org/10.1080/15287390490493475

Benedict KM, Reses H, Vigar M, Roth DM, Roberts VA, Mattioli M, Cooley LA, Hilborn ED, Wade TJ, Fullerton KE, Yoder JS, Hill VR (2017) Surveillance for waterborne disease outbreaks associated with drinking water—United States, 2013-2014. MMWR Morb Mortal Wkly Rep 66:1216–1221. https://doi.org/10.15585/mmwr.mm6644a3

Bouvy M, Falcão D, Marinho M, Pagano M, Moura A (2000) Occurrence of *Cylindrospermopsis* (cyanobacteria) in 39 Brazilian tropical reservoirs during the 1998 drought. Aquat Microb Ecol 23:13–27

Bullerjahn GS, McKay RM, Davis TW, Baker DB, Boyer GL, D'Anglada LVD, Doucette GJ, Ho JC, Irwin EG, Kling CL, Kudela RM, Kurmayer R, Michalak AM, Ortiz JD, Otten TG, Paerl HW, Qin B, Sohngen BL, Stumpf RP, Visser PM, Wilhelm SW (2016) Global solutions to regional problems: collecting global expertise to address the problem of harmful cyanobacterial blooms. A Lake Erie case study. Harmful Algae 54:223–238. https://doi.org/10.1016/j.hal.2016.01.003

Canadian Council of Ministers of the Environment (CCME) (2002) From source to tap: the multi-barrier approach to safe drinking water. CCME, Ottawa, Canada. https://www.canada.ca/content/dam/hc-sc/migration/hc-sc/ewh-semt/alt_formats/hecs-sesc/pdf/water-eau/tap-source-robinet/tap-source-robinet-eng.pdf. Accessed 21 May 2024

Chakraborty R, Khan KN, Dibaba DT, Khan MA, Ahmed A, Islam MZ (2019) Health implications of drinking water salinity in coastal areas of Bangladesh. Int J Environ Res Pub Health 16:3746. https://doi.org/10.3390/ijerph1619374

Cissé G (2019) Food-borne and water-borne diseases under climate change in low- and middle-income countries: further efforts needed for reducing environmental health exposure risks. Acta Trop 194:181–188. https://doi.org/10.1016/j.actatropica.2019.03.012

City of Toledo. Department of Public Utilities (2014) Microcystin event preliminary summary. https://www.toledoblade.com/attachment/2014/08/04/72-page-preliminary-study-from-the-City-of-Toledo-on-water-crisis.pdf. Accessed 24 June 2024

Del Giudice D, Fang S, Scavia D, Davis TW, Evans MA, Obenour DR (2021) Elucidating controls on cyanobacteria bloom timing and intensity via Bayesian mechanistic modeling. Sci Total Environ 755:142487. https://doi.org/10.1016/j.scitotenv.2020.142487

Department of Public Health Engineering, Bangladesh (DPHE, B) (2019) Water quality parameters. DPHE, B, Dhaka, Bangladesh. http://dphe.gov.bd/site/page/15fa0d7b-11f1-45c0-a684-10a543376873/Water-Quality-Parameters-. Accessed 4 June 2021

Effler P, Isaäcson M, Arntzen L, Heenan R, Canter P, Barrett T, Lee L, Mambo C, Levine W, Zaidi A, Griffin PM (2001) Factors contributing to the emergence of *Escherichia coli* O157 in Africa. Emerg Infect Dis 7(5):812–819. https://doi.org/10.3201/eid0705.017507

Fitzsimmons EG. Tap water ban for Toledo residents (2014) New York Times. https://www.nytimes.com/2014/08/04/us/toledo-faces-second-day-of-water-ban.html. Accessed 24 June 2024

Fox KR, Lytle DA (1996) Milwaukee's crypto outbreak: investigation and recommendations. J Am Water Works Assoc 88(9):87–94. https://doi.org/10.1002/j.1551-8833.1996.tb06615.x

Frisbie SH, Mitchell EJ, Molla AR (2024) Sea level rise from climate change is expected to increase the release of arsenic into Bangladesh's drinking well water by reduction and by the salt effect. PLoS ONE 19(1):e0295172. https://doi.org/10.1371/journal.pone.0295172

Gersony L Lake Erie's failed algae strategy hurts poor communities the most (2022) Michigan public. https://www.michiganpublic.org/environment-climate-change/2022-09-20/lake-eries-failed-algae-strategy-hurts-poor-communities-the-most. Accessed 24 June 2024

Greenwood KP, Reid SA (2020) Clustering of cryptosporidiosis in Queensland, Australia, is not defined temporally or by spatial diversity. Int J Parasitol 50(3):209–216. https://doi.org/10.1016/j.ijpara.2019.11.010

Hashizume M, Armstrong B, Wagatsuma Y, Faruque ASG, Hayashi T, Sack DA (2008) Rotavirus infections and climate variability in Dhaka, Bangladesh: a time-series analysis. Epidemiol Infect 136:1281–1289. https://doi.org/10.1017/S0950268807009776

Hashizume M, Faruque ASG, Terae T, Yunus M, Streatfield K, Yamamoto T, Moji K (2011) The Indian Ocean dipole and cholera incidence in Bangladesh: a time-series analysis. Environ Health Persp 119(2):239–244. https://doi.org/10.1289/ehp.1002302

Hrudy SE, Payment P, Huck PM, Gillham RW, Hrudey EJ (2003) A fatal waterborne disease epidemic in Walkerton, Ontario: comparison with other waterborne outbreaks in the developed world. Water Sci Tech 43(3):7–14. https://doi.org/10.2166/wst.2003.0146

Hu W, Mengersen K, Tong S (2010) Risk factor analysis and spatiotemporal CART model of cryptosporidiosis in Queensland, Australia. BMC Infect Dis 10:311. https://doi.org/10.1186/1471-2334-10-311

Intergovernmental Panel on Climate Change (IPCC) (2022) Climate change 2022: impacts, adaptation and vulnerability. IPCC, Cambridge University Press, Cambridge, United Kingdom. https://report.ipcc.ch/ar6/wg2/IPCC_AR6_WGII_FullReport.pdf. Accessed 23 May 2024

Jones FK, Ko AI, Becha C, Joshua C, Musto J, Thomas S, Ronsse A, Kirkwood CD, Sio A, Aumua A, Nilles EJ (2016) increased rotavirus prevalence in diarrheal outbreak precipitated by localized flooding, Solomon Islands, 2014. Emerg Infect Dis 22(5):875–879. https://doi.org/10.3201/eid2205.151743

Joseph C, Hamilton G, O'Connor M, Nicholas S, Marshall R, Stanwell-Smith R, Sims R, Ndawula E, Casemore U, Gallagher P, Harnett P (1991) Cryptosporidiosis in the Isle of Thanet; an outbreak associated with local drinking water. Epidemiol Infect 107:509–519. https://doi.org/10.1017/s0950268800049207

Khan AE, Ireson A, Kovats S, Mojumder SK, Khusru A, Rahman A, Vineis P (2011) Drinking water salinity and maternal health in coastal Bangladesh: implications of climate change. Env Health Persp 119(9):1328–1332. https://doi.org/10.1289/ehp.1002804

Khan AE, Scheelbeek PFD, Shilpi AB, Chan Q, Mojumder SK, Rahman A, Haines A, Vineis P (2014) Salinity in drinking water and the risk of (pre)eclampsia and gestational hypertension in coastal Bangladesh: a case control study. PLoSONE 9(9):e108715. https://doi.org/10.1371/journal.pone.0108715

Lad A, Breidenbach JD, Su RC, Murray J, Kuang R, Mascarenhas A, Najjar J, Patel S, Hegde P, Youssef M, Breuler J, Kleinhenz AL, Ault AP, Westrick JA, Modyanov NN, Kennedy DJ, Haller ST (2022) As we drink and breath: adverse health effects of microcystins and other harmful algal bloom toxins in the liver, gut, lungs and beyond. Life 12:418. https://doi.org/10.3390/life12030418

Lal A, Cornish LM, Fearnley E, Glass K, Kirk M (2015) Cryptosporidiosis: a disease of tropical and remote areas in Australia. PLoS Neg Trop Dis 9(9):e0004078. https://doi.org/10.1371/journal.pntd.0004078

Luque Fernádez MÁ, Bauernfeind A, Díaz Jiménez J, Linares Gil C, El Omeiri N, Herrera Guibert D (2008) Influence of temperature and rainfall on the evolution of cholera epidemics in Lusaka, Zambia,2003—2006: Analysis of a time series. Trans R Soc Trop Med Hyg 103(2):137–143. https://doi.org/10.1016/j.trstmh.2008.07.017

Mero S, Lääveri T, Ursing J, Rombo L, Kofoed P-E, Kantele A (2023) Seasonal variation of diarrhoeal pathogens among Guinea-Bissauan children under 5 years of age. PLoS Negl Trop Dis 17(3):e0011179. https://doi.org/10.1371/journal.pntd.0011179

Mitchell EJ, Frisbie SH (2023) A comprehensive survey and analysis of international drinking water regulations for inorganic chemicals with comparisons to the World Health Organization's drinking-water guidelines. PLoS ONE 18(11):e0287937. https://doi.org/10.1371/journal.pone.0287937

New Humanitarian (2007) Rising sea levels threaten agriculture. The New Humanitarian. https://www.thenewhumanitarian.org/report/75094/bangladesh-rising-sea-levels-threaten-agriculture. Accessed 30 June 2024

Peterson A (2022) Toledo's water issue is an affordability issue, too. Midstory. https://storymaps.arcgis.com/stories/2624f25f060f4248b02a0fc7ae0174d9/print. Accessed 24 June 2024

Rasheed S, Siddique AK, Sharmin T, Hasan AMR, Hanifi SMA, Iqbal M, Bhuiya A (2016) Salt intake and health risk in climate change vulnerable coastal Bangladesh: what role do beliefs and practices play? PLoS ONE 11(4):e0152783. https://doi.org/10.1371/journal.pone.0152783

Rose AK, Fabbro L, Kinnear S (2018) Cyanobacteria breakthrough: effects of *Limnothrix redekei* contamination in an artificial bank filtration on a regional water supply. Harmful Algae 76:1–10. https://doi.org/10.1016/j.hal.2018.04.010

Rose JB, Daeschner S, Easterling DR, Curriero FC, Subhash L, Patz JA (2000) Climate and waterborne disease outbreaks. J Am Water Works Assoc 92(9):77–87. https://doi.org/10.1002/j.1551-8833.2000.tb09006.x

Scheelbeek PFD, Chowdhury MAH, Haines A, Alam DS, Hoque MA, Butler AP, Khan AE, Mojumder SK, Blangiardo MAG, Elliott P, Vineis P (2017) Drinking water salinity and raised blood pressure: evidence from a cohort study in coastal Bangladesh. Env Health Persp 125(5):057007. https://doi.org/10.1289/EHP659

Scheelbeek PFD, Khan AE, Mojumder S, Elliott P, Vineis P (2016) Drinking water sodium and elevated blood pressure of health pregnant women in salinity-affected coastal areas. Hypertension 68(2):464–470. https://doi.org/10.1161/HYPERTENSIONAHA.116.07743

Sharma S, Patel P, Kulkarni SV, Deoshatwar A, Yada R, Tanwar S, Manohar K, Dolla JR, Jain SK, Singh SK, Dikid T (2023) An outbreak of acute neurological illness associated with drinking water source following a cyclone in Eluru, west Godavari district, Andhra Pradesh, India, December 2020. Clin Epidemiol Glob Health 20:101261. https://doi.org/10.1016/j.cegh.2023.101261

Steffen MM, Davis TW, McKay RML, Bullerjahn GS, Krausfeldt LE, Stough JMA, Neitzey ML, Gilbert NE, Boyer GL, Johengen TH, Gossiaux DC, Burtner AM, Palladino D, Rowe MD, Dick GJ, Meyer KA, Levy S, Boone BE, Stumpf RP, Wynne TT, Zimba PV, Gutierrez D, Wilhelm SW (2017) Ecophysiological examination of the Lake Erie *Microcystis* bloom in 2014: linkages between biology and the water supply shutdown of Toledo, OH. Environ Sci Technol 51(12):6745–6755. https://doi.org/10.1021/acs.est.7b00856

Talukder MRR, Rutherford S, Phung D, Islam MZ, Chu C (2016) The effect of drinking water salinity on blood pressure in young adults of coastal Bangladesh. Env Poll 214:248–254. https://doi.org/10.1016/j.envpol.2016.03.074

Van de Vuurst P, Escobar LE (2023) Climate change and infectious disease: a review of evidence and research trends. Infect Dis Poverty 12:51. https://doi.org/10.1186/s40249-023-01102-2

Weinberg HS, Krasner SW (2022) The occurrence of disinfection by-products (DBPs) of health concern in drinking water: results of a national DBP occurrence study. United States Environmental Protection Agency (U.S. EPA), Athens, Georgia. https://www.researchgate.net/profile/Susan-Richardson-5/publication/266336355_The_Occurrence_of_Disinfection_By-Products_DBPs_of_Health_Concern_in_Drinking_Water_Results_of_a_Nationwide_DBP_Occurrence_Study/links/5efe1c8292851c52d61145c7/The-Occurrence-of-Disinfection-By-Products-DBPs-of-Health-Concern-in-Drinking-Water-Results-of-a-Nationwide-DBP-Occurrence-Study.pdf. Accessed 7 June 2024

Wines M (2014) Behind Toledo's water crisis, a long-troubled Lake Erie. New York Times. https://www.nytimes.com/2014/08/05/us/lifting-ban-toledo-says-its-water-is-safe-to-drink-again.html. Accessed 24 Aug 2024

Woodall CJ (2009) Waterborne diseases—what are the primary killers? Desalination 243(1–3):616–621. https://doi.org/10.1016/j.desal.2008.05.110

World Health Organization (WHO) (2012) Guideline: sodium intake for adults and children. WHO, Geneva, Switzerland. https://www.who.int/publications/i/item/9789241504836. Accessed 30 June 2024

World Health Organization (WHO) (2022) Guidelines for drinking-water quality, 4th edn incorporating the first and second addenda. WHO, Geneva, Switzerland. https://apps.who.int/iris/rest/bitstreams/1414381/retrieve. Accessed 30 June 2024

World Health Organization (WHO) (2003) Sodium in drinking-water: background document for development of WHO guidelines for drinking-water quality. WHO, Geneva, Switzerland. https://www.who.int/water_sanitation_health/dwq/chemicals/sodium.pdf. Accessed 30 June 2024

Chapter 20
Harnessing Artificial Intelligence to Transform Drinking Water Safety

Holger Janes and Boris Faybishenko

1 Learning Objectives

1. Identify the goal of developing and using intelligent AI systems in the water industry.
2. Identify key AI technologies and data-driven approaches to enhance drinking water safety, including predictive modeling, real-time monitoring, and contamination risk assessment.
3. Compare traditional and AI-driven methodologies in water safety management, assessing their effectiveness, scalability, and potential for mitigating chemical and microbial threats.
4. Critically assess AI-driven water safety solutions' ethical, regulatory, and policy considerations, exploring pathways for interdisciplinary collaboration and equitable implementation, and provide examples of using the AI-fuzzy systems modeling of arsenic contaminated water.

2 Introduction: The Universal Quest for Safe Water

Access to safe drinking water is a universal human right, yet millions worldwide face daily challenges in securing this basic necessity. Waterborne infections cause devastation in developing and undeveloped countries. The World Health Organization (WHO) assessed that 3.6 million people worldwide die from waterborne illnesses,

H. Janes (✉)
Better Life Laboratories, Inc., Tallinn, Estonia

B. Faybishenko
Lawrence Berkeley National Laboratory, Berkeley, CA, USA
e-mail: bafaybishenko@lbl.gov

A. R. Molla et al. (eds.), *Climate Change, Drinking Water Security, and Public Health*, https://doi.org/10.1007/978-3-032-12074-8_20

among which around 2.2 million children. Waterborne diseases are brought on by drinking water infected with pathogens like dangerous bacteria, viruses, protozoa, etc. (WHO 2023).

One of the most toxic chemicals in drinking water is arsenic that ranks twentieth among the elements in abundance in the earth's crust (National Research Council 1977). The abundance of arsenic in the earth's continental crust is generally given as 1.5 to 2 micrograms per kilogram (mg/kg, ppm). Thus, it is relatively scarce. Nevertheless, it is a major constituent of no fewer than 245 mineral species. Inorganic arsenic is naturally present at high levels in the groundwater of several countries, including Argentina, Bangladesh, Cambodia, Chile, China, India, Mexico, Pakistan, the United States of America (U.S.), and Viet Nam. The United States Geological Survey (2017) described the distribution of arsenic concentrations from over 2000 domestic wells in Connecticut and identified the areas of high and low potential exposure (Fig. 20.1).

In rural Bangladesh, villagers who once celebrated the advent of tubewells (i.e., boreholes) to combat deadly waterborne diseases now face a silent crisis: chronic exposure to arsenic-contaminated water. Meanwhile, in the United States, residents of underserved communities struggle with aging infrastructure and chemical contamination, such as lead (Pb) and perfluoroalkyl and polyfluoroalkyl substances (PFAS), highlighting persistent inequities in public health protections (Berthold et al. 2023; Frank et al. 2019). Despite differing circumstances, these examples underscore a shared truth—the quest for safe drinking water transcends geography and socio-economic boundaries.

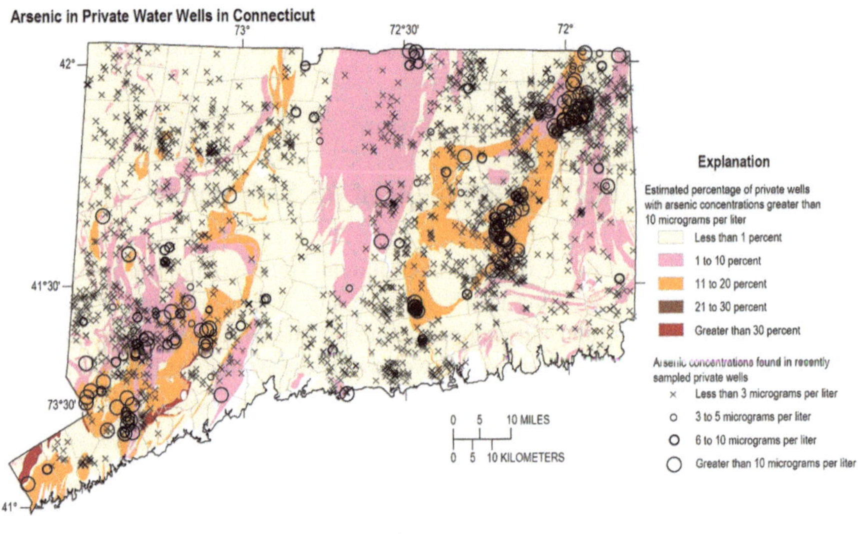

Fig. 20.1 Estimated percentages (< 1%, 1% to 10%, 11% to 20%, 21% to 30%, >30%) of private wells with arsenic concentrations greater than 10 micrograms per liter (μg/L). (Reproduced from The United States Geological Survey 2017)

While traditional solutions have addressed specific aspects of water safety, they have often failed to account for the dynamic and interconnected nature of the problem. Climate change introduces new risks, such as saltwater intrusion and mobilization of toxic elements, while scientific advancements reveal synergistic health effects of co-exposures to multiple contaminants (Frisbie et al. 2024). Addressing these complex challenges requires a paradigm shift toward innovative, data-driven approaches integrating environmental, health, and socio-economic factors. To successfully address this worldwide threat, optimal solutions built on cutting-edge deep learning and machine learning developments can be deployed (Joy et al. 2023). The goal of developing the field of artificial intelligence (AI) is to create autonomous, intelligent systems to assess and manage drinking water optimal solutions.

This chapter presents a conceptual understanding of how AI can revolutionize our approach to drinking water safety by offering scalable and equitable solutions. With its ability to process vast amounts of data, predict risks, and optimize interventions, AI emerges as a vital tool to address water safety's complex and evolving challenges of the assessment and prediction of drinking water quality. Drawing on examples from Bangladesh, the United States, and other regions, we analyze the evolution of drinking water issues over time, the tradeoffs of various interventions, and the transformative potential of AI to ensure safe water access for all.

We begin with a historical overview, tracing the shifts in water safety concerns and the emergence of new threats. From there, we delve into three critical global challenges and explore promising AI-driven solutions to tackle them. Through this narrative, we aim to show that with the right tools and strategies, we can effectively address today's challenges while anticipating and mitigating those of the future.

3 Journey through Time: From Manually Pumped Wells to Smart Technical Systems

3.1 Historical Context

For much of human history, securing clean water was a localized challenge, primarily focused on avoiding immediate threats like visible contamination and microbial diseases. In the nineteenth and early twentieth centuries, cholera and typhoid outbreaks were rampant, fueled by inadequate sanitation and reliance on untreated surface water sources. Advances in public health, including the chlorination of municipal water supplies and the construction of centralized treatment systems, marked a turning point in mitigating these microbial threats.

However, as surface water contamination persisted, many regions turned to groundwater as a safer alternative. In Bangladesh, the installation of millions of tubewells in the latter half of the twentieth century was a landmark intervention (Frisbie et al. 2002). Tubewells dramatically reduced waterborne diseases like cholera, saving countless lives. Yet, this progress came at a cost. By the 1990s, it became evident that groundwater in many areas was contaminated with arsenic, a naturally occurring toxic element. The reliance on these wells led to widespread chronic exposure, resulting in serious health impacts such as skin lesions and cancers.

In the United States, industrialization brought its own set of challenges. While municipal systems improved water safety for urban populations, industrial waste introduced chemical contaminants like lead, mercury, and, more recently, PFAS (Burger and Gochfeld 2021). These contaminants, often invisible and long-lasting, highlighted the limitations of existing regulatory frameworks and detection methods.

3.2 Present Challenges

Today, water safety challenges are more interconnected and complex than ever before. Climate change intensifies these issues by altering water availability, increasing the frequency of extreme weather events, and promoting the mobilization of toxic elements in aquifers (Bamal et al. 2024; Jasechko et al. 2024). In regions like Bangladesh, rising sea levels introduce saltwater intrusion into freshwater aquifers, further compounding arsenic contamination (Faybishenko and Frisbie 2025; Friedman et al. 2024; Frisbie et al. 2024). Meanwhile, in the United States, aging infrastructure and disparities in resource allocation leave vulnerable communities disproportionately exposed to contaminants like lead and manganese (Mn; Jamei et al. 2022; Levin et al. 2024).

Compounding these issues is the phenomenon of co-exposure, where individuals are simultaneously exposed to multiple contaminants. Research has shown that arsenic and manganese, when present together, can have synergistic neurotoxic effects, particularly in children (Heng et al. 2022). Traditional monitoring systems are ill-equipped to address these layered risks, creating an urgent need for innovative solutions.

3.3 Future Threats

Looking ahead, threats to drinking water safety will only grow. Emerging contaminants, such as microplastics and pharmaceutical residues, present new challenges that current technologies are unprepared to address (Nathanael et al. 2024). Additionally, the socio-economic impacts of climate change—including population displacement and increased pressure on water resources—will strain already fragile water systems.

AI is changing the water industry (Autodesk 2022), because it offers a pathway to tackle future threats. Predictive modeling can forecast contamination risks under various climate scenarios, helping communities prepare and adapt (Huntingford et al. 2019). Machine learning algorithms can analyze complex datasets to identify emerging patterns, such as new contaminants or shifting water quality trends. These tools will be essential in building resilient water systems that can withstand the pressures of a rapidly changing world.

3.4 AI's Role in the Journey

Throughout this journey from historical to future challenges, AI has the potential to redefine how we approach water safety. In the past, limited data and reactive policies hindered progress. Today, AI enables real-time monitoring and predictive analytics, empowering stakeholders to make informed decisions (Joha et al. 2024). For example, Access to safe drinking water, once a localized and manageable challenge, has evolved into a complex global issue, shaped by the interplay of natural, anthropogenic, and climatic factors. From the arsenic-driven aquifers of Bangladesh to the lead-contaminated pipes of urban America, the burden of unsafe water affects millions. Addressing this burden demands more than incremental progress but transformative innovation—an endeavor for which AI is uniquely positioned. With its unparalleled capacity to analyze vast and multidimensional datasets, predict risks, and optimize interventions (Himeur et al. 2023), AI offers the opportunity to redefine how humanity safeguards one of its most fundamental resources: drinking water.

As we move toward a future where water safety challenges grow increasingly dynamic and interconnected, AI emerges as a transformative tool, bridging the gaps between science, policy, and human well-being (Bibri et al. 2024). This chapter delves into AI's potential to address global water safety challenges and highlights the global applicability of AI-driven methodological solutions for managing chemical contamination in drinking water. Through predictive modeling, advanced detection techniques, and data integration, AI can anticipate emerging threats, optimize resource allocation, and build resilient water systems (Dwivedi et al. 2021). Lai et al. (2022) used the AI models to analyze historical data, rainfall patterns, and water demand to optimize reservoir operations and maximize water storage and release timing. Guo et al. (2023) applied AI algorithms to assess water demand, availability, and environmental constraints to optimize water allocation schemes, minimizing conflicts and maximizing efficiency. Adikari et al. (2021) performed AI-based modeling to analyze rainfall patterns, soil moisture levels, and climate data to predict and mitigate drought events, which is critical for early warning systems and water management. Krishnan et al. (2022) showed that AI-driven systems can integrate data from multiple sources, including weather forecasts, river flows, and water usage. AI models can be used to assess the environmental impact of water resource management activities, considering water quality, ecosystem health, and habitat preservation factors (Xiang et al. 2021). AI-powered sensors and data analytics are essential for real-time monitoring of water parameters and automatic control of water systems, optimizing water allocation and usage in response to changing conditions (Chowdury et al. 2019).

4 Redefining Risk Assessment: AI-Powered Solutions for Arsenic Contamination in Groundwater

4.1 *Introduction: A Global Challenge*

Arsenic contamination in groundwater poses a severe threat to the health and safety of millions worldwide (Sanyal et al. 2020). A known example originates from Bangladesh, where nearly half the land area has groundwater arsenic levels exceeding the World Health Organization guideline of 10 micrograms per liter (μg/L). Prolonged exposure to arsenic has been linked to cancer, cardiovascular diseases, and cognitive impairments, creating a silent yet devastating public health emergency. However, this issue extends far beyond Bangladesh, with communities across South Asia, Africa, and even parts of North America facing similar threats from arsenic-laden water supplies.

Efforts to combat arsenic contamination have historically been hindered by significant challenges, particularly the limitations of early groundwater datasets. Detection methods often lacked the precision to measure ultra-low arsenic concentrations, frequently reporting values below 2 μg/L as "<2 μg/L," obscuring critical details about contamination levels. These gaps in data have impeded effective mitigation, leaving millions exposed to chronic health risks.

Today, AI offers an innovative pathway to overcome these obstacles. By reanalyzing historical data with advanced algorithms, AI not only fills the gaps in our understanding but also enables the prediction and prevention of future contamination events. This section explores how AI can revolutionize groundwater safety, transforming historical limitations into actionable solutions for a safer, more sustainable future.

4.2 *Reimagining Historical Data*

The arsenic crisis, as we know it, is partly yet deeply rooted in the limitations of historical data. Early surveys in Bangladesh, conducted with rudimentary tools, were foundational for understanding the extent of contamination. However, these datasets fell short of capturing arsenic levels with sufficient granularity. Imagine trying to navigate a vast landscape with a map that erases all but the largest landmarks—the finer details of the terrain remain hidden, leaving travelers vulnerable to unseen hazards. Similarly, the absence of precise data concealed contamination hotspots that silently endangered entire communities.

Here, AI enters as both a magnifying lens and a cartographer. By training algorithms on high-resolution datasets—such as those generated through inductively coupled plasma mass spectrometry (ICP-MS)—researchers can revisit and reinterpret historical records. These AI models use environmental covariates, like aquifer geology and pH levels, to infer arsenic concentrations with unprecedented accuracy.

Crucially, transfer learning enables these models to extend their insights to regions where data is sparse, adapting knowledge from well-studied areas to fill critical gaps.

The result is nothing short of transformative. By detecting hidden hotspots of contamination, AI empowers policymakers to direct interventions where they are most urgently needed. This newfound clarity transforms arsenic risk assessment from a reactive endeavor to a proactive strategy, marking the first step toward a safer future.

4.3 Navigating a Dynamic Landscape

Arsenic contamination is not static. It is a dynamic and evolving challenge shaped by driving forces such as climate change, land use patterns, and shifting hydrological conditions. In coastal Bangladesh, for example, rising sea levels drive saltwater intrusion into freshwater aquifers, altering the chemical balance and mobilizing arsenic previously locked away in sediment. These changes complicate efforts to contain contamination, as areas once considered safe become vulnerable over time.

To address this, AI provides a compass for navigating this uncertain terrain. AI-powered models can simulate how contamination patterns might evolve under various scenarios by integrating real-time environmental data with long-term climate projections. These simulations take into account factors like rainfall variability, aquifer salinity, and groundwater extraction rates, painting a vivid picture of future risks.

This forward-looking approach transforms the way we respond to contamination. Instead of waiting for crises to unfold, we can identify at-risk areas before contamination reaches critical levels. Communities can be equipped with early-warning systems, testing campaigns can target emerging hotspots, and infrastructure upgrades can be implemented preemptively. AI shifts the narrative from crisis management to crisis prevention, ensuring that the future does not need to mirror the past.

4.4 Enhancing Detection at the Frontlines

Resource limitations present a formidable barrier to effective intervention in many regions affected by arsenic contamination. Low-cost sensors, while widely available, often lack the sensitivity needed to detect arsenic at ultra-low concentrations. The challenge is akin to trying to detect a whisper in a crowded room—the tools at hand are simply not equipped for the task.

AI, however, has the power to amplify these whispers. By calibrating low-cost sensors against high-precision instruments, AI algorithms can correct for inaccuracies and enhance detection capabilities. For instance, dynamic signal amplification techniques enable these sensors to identify trace amounts of arsenic even in challenging environmental conditions. This innovation brings high-level precision to settings where resources are scarce, democratizing access to reliable water quality monitoring.

The impact of this technology is profound. Portable, AI-enhanced sensors empower local communities to monitor their water in real-time, reducing reliance on centralized laboratories and accelerating response times. By placing the detection tools directly in the hands of those most affected, AI fosters a new era of decentralized water safety.

4.5 A Holistic Vision for Risk Assessment

Arsenic contamination is not just a chemical problem—it is a human one, deeply intertwined with socio-economic vulnerabilities and public health outcomes. Addressing this complexity requires a holistic approach that transcends traditional risk assessment.

AI excels in synthesizing diverse datasets, weaving together information on arsenic concentrations, health records, socio-economic indices, and hydrological models. This integrative capability reveals patterns that might otherwise remain invisible. For example, co-exposure to arsenic and manganese has been shown to intensify neurotoxic effects, highlighting the need for multi-dimensional risk assessments. Similarly, data on socio-economic factors can illuminate disparities in exposure and access to mitigation resources, guiding equitable policy reforms.

By adopting this holistic perspective, AI enhances our understanding of arsenic risks and provides a roadmap for addressing them in a way that prioritizes justice and sustainability. The intersection of data and humanity becomes the foundation for solutions that leave no one behind.

4.6 Conclusion: Toward a Safer Future

The story of arsenic contamination in groundwater is one of complexity and resilience. It is a challenge that has persisted for decades, defying simple solutions and demanding innovation at every turn. Today, AI offers a new chapter in this narrative. By reinterpreting historical data, predicting future risks, enhancing detection technologies, and fostering holistic assessments, AI redefines what is possible in the realm of water safety.

This is not merely a story of technology; it is a story of empowerment. AI equips us with the tools to see more clearly, act more decisively, and build a future where access to safe drinking water is a universal right, not a privilege. As we confront the evolving landscape of arsenic contamination, let this be our guiding principle: the solutions we create today must illuminate the path to a safer, more equitable tomorrow.

5 Understanding and Addressing Co-Exposures: Leveraging AI to Navigate Complexity

5.1 Introduction: The Challenge of Co-Exposures

In the intricate web of drinking water safety, co-exposures—the simultaneous exposure to multiple contaminants—emerge as an urgent frontier (WHO 2017). Unlike the relatively straightforward risks of single-contaminant exposure, co-exposures introduce a labyrinth of interactions that are non-linear, context-dependent, and difficult to predict. These interactions can either amplify harm or, in rare instances, mitigate it. For example, co-exposure to manganese and lead amplifies neurotoxic effects, particularly in children, leading to significant cognitive impairments. Conversely, magnesium (Mg) in groundwater can possibly offset some cardiovascular risks of arsenic, but only under specific environmental conditions (Bhattacharya et al. 2002; Shlezinger et al. 2018).

These complex dynamics demand a new generation of tools that go beyond traditional risk assessment methodologies. AI emerges as a transformative force, capable of integrating vast and diverse datasets, modeling intricate chemical interactions, and forecasting risks under dynamic conditions. This section explores how AI can unravel the complexity of co-exposures, offering innovative pathways to ensure water safety.

5.2 Unraveling the Double-Edged Nature of Co-Exposures

The dual nature of co-exposures—where contaminants can act synergistically, antagonistically, or in entirely novel ways—requires a nuanced understanding of their interplay. Consider three pivotal scenarios:

5.2.1 Compounding Toxicities

In Bangladesh, overlapping contamination of manganese, lead, and arsenic in shallow aquifers compounds neurological and cognitive risks for children. Similarly, in parts of India, the combined exposure to arsenic and fluoride worsens skeletal fluorosis, intensifying bone health issues. Such examples highlight how contaminants can amplify harm when co-occurring.

5.2.2 Balancing Effects

On the other hand, magnesium in groundwater can mitigate arsenic-induced cardiovascular toxicity by supporting vascular health and reducing oxidative stress. However, this protective effect depends on localized factors such as water pH, redox potential, and Mg concentration, emphasizing the need for context-specific analysis.

5.2.3 Emerging Challenges

The rise of contaminants like PFAS, often termed "forever chemicals," further complicates co-exposure assessments. When combined with nitrates, as seen in agricultural regions, PFAS can disrupt endocrine systems in ways that remain poorly understood. These emergent risks underscore the urgency for advanced predictive models.

Traditional approaches to risk assessment struggle to capture these complexities, particularly when interactions defy linear models. AI, however, thrives on complexity, offering the adaptability and precision needed to untangle the web of co-exposures which will be outlined step-by-step in sub-sections 5.3 to 5.6.

5.3 Mapping Co-Exposure Hotspots

5.3.1 The Challenge

Identifying regions where multiple contaminants co-occur at hazardous levels requires the integration of geospatial, chemical, and socio-economic data—a daunting task for conventional methods.

5.3.2 The AI Solution

Advanced clustering algorithms, such as k-means and DBSCAN, paired with geospatial analysis, process vast datasets to pinpoint co-exposure hotspots. In Bangladesh, for instance, AI could map areas where manganese (Mn), lead, and arsenic (As) concentrations exceed toxic thresholds while overlaying demographic data to identify vulnerable populations. Similarly, in agricultural regions of the United States, AI can highlight intersections of PFAS and nitrate contamination, offering actionable insights for remediation.

5.3.3 The Impact

High-resolution hotspot maps enable policymakers to prioritize interventions, such as replacing contaminated wells, installing water treatment systems, or launching targeted community health monitoring. These insights ensure that resources are directed where they are most needed, maximizing public health benefits.

5.4 Modeling Biogeochemical Interactions

5.4.1 Modeling Approach and Challenges

Hu et al. (2023) summarized four common steps used in predictive modeling studies: (1) data sources, (2) feature engineering, (3) model training, and (4) presentation of model results, which are shown in Fig. 20.2.

The biogeochemical processes governing the dynamics and quality of drinking water are generally nonlinear. However, we usually do not have sufficient information (see the box "Data source" in Fig. 20.2) to develop reliable deterministic or probabilistic mathematical models to describe complex nonlinear biogeochemical processes. An alternative approach to solving this problem is the use of AI methods. One of the modern AI approaches for modeling and predicting water quality to control water pollution and develop an efficient operation of monitoring drinking water is applying fuzzy logic and fuzzy system modeling methods with the application of fuzzy logic based on AI software functions. Using fuzzy logic, AI is analogous to human thinking and answers conditional questions. The effective adaptive neuro-fuzzy inference system (ANFIS) algorithms are available in Matlab (https://www.mathworks.com/help/fuzzy/neuro-adaptive-learning-and-anfis.html), R (R library *frbs*) and Python's ANFIS package that is essentially a Python refactoring of the R code.

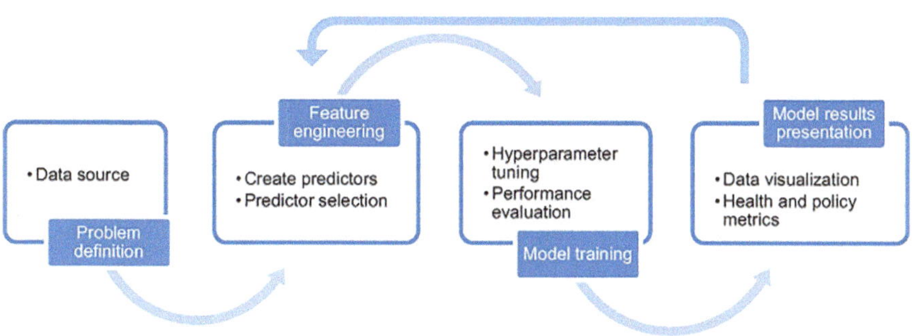

Fig. 20.2 Schematic of workflow for developing a statistical model for predicting drinking water quality. (Hu et al. 2023) (Reproduced from Hu et al. 2023, Figure 1. https://doi.org/10.1007/s40572-022-00389-x, licensed under the terms of the Creative Commons Attribution 4.0 International License (http://creativecommons.org/licenses/by/4.0/))

For example, Al-Adhaileh and Alsaade (2021) developed a method to predict the water quality index (WQI) using the ANFIS algorithm. They implemented the feed-forward neural network (FFNN) and K-nearest neighbors to classify water quality using the dataset with eight significant parameters. Based on the FFNN algorithm, the authors achieved the highest accuracy (100%) for water quality classification (WQC) and accurately predicted water quality index.

Faybishenko et al. (2024) employed fuzzy regression and fuzzy multivariate clustering techniques to analyze arsenic-polluted water samples originating from acid rock drainage in waste rock dumps. This research focuses on understanding the complex relationships between variables associated with arsenic contamination, such as water arsenic concentration, pH levels, and soil characteristics. Based on the experimental data from the paper by Frisbie et al. (2024), Chap. 11 of this book shows the application of fuzzy regression and fuzzy clustering of arsenic in Bangladesh.

One of the challenging problems for modeling of quality of drinking water is that there are no well-established methods of modeling how contaminants interact within biological systems. Environmental contexts are critical for developing purpose-oriented AI solutions.

5.4.2 The AI Solutions

Machine learning models, such as Random Forest, Gradient Boosting, and Bayesian Networks, simulate non-linear interactions between contaminants. These models predict health outcomes more precisely by incorporating variables like temperature, pH, and aquifer geochemistry. For example, AI can model the synergistic neurotoxicity of manganese and lead (Neal and Guilarte 2013), helping to establish safe exposure thresholds. Similarly, it can simulate magnesium's likely protective effects against arsenic-induced cardiovascular risks, guiding efforts to increase magnesium supplementation in affected regions (Bhattacharya et al. 2002; Shlezinger et al. 2018).

5.4.3 The Impact

By uncovering the conditions under which contaminants interact, AI facilitates more precise risk assessments and informs policies tailored to local environmental and socio-economic contexts. This empowers decision-makers with the data needed to address specific challenges effectively.

5.5 Real-Time Monitoring with AI-Enhanced Sensors

5.5.1 The Challenge

Traditional sensors often detect only single contaminants, limiting their capacity to assess co-exposure risks in real-time.

5.5.2 The AI Solution

AI-optimized Internet of Things (IoT) sensors revolutionize monitoring by simultaneously measuring multiple contaminants and dynamically adjusting for environmental variables like salinity and turbidity. In coastal Bangladesh, for example, these sensors could track manganese, lead, and arsenic concentrations in real-time, providing communities with immediate warnings of unsafe water. In urban areas, they can monitor lead and PFAS levels in municipal water supplies, enabling swift responses to contamination events.

5.5.3 The Impact

Real-time, AI-driven monitoring democratizes access to water quality data, reducing reliance on centralized laboratories and accelerating mitigation efforts, especially in remote or underserved areas. This ensures that communities could have online geoportal tools to safeguard their health proactively.

5.6 Predictive Modeling for Future Risks

5.6.1 The Challenge

Climate change and anthropogenic activities are altering contaminant dynamics, complicating efforts to predict future co-exposure risks.

5.6.2 The AI Solution

By integrating climate projections, land use data, and hydrological models, AI can simulate how co-exposure risks evolve under various scenarios. For instance, rising sea levels in coastal Bangladesh may increase arsenic and manganese mobilization through saltwater intrusion. AI models can predict these shifts and guide preemptive strategies. Similarly, in drought-prone regions of Sub-Saharan Africa, AI could forecast how declining water tables might concentrate contaminants, amplifying co-exposure risks.

5.6.3 The Impact

Predictive modeling equips policymakers with the foresight needed to anticipate and mitigate emerging risks. This ensures that communities remain resilient in the face of changing environmental conditions.

5.7 Conclusion: Charting a Path Forward

Co-exposures present a multifaceted challenge that demands equally multifaceted solutions. AI, with its unparalleled capacity to process complexity, offers a transformative framework for addressing these risks. From mapping hotspots to modeling interactions and predicting future dynamics, AI redefines what is possible in the realm of drinking water safety.

As we navigate the complexities of co-exposures, the integration of AI-driven methodologies provides more than just technical innovation—it offers hope. Hope for communities grappling with contaminated water, for policymakers seeking actionable insights, and for a future where safe drinking water is a reality for all. This is the promise of AI: to pave the path forward, even in the face of the most intricate challenges.

6 Toward Global Standards for Groundwater Safety

6.1 Introduction: Bridging the Gap in Groundwater Guidelines

Groundwater, the lifeline for billions worldwide, is now the primary source of drinking water for much of the global population. However, its assessment and management lack universally accepted standards. This gap leaves populations vulnerable to inequitable access to clean water, fragmented evaluation practices, and insufficient responses to contamination risks.

This section proposes establishing standardized guidelines to address this gap. Using a preliminary framework of key indicators presented in the accompanying table (see Table 20.1), we aim to highlight how clear, actionable benchmarks can enhance groundwater safety. By exploring its historical context, risks, and global dependence, we make the case for why establishing such guidelines is imperative for water security.

Table 20.1 Key indicators and mitigation strategies for groundwater safety

Indicator	Arsenic (As)	Fluoride (F^-)	Nitrate (NO_3^-)
Thresholds (WHO; Bangladesh; Mitchell and Frisbie 2023)	WHO: 10 µg/L (WHO 2022); Bangladesh: 50 µg/L (People's Republic of Bangladesh 2025)	WHO: 1500 µg/L (WHO 2022); Bangladesh: 1000 µg/L (People's Republic of Bangladesh 2025)	WHO: 50,000 µg/L (WHO 2022); Bangladesh: 10,000 µg/L (People's Republic of Bangladesh 2025)
Health risk score	High	Medium	High
Mitigation strategies	Testing to identify safer wells, switch to deep aquifers, point-of-use filters	Community defluoridation plants, groundwater blending	Precision agriculture, buffer zones, wetlands
Remote sensing capacity	No	Yes	Yes
Measurable indicators	Oxidation-reduction potential (ORP), pH, conductivity	Alkalinity, specific conductance	Land use maps
Rationale of the indicator	Identifies toxicity risk due to natural arsenic mobilization under reducing conditions	Detects high fluoride levels due to natural alkalinity and evaporation	Tracks agricultural runoff and leaching into groundwater systems
Description of impact	Chronic exposure causes cancers, cardiovascular diseases, and neurological impairments	Chronic exposure causes skeletal and dental fluorosis	Causes methemoglobinemia (blue baby syndrome) in infants and contributes to eutrophication risks
Methods of direct measure	Arsenic by inductively coupled plasma mass spectrometry (ICP-MS), graphite furnace atomic absorption spectroscopy (GFAAS), and spectrophotometry	Fluoride by ion selective electrode (ISE), ion chromatography (IC), and spectrophotometry	Nitrate by ion selective electrode (ISE), ion chromatography (IC), and spectrophotometry
Estimated cost of sample analysis (USD)	\$2000–\$5000	\$500–\$2000	\$1000–\$3000
Policy relevance	Critical for guiding water safety regulations and rural health interventions	Relevant for arid regions with endemic fluorosis issues	Supports nitrate reduction strategies and sustainable agriculture policies

6.2 *The Relatively New Reliance on Groundwater*

Historically, human populations relied on surface water—such as freshwater streams—for drinking water. Drinking directly from streams was common practice, with surface water undergoing natural purification processes as it flowed through the environment. However, the mid-1800s marked a paradigm shift when vast underground freshwater reservoirs were discovered and tapped. Groundwater bypasses the surface purification process, reaching users directly through pumping systems via metal pipes. While this shift improved access to water, it also introduced significant risks. Groundwater contains chemical components from the Earth's crust that are typically filtered or oxidized as it moves naturally toward surface layers. However, when groundwater is pumped directly from the ground, these natural purification processes are bypassed, allowing harmful chemicals to remain in the water. This shift has introduced significant risks, as groundwater interacts with subsurface geology, mobilizing toxic elements such as arsenic, manganese, and fluoride. Unlike surface water, which is visibly susceptible to pollution, groundwater contamination is invisible, complicating detection and management. These challenges highlight the pressing need for comprehensive, standardized approaches to ensure groundwater safety.

6.3 The Case for Clear Guidelines

Without clear and consistent guidelines, groundwater safety assessments often rely on fragmented local practices that vary widely in scope and rigor. The absence of international standards creates gaps in our ability to evaluate contamination comprehensively and address disparities in water quality across regions.

Table 20.1 highlights preliminary steps on how the new standards could be created with AI by selecting relevant indicators for assessment chosen for their prevalence and severe health impacts. Beyond identifying these risks, the table illustrates the need to standardize the measurable components, methodologies, and policy implications associated with groundwater safety. This structured approach ensures assessments are scientifically rigorous and also practical and actionable across diverse global contexts.

While this table is not yet exhaustive, it illustrates the necessity of developing a structured framework to guide groundwater assessments. Such a framework would enable decision-makers to standardize data collection, compare risks across regions, and prioritize interventions effectively. The logic of this proposition leverages the recent work where AI was used to develop a novel indicator database for assessing the ecological impacts of armed conflicts globally.

6.4 The Importance of Groundwater Safety Guidelines

Groundwater's critical role as the main source of drinking water for billions makes the development of safety guidelines a global imperative. Consider these key factors:

- **Global Dependence:** Groundwater accounts for nearly half of the world's drinking water and sustains essential agricultural and industrial activities.
- **Health Risks:** Contaminants like arsenic, fluoride, and nitrates pose severe health threats, from cancer and neurological impairments to skeletal fluorosis and methemoglobinemia.
- **Equity Challenges:** Vulnerable populations in low-income regions often bear the brunt of groundwater contamination, exacerbating existing socio-economic inequalities.
- **Dynamic Risks:** Climate change, land use alterations, and industrial activities continually reshape contamination profiles, requiring adaptive and forward-looking strategies.

Establishing clear groundwater guidelines ensures that global communities can proactively address these challenges. Such standards will harmonize efforts, improve health outcomes, and enable resource-strapped regions to respond more effectively to the dynamic risks of contamination and climate change.

7 Conclusion: The Path Forward for AI to Establish Drinking Water Safety

Access to safe drinking water is not just a necessity; it is a fundamental human right and a cornerstone of public health and global equity. Yet, the challenges surrounding water safety remain among our time's most complex and pressing issues. Across continents, from the arsenic-contaminated aquifers of Bangladesh (Frisbie et al. 2002; Jamei et al. 2022) to the aging infrastructure in the United States (Levin et al. 2024), the story is strikingly similar: communities struggling against invisible contaminants, systemic inequities, and environmental changes that exacerbate the risks (Jasechko et al. 2024). This shared plight calls for an urgent and transformative response that bridges the gaps between technology, policy, and humanity.

AI has emerged as a beacon of hope in this arena, a tool capable of addressing the multifaceted nature of water safety challenges. It offers the ability to reimagine how we collect, analyze, and act on data, turning reactive solutions into proactive strategies (Huntingford et al. 2019). Through predictive modeling, hotspot identification, and real-time monitoring, AI provides unparalleled insights that were previously out of reach (Bibri et al. 2024). By integrating vast and diverse datasets, AI not only helps uncover hidden risks but also fosters holistic approaches that account for the socio-economic, environmental, and health dimensions of water safety (Himeur et al. 2023). The potential is profound, but the journey forward requires careful navigation.

One of the most critical steps is the establishment of global guidelines for groundwater safety. Groundwater has become the primary source of drinking water for billions. Yet, its assessment and management remain inconsistent, fragmented, and often inadequate (Frisbie et al. 2024). The absence of universal standards perpetuates inequalities, leaving the most vulnerable populations disproportionately exposed to unsafe water (WHO 2017). The proposed table in this work offers a glimpse into what such guidelines might entail, identifying key indicators like arsenic, fluoride, and nitrate levels while emphasizing measurable components, mitigation strategies, and policy relevance. However, this is merely a starting point. The table must evolve through interdisciplinary collaboration, drawing on the expertise of hydrologists, AI researchers, public health officials, and policymakers to create a robust, adaptable framework that serves diverse global contexts.

While the promise of AI is undeniable, it is crucial to acknowledge its limitations. AI-driven solutions rely heavily on data availability and quality, yet many regions affected by water contamination lack the infrastructure for comprehensive data collection (Joha et al. 2024). In rural and resource-poor areas, gaps in digital literacy, access to technology, and financial resources could hinder the adoption of AI (Bamal et al. 2024). Moreover, the initial costs of deploying AI systems—including sensor networks, high-performance computing, and skilled personnel—may be prohibitive for many communities (Friedman et al. 2024). These barriers highlight the importance of equitable investments and capacity-building initiatives to ensure that AI's benefits are not confined to wealthy nations or urban centers. Bridging the digital divide is as critical as the technological innovations themselves.

Ethical considerations must also remain at the forefront. Transparency in AI algorithms, community involvement in decision-making, and safeguards against misuse are essential to building trust and ensuring that solutions align with the needs of those they are meant to serve (Dwivedi et al. 2021). AI is not a panacea; it is a tool that must be wielded with care, guided by principles of equity, sustainability, and inclusivity.

Despite these challenges, the potential rewards far outweigh the costs. Imagine a future where real-time water quality data is accessible to every community, enabling early warnings for contamination events and empowering local stakeholders to take immediate action (Levin et al. 2024). Picture a world where predictive models guide policymakers in allocating resources to the areas of greatest need, preventing crises before they escalate (Jasechko et al. 2024). Envision a global collaboration where data from rural Bangladesh informs strategies for drought-prone regions in Sub-Saharan Africa, and innovations from Silicon Valley adapt to the unique challenges of remote Himalayan villages. This is not a distant dream; it is a future within our grasp if we commit to the thoughtful integration of AI into water safety efforts.

Before applying AI to address water-related problems, baseline studies are essential to assess the implementation capabilities, return on investment, and impact of the intervention (Maroju et al. 2023; Maskrey et al. 2016). Policymakers and stakeholders must thoroughly examine social, economic, and cultural issues before implementing AI in the water sector. The application of AI can remarkably reduce water supply and sanitation systems costs and help ensure compliance with the quality of drinking and wastewater treatment.

The path forward will require more than technological advancements. It will demand bold leadership, interdisciplinary collaboration, and a shared commitment to equity and sustainability (Sanyal et al. 2020). International organizations, governments, private sector innovators, and grassroots communities must come together to chart this course. The solutions must be as dynamic and interconnected as the challenges they aim to address.

As we stand at this crossroads, the urgency to act has never been greater. The threats to water safety are growing—driven by climate change, industrialization, and population growth—but so too are the opportunities to create lasting change. AI offers a transformative lens through which we can address these challenges, but we must ensure that the technology serves humanity's best interests. Our choices today will shape the quality of life for generations, determining whether safe drinking water remains an aspiration or has become a universal reality.

This journey is not merely about safeguarding a resource but about upholding dignity, fostering resilience, and empowering communities worldwide. With AI as our ally, we have the tools to illuminate the path forward, turning the dream of safe drinking water for all into an enduring legacy of progress, equity, and hope.

8 Discussion Questions

1. **Bridging Science and Policy:** AI-driven solutions have the potential to revolutionize drinking water safety by providing predictive insights, improving monitoring systems, and informing policy decisions. However, regulatory frameworks often lag behind technological advancements, and AI adoption varies globally. What are the key barriers to integrating AI-based water safety solutions into national and international policy frameworks? Can you identify examples where AI-driven environmental policies have succeeded or failed? What interdisciplinary strategies—across science, governance, and industry—could accelerate the responsible implementation of AI in water safety?
2. **Ethics, Equity, and AI in Water Safety:** AI offers powerful tools to monitor and mitigate water contamination risks, but access to such technology remains uneven, favoring wealthier nations and urban centers. Who should be responsible for ensuring AI-driven water safety initiatives reach the most vulnerable populations—governments, NGOs, private companies, or international agencies? What are the practical trade-offs between cost, scalability, and ethical obligations? How can we prevent AI from becoming an elite-driven tool that widens the global water safety divide rather than closing it?
3. **The Future of Water Security in a Changing Climate:** Climate change is intensifying global water security challenges, from sea level rise increasing arsenic contamination to extreme droughts straining freshwater supplies. Beyond monitoring, how can AI actively contribute to climate adaptation strategies for drinking water safety? Can AI-driven models help communities preemptively manage climate-related water crises? What current limitations in AI technology must be addressed before it can be fully leveraged for long-term water security?

Competing Interests This study was carried out without external funding and thereby no competing interests exist.

References

Al-Adhaileh MH, Alsaade FW (2021) Modelling and prediction of water quality by using artificial intelligence. Sustain 13(8):4259. https://doi.org/10.3390/su13084259

Autodesk (2022) AI for water: 10 ways AI is changing the water industry. https://blogs.autodesk.com/innovyze/2022/05/01/ai-in-water-10-ways-ai-is-changing-the-water-industry/. Accessed 1 Feb 2025

Bamal A, Uddin MG, Olbert AI (2024) Harnessing machine learning for assessing climate change influences on groundwater resources: a comprehensive review. Heliyon 10:e37073. https://doi.org/10.1016/j.heliyon.2024.e37073

Berthold TA, McCrary A, deVilleneuve S, Schramm M (2023) Let's talk about PFAS: inconsistent public awareness about PFAS and its sources in the United States. PLoS One 18(11):e0294134. https://doi.org/10.1371/journal.pone.0294134

Bhattacharya P, Jacks G, Frisbie SH, Smith E, Naidu R, Sarkar B (2002) Arsenic in the environment: a global perspective. In: Sarkar B (ed) Heavy metals in the environment. Marcel Dekker, New York, pp 147–215

Bibri SE, Krogstie J, Alahi A, Kaboli A (2024) Smarter eco-cities and their leading-edge artificial intelligence of things solutions for environmental sustainability: a comprehensive systematic review. Environ Sci Ecotechnol 19:100330. https://doi.org/10.1016/j.ese.2023.100330

Burger J, Gochfeld M (2021) Biomonitoring selenium, mercury, and selenium: mercury molar ratios in selected species in northeastern United States estuaries: risk to biota and humans. Environ Sci Pollut Res Int 28(15):18392–18406. https://doi.org/10.1007/s11356-020-12175-z

Chowdury MSU, Emran TB, Ghosh S, Pathak A, Alam MM, Absar N, Andersson K, Hossain MS (2019) IoT based real-time river water quality monitoring system. Procedia Comput Sci 155:161–168. https://www.sciencedirect.com/science/article/pii/S1877050919309391/pdf

Dwivedi YK, Hughes L, Ismagilova E, Aarts G, Coombs C, Crick T, Duan Y, Dwivedi R, Edwards J, Eirug A, Galanos V, Ilavarasan PV, Janssen M, Jones P, Kar AK, Kizgin H, Kronemann B, Lal B, Lucini B, Medaglia R, Meunier-FitzHugh KL, Meunier-FitzHugh Leslie CL, Misra S, Mogaji E, Sharma SK, Singh JB, Raghavan V, Raman R, Rana NP, Samothrakis S, Spencer J, Tamilmani K, Tubadji A, Walton P, Williams MD (2021) Artificial intelligence (AI): multidisciplinary perspectives on emerging challenges, opportunities, and agenda for research, practice, and policy. Int J Inf Manag 57:101994. https://doi.org/10.1016/j.ijinfomgt.2019.08.002

Faybishenko B, Bakhtavar E, Hewage K, Sadiq R (2024) Chemical composition of arsenic-based acid mine drainage in the downstream of a gold mine: fuzzy regression and clustering analysis. J Hazard Mater 465:133250. https://doi.org/10.1016/j.jhazmat.2023.133250

Faybishenko B, Frisbie SH (2025) On the application of the statistical and fuzzy systems regression and clustering to analyze the multivariate chemical composition of groundwater: case study of arsenic-contaminated groundwater in Bangladesh. In: Molla AR, Frisbie SH, Sherif M, Singh VP (eds) Climate change, drinking water security, and public health. Springer Nature Switzerland AG, Gewerbestrasse 11, 6330 Cham, Switzerland

Frank JJ, Poulakos AG, Tornero-Velez R, Xue J (2019) Systematic review and meta-analyses of lead (Pb) concentrations in environmental media (soil, dust, water, food, and air) reported in the United States from 1996 to 2016. Sci Total Environ 694:133489. https://doi.org/10.1016/j.scitotenv.2019.07.295

Friedman A, Boselli E, Ogneva-Himmelberger Y, Heiger-Bernays W, Brochu P, Burgess M, Schildroth S, Denehy A, Downs T, Papautsky I, Claus Henn B (2024) Manganese in residential

drinking water from a community-initiated case study in Massachusetts. J Expo Sci Environ Epidemiol 34:58–67. https://doi.org/10.1038/s41370-023-00563-9
Frisbie SH, Mitchell EJ, Molla AR (2024) Sea level rise from climate change is expected to increase the release of arsenic into Bangladesh's drinking well water by reduction and by the salt effect. PLoS One 19(1):e0295172. https://doi.org/10.1371/journal.pone.0295172
Frisbie SH, Ortega R, Maynard DM, Sarkar B (2002) The concentrations of arsenic and other toxic elements in Bangladesh's drinking water. Environ Health Perspect 110(11):1147–1153. https://doi.org/10.1289/ehp.021101147
Guo L, Xie X, Zeng J, An N, Wang Z, Gao L, Wang Y, Yang J (2023) Optimization model of water resources allocation in coal mine area based on ecological environment priority. Water 15(6):1205. https://doi.org/10.3390/w15061205
Heng YY, Asad I, Coleman B, Menard L, Benki-Nugent S, Hussein Were F, Karr CJ, McHenry MS (2022) Heavy metals and neurodevelopment of children in low and middle-income countries: a systematic review. PLoS One 17(3):e0265536. https://doi.org/10.1371/journal.pone.0265536
Himeur Y, Elnour M, Fadli F, Meskin N, Petri I, Rezgui Y, Bensaali F, Amira A (2023) AI-big data analytics for building automation and management systems: a survey, actual challenges and future perspectives. Artif Intell Rev 56:4929–5021. https://doi.org/10.1007/s10462-022-10286-2
Hu XC, Dai M, Sun JM, Sunderland EM (2023) The utility of machine learning models for predicting chemical contaminants in drinking water: promise, challenges, and opportunities. Curr Envir Health Rpt 10:45–60. https://doi.org/10.1007/s40572-022-00389-x
Huntingford C, Jeffers ES, Bonsall MB, Christensen HM, Lees T, Yang H (2019) Machine learning and artificial intelligence to aid climate change research and preparedness. Environ Res Lett 14:124007. https://doi.org/10.1088/1748-9326/ab4e55
Jamei M, Karbasi M, Malik A, Abualigah L, Towfiqul Islam ARM, Yaseen ZM (2022) Computational assessment of groundwater salinity distribution within coastal multi-aquifers of Bangladesh. Sci Rep 12:11165. https://doi.org/10.1038/s41598-022-15104-x
Jasechko S, Seybold H, Perrone D, Fan Y, Shamsudduha M, Taylor RG, Fallatah O, Kirchner JW (2024) Rapid groundwater decline and some cases of recovery in aquifers globally. Nature 625:715–721. https://doi.org/10.1038/s41586-023-06879-8
Joha MI, Rahman MM, Nazim MS, Jang YM (2024) A secure IIoT environment that integrates AI-driven real-time short-term active and reactive load forecasting with anomaly detection: a real-world application. Sensors 24:7440. https://doi.org/10.3390/s24237440
Joy C, Sundar GN, Narmadha D (2023) Artificial intelligence based test systems to resist water-borne diseases by early and rapid identification of pathogens: a review. Comput Sci 4:180. https://doi.org/10.1007/s42979-022-01620-0
Krishnan SR, Nallakaruppan MK, Chengoden R, Koppu S, Iyapparaja M, Sadhasivam J, Sethuraman S (2022) Smart water resource management using artificial intelligence—a review. Sustainability 14(20):13384. https://doi.org/10.3390/su142013384
Lai V, Huang YF, Koo CH, Ahmed AN, El-Shafie A. Arch (2022) A review of reservoir operation optimisations: from traditional models to metaheuristic algorithms. Comput Methods Eng 29:3435–3457. doi:https://doi.org/10.1007/s11831-021-09701-8
Levin R, Villanueva CM, Beene D, Cradock AL, Donat-Vargas C, Lewis J, Martinez-Morata I, Minovi D, Nigra AE, Olson ED, Schaider LA, Ward MH, Deziel NC (2024) US drinking water quality: exposure risk profiles for seven legacy and emerging contaminants. J Expo Sci Environ Epidemiol 34:3–22. https://doi.org/10.1038/s41370-023-00597-z
Maroju RG, Choudhari SG, Shaikh MK, Borkar SK, Mendhe H (2023) Application of artificial intelligence in the management of drinking water: a narrative review. Cureus 15(11):e49344. https://doi.org/10.7759/cureus.49344
Maskrey SA, Mount NJ, Thorne CR, Dryden I (2016) Participatory modelling for stakeholder involvement in the development of flood risk management intervention options. Environ Model Softw 82:275–294. https://doi.org/10.1016/j.envsoft.2016.04.027

Mitchell EJ, Frisbie SH (2023) A comprehensive survey and analysis of international drinking water regulations for inorganic chemicals with comparisons to the World Health Organization's drinking-water guidelines. PLoS One 18(11):e0287937. https://doi.org/10.1371/journal.pone.0287937

Nathanael RJ, Adyanis LN, Oginawati K (2024) The last decade epidemiologic concern of drinking water contaminants of emerging concern (CECs) in Asian countries: a scoping review. Heliyon 10:e39236. https://doi.org/10.1016/j.heliyon.2024.e39236

National Research Council (US) Committee on Medical and Biological Effects of Environmental Pollutants (1977) Arsenic: medical and biologic effects of environmental pollutants. National Academies Press, Washington, DC. 3, Distribution of arsenic in the environment. https://www.ncbi.nlm.nih.gov/books/NBK231016/

Neal AP, Guilarte TR (2013) Mechanisms of lead and manganese neurotoxicity. Toxicol Res (Camb) 2(2):99–114. https://doi.org/10.1039/C2TX20064C

People's Republic of Bangladesh (2025) Water quality parameters. http://dphe.gov.bd/site/page/15fa0d7b-11f1-45c0-a684-10a543376873/Water-Quality-Parameters-. Accessed 12 July 2025

Sanyal T, Bhattacharjee P, Paul S, Bhattacharjee P (2020) Recent advances in arsenic research: significance of differential susceptibility and sustainable strategies for mitigation. Front Public Health 8:464. https://doi.org/10.3389/fpubh.2020.00464

Shlezinger M, Amitai Y, Akriv A, Gabay H, Shechter M, Leventer-Roberts M (2018) Association between exposure to desalinated sea water and ischemic heart disease, diabetes mellitus and colorectal cancer; a population-based study in Israel. Environ Res 166:620–627. https://doi.org/10.1016/j.envres.2018.06.053

United States Geological Survey (USGS) (2017) Arsenic in private wells in Connecticut. https://www.usgs.gov/media/images/arsenic-private-wells-connecticut. Accessed 11 June 2025

World Health Organization (WHO) (2017) Chemical mixtures in source water and drinking-water. World Health Organization, Geneva

World Health Organization (WHO) (2022) Guidelines for drinking-water quality, 4th edn. WHO, Geneva, Switzerland, p 340–343, 402–405, 438–444, 525–527. https://www.who.int/publications/i/item/9789240045064. Accessed 12 July 2025

World Health Organization (WHO) (2023) Drinking-water. https://www.who.int/news-room/fact-sheets/detail/drinking-water. Accessed 3 Feb 2025

Xiang X, Li Q, Khan S, Khalaf OI (2021) Urban water resource management for sustainable environment planning using artificial intelligence techniques. Environ Impact Assess Rev 86:106515. https://doi.org/10.1016/j.eiar.2020.106515

Index

A. R. Molla et al. (eds.), *Climate Change, Drinking Water Security, and Public Health*, https://doi.org/10.1007/978-3-032-12074-8

FSC
www.fsc.org
MIX
Papier aus verantwortungsvollen Quellen
Paper from responsible sources
FSC® C105338

If you have any concerns about our products,
you can contact us on
ProductSafety@springernature.com

In case Publisher is established outside the EU,
the EU authorized representative is:
Springer Nature Customer Service Center GmbH
Europaplatz 3, 69115 Heidelberg, Germany

Printed by Libri Plureos GmbH
in Hamburg, Germany